Nietzsche on Time and History

Nietzsche on Time and History

Edited by

Manuel Dries

Walter de Gruyter · Berlin · New York

♾ Printed on acid-free paper which falls within the guidelines
of the ANSI to ensure permanence and durability.

Library of Congress Cataloging-in-Publication Data

A CIP catalogue record for this book is available from the Library of Congress.

ISBN 978-3-11-019009-0

Bibliographic information published by the Deutsche Nationalbibliothek

The Deutsche Nationalbibliothek lists this publication in the Deutsche Nationalbibliografie;
detailed bibliographic data are available in the Internet at http://dnb.d-nb.de.

Printed in Germany
Cover design: Martin Zech, Bremen.
Printing and binding: Hubert & Co GmbH & Co KG, Göttingen.

If there is no goal in the whole of history of man's lot, then we must put one in: assuming, on the one hand, that we have need of a goal, and on the other that we've come to see through the illusion of an immanent goal and purpose. And the reason we have need of goals is that we have need of a will—which is the spine of us. 'Will' as the compensation of lost 'belief', i.e., for the idea that there is a divine will, one which has plans for us.

Friedrich Nietzsche, Nachlaß Summer 1886–Spring 1887, KSA 12, 6[9]

We are still growing continually, our sense of time and place, etc., is still developing.

Friedrich Nietzsche, Nachlaß April–June 1885, KSA 11, 34[124]

'Timeless' to be rejected. At a particular moment of a force, an absolute conditionality of the redistribution of all forces is given: it cannot stand still. 'Change' is part of the essence, and therefore so is temporality—which, however, just amounts to one more conceptual positing of the necessity of change.

Friedrich Nietzsche, Nachlaß May–July 1885, KSA 11, 35[55]

Acknowledgements

The essays in this volume were first presented at the 15th International Conference of the Friedrich Nietzsche Society of Great Britain and Ireland, held at Peterhouse, Cambridge, in September 2005. I wish to take this opportunity to thank delegates and all those who participated in this event for the debates and critical discussions that shaped the research presented here.

I would further like to thank the Faculty of Philosophy and the Department of German at Cambridge for providing generous financial support. I am much indebted to Raymond Geuss, Simon Blackburn, Margaret Clare Ryan, and the Executive Committee of the Friedrich Nietzsche Society for their advice during the various stages of the editing of this volume.

Two of the contributions in this book are either drawn from or have appeared in full elsewhere, and this material appears with permission and my thanks. Raymond Geuss' article was previously published in his collection of essays *Outside Ethics* (Princeton: Princeton University Press, 2005). It is reprinted here with permission of Princeton University Press. Lawrence J. Hatab's article is drawn in parts from the text of his book *Nietzsche's Life Sentence: Coming to Terms with Eternal Recurrence* (New York: Routledge University Press, 2005). I thank Routledge for permission to publish the essay.

The four excerpts of printed music of Wagner, Bizet, and Stravinsky in Jonathan R. Cohen's essay appear here with permission of Dover Publishing, Chester Music Limited (Music Sales) and Schott Music GmbH & Co. KG. Every effort has been made to trace and contact copyright holders. If there are any inadvertent omissions I apologize to those concerned and undertake to include suitable acknowledgements in future editions.

Finally, I would like to thank Walter de Gruyter Publishers for taking on this volume, Gertrud Grünkorn, Christoph Schirmer, and Jana Pokorny for their patient support, and Angela Blackburn for copyediting the final manuscript.

Contents

Part IV: Nietzsche's Contemporaries

Part V: Tragic and Musical Time

Notes on Contributors

TINNEKE BEECKMAN is postdoctoral researcher for the Fund of Scientific Research, Flanders. She works for the Department of Philosophy, University of Brussels, Belgium.

THOMAS H. BROBJER is Associate Professor in the Department of the History of Science and Ideas at Uppsala University, Sweden.

JONATHAN R. COHEN is Professor of Philosophy at the University of Maine in Farmington, USA.

MANUEL DRIES is Research Fellow at Wolfson College and the Faculty of Philosophy, University of Oxford, UK.

RAYMOND GEUSS is Professor of Philosophy at the University of Cambridge, UK.

KATHERINE C. HARLOE is a Career Development Fellow in Classics at St. Anne's College, University of Oxford, UK.

LAWRENCE J. HATAB is Louis I. Jaffe Professor of Philosophy at Old Dominion University, Virginia, USA.

R. KEVIN HILL is Associate Professor of Philosophy at Portland State University, USA.

ANTHONY K. JENSEN is a Visiting Assistant Professor in the Department of Philosophy at Xavier University in Cincinnati, Ohio, USA.

PAUL S. LOEB is Professor of Philosophy at the University of Puget Sound, USA.

ANDREA ORSUCCI is Professor of Philosophy at the University of Cagliari, Italy.

JOHN RICHARDSON is Professor of Philosophy at New York University, USA.

MARTIN A. RUEHL is University Lecturer of German at Sidney Sussex College, University of Cambridge, UK.

HERMAN SIEMENS is University Lecturer in Philosophy at the University of Leiden, The Netherlands.

Abbreviations and Translations

Friedrich Nietzsche's published and unpublished writings (Nachlaß) are quoted according to the following abbreviations:

A	*The Anti-Christ*, cited by section number.
AOM	'Assorted Opinions and Maxims' (vol. 2, pt 1, of *Human, All Too Human*), cited by section number.
BAW	*Historisch-kritische Gesamtausgabe. Werke*, ed. Hans Joachim Mette, 5 vols. (Munich: C. H. Beck, 1933–1940), cited by volume and page number.
BAB	*Historisch-kritische Gesamtausgabe. Briefe*, ed. Hans Joachim Mette, 4 vols. (Munich: C. H. Beck, 1933–1940), cited by volume and page number.
BGE	*Beyond Good and Evil*, cited by section number.
BT	*The Birth of Tragedy*, cited by section number and KSA page number.
CV	'Five Prologues to Five Unwritten Books', cited by number and KSA page number.
CW	*The Case of Wagner*, cited by section number.
D	*Daybreak*, cited by section number.
EH	*Ecce Homo*, cited by section heading and (when applicable) number.
EI	'On the Future of Our Educational Institutions', cited by section number.
GM	*On the Genealogy of Morality*, cited by essay and section number.
GS	*The Gay Science*, cited by section number.
HA	*Human, All Too Human*, cited by volume and section number.
CV	'Five Prefaces to Five Unwritten Books', cited by preface number and KSA page number.
KGB	*Briefwechsel. Kritische Gesamtausgabe*, ed. Giorgio Colli and Mazzino Montinari (Berlin: Walter de Gruyter, 1975–), cited by volume and page number.
KGW	*Werke. Kritische Gesamtausgabe*, established by Giorgio Colli and Mazzino Montinari, ed. Wolfgang Müller-Lauter and Karl Pestalozzi (Berlin: Walter de Gruyter, 1967–), cited by volume, part, and page number.

KSA *Sämtliche Werke. Kritische Studienausgabe*, ed. Giorgio Colli and Mazzino Montinari (Berlin: Walter de Gruyter, 1967–), cited by volume and page number. The Nachlaß is cited by date, KSA volume, notebook section, and fragment number.

KSB *Sämtliche Briefe. Kritische Studienausgabe Briefe*, ed. Giorgio Colli and Mazzino Montinari (Berlin: Walter de Gruyter, 1986–), cited by volume and page number.

NCW *Nietzsche contra Wagner*, cited by section heading.

OTL 'On Truth and Lying in a Non-Moral Sense', cited by KSA page number.

PTAG 'Philosophy in the Tragic Age of the Greeks', cited by section number.

TI *Twilight of the Idols*, cited by section heading and number.

UM *Untimely Meditations*, cited by part and section number, and (when applicable) KSA page number.

Z *Thus Spoke Zarathustra*, cited by part, section heading, and (when applicable) number.

Note on Translations of Nietzsche's Works

The contributors to this volume have used different translations of Nietzsche's texts, often modified by the individual contributor. At the end of each essay the reader will find a list of the translations used. Where no such list has been provided the contributor has relied exclusively on his or her own translations. All translations from Nietzsche's Nachlaß are usually by the individual contributors, although other translations have been consulted whenever possible, notably *The Will to Power*, trans. Walter Kaufmann and R. J. Hollingdale (New York: Random House, 1967), and *Writings from the Late Notebooks*, ed. Rüdiger Bittner, trans. Kate Sturge (Cambridge: Cambridge University Press, 2003).

Nietzsche's Critique of Staticism
Introduction to Nietzsche on Time and History

Manuel Dries

> Motion must first disappear, i.e. lead to a static effect before it appears to our feeling. Feeling is the sign of a motion that has been made statically perceptible, i.e. a contained and annihilated motion. (Nachlaß Summer 1875, KSA 8, 9[1])
>
> Every thing is a sum of judgements (fears, hopes, some inspire confidence, others do not). Now, the better we know physics the less phantasmal this sum of judgements becomes ... Finally we understand: a thing is a sum of excitations within us: however, since we are nothing fixed [*Festes*] a thing is also not a fixed sum. And the more stability we attribute to things, – – – (Nachlaß Spring 1880–Spring 1881, KSA 9, 10[F100])
>
> If there is no goal in the whole of history of man's lot, then we must put one in: assuming, on the one hand, that we have *need* of a goal, and on the other that we've come to see through the illusion of an immanent goal and purpose. (Nachlaß Summer 1886–Spring 1887, KSA 12, 6[9])

Why are we still intrigued by Nietzsche? One might think of a number of answers to this question: the variety of his interests, his entertaining and accessible style, perhaps his aphoristic ambiguity that leaves so much more work to the interpreter. I am not convinced that this suffices to explain the sustained interest in Nietzsche's philosophy. What I will argue in this introduction is that this sustained interest stems from Nietzsche's challenge to what I will call the 'staticism' inherent in our ordinary experience. By 'staticism' I mean, roughly speaking and in general, the view that the world is a collection of enduring, re-identifiable objects that change only very gradually and according to determinate laws. The claim I wish to make is simple: as long as human beings subscribe to the 'staticist picture' Nietzsche will remain of interest. Why is this so? In short: because ordinary experience is 'what is the case' (for most of us) and it is also quite clearly not the case. Should it turn out that staticism is a kind of anthropological constant that each generation of philosophers eventually has to face critically then Nietzsche will remain of interest at least until someone else provides a more comprehensive examination and critique of it.

What do I mean more precisely by the term 'staticism'? The staticism Nietzsche is already suspicious of very early on in writings such as *On*

Becoming in History, and then questions almost obsessively in his later works and notebooks, is usually a variant of the complex view comprised of the conjunction of the following three propositions:

(i) The world is best conceived as a world of relatively easily distinguishable, property-instantiating objects that remain sufficiently identical over time to be named, referred to and remembered.
(ii) The collection of objects called 'the world' is governed by laws that are sufficiently determinate to prevent chaos from ensuing, and to allow humans, objects with special properties, to make some predictions about what will happen in the future.
(iii) The existence of this deterministic world of objects is somehow compatible with the possibility of actual choice and voluntary action.

I will call this the staticist worldview. Nietzsche's emphasis on time and history is usually both a critique of the staticist worldview and, less often so, his attempt to develop an alternative worldview, an alternative that is, however, not simply a negation of the staticist worldview. It is for this reason that I wish to preface *Nietzsche on Time and History* with a few remarks on Nietzsche's critique of staticism.

I will first discuss Nietzsche's rejection of the remnants of staticism in Hegel and Schopenhauer (both of whom, he holds, remain fundamentally opposed to the very existence of time and history *proper*). I will then briefly outline why Nietzsche deems the belief in any variant of the staticist picture as problematic. Finally, I will examine what I believe is Nietzsche's adualistic-dialetheic stance towards the staticist worldview. In the final section, I will comment on the different ways these issues are addressed in *Nietzsche on Time and History.*

Nietzsche's Predecessors: Schopenhauer and Hegel

Nietzsche believed most if not all *homines mensurae* to be in thrall to the staticist, ordinary standpoint. It is the way the world first seems or appears to them. After a relatively short period of discipleship, Nietzsche realized that Schopenhauer's philosophy remained, despite its subversive metaphysical and critical aspects, firmly embedded within a philosophical tradition that hypostatizes the atemporal, thereby tacitly supporting the staticist picture. Schopenhauer distinguishes between a reality as it is in and for itself, a metaphysical will that is *not* (still a common misconception) the thing in itself but nevertheless 'the nearest and clearest *phenomenon* of the thing-in-itself' (WWR II 18), and an illusory actuality of becoming that has the ontological status of a problematic, *mere* appearance (*Schein*). Commitment to a number of Kantian dualisms leads Schopenhauer to attack in *The World as Will and Representation* any philosophy focussed on time and history. Philosophy, he writes in 'On history'

> should not consider ... that which is always becoming and never is … On the contrary, it should keep in view that which always is, and never becomes and passes away ... The true philosophy of history consists in the insight that, in spite of all these endless changes and their chaos and confusion, we yet always have before us the same, identical, unchangeable essence, acting in the same way today as it did yesterday and always. (Schopenhauer 1969, vol. 2, p. 444)

For Schopenhauer, then, accepting temporality, and its appearance for human beings *as* history, as essential is fundamentally misguided. Nietzsche realized that such privileging of ideas devoid of change ultimately leads to a non-Christian but equally world-negating pessimism: if philosophy is supposed to contemplate that which is permanent and unchanging, then the confrontation with impermanence poses a real problem. If that which is permanent has added value, and binary thinking demands a necessary choice or exclusive disjunction, it follows that the value of becoming approaches zero. As Schopenhauer puts it in his 'Additional Remarks on the Doctrine of the Vanity of Existence':

> This vanity [of existence] finds its expression ... in constant becoming without being; in constant desire without satisfaction ... *Time* is that by virtue whereof at every moment all things in our hands come to naught and thereby lose all true value. (Schopenhauer 1974, vol. 2, p. 283)

Nietzsche would ultimately reject Schopenhauer's 'chronophobic' evaluation of existence but primarily because it thereby tacitly supports the staticist worldview.

The same holds, in Nietzsche's evaluation, for Hegel. Hegel had drawn attention to the *concept* of becoming (cf. GS 357) but did so within a macro-teleological, systematic philosophy. 'Becoming' denotes not only the original, restless-creative oscillation between determination and indeterminacy—being and nothing—that gets the micro-teleological dialectical becoming under way, but more importantly the macro-teleological, necessary *autopoesis* of a mind-like absolute substance. Nietzsche's Hegel is the optimistic (cf. DS 6, KSA 1, p. 191; also KSA 8, p. 56) panlogicist (UM II 8, KSA 1, p. 309) who imbues 'the whole' with meaning only by attributing to it an organic, macro-teleological, rational and thereby stable composition. In *Untimely Meditation II* Nietzsche cautions against such an uncritical view of becoming since it still contains all the attributes of necessary *Sein*, 'being'—the staticist concept *par excellence* for Nietzsche:

> If every success is in itself a rational necessity, if every event is the victory of the logical or the 'idea' — then quickly down on your knees and hold in reverence the entire stepladder of 'successes'. (UM II 8, KSA 1, p. 309)

The Hegelian system both presupposes and culminates in the *unsinnige* absolute Idea that 'alone has *Being*, imperishable life, *truth known to itself*, and is *all truth* ... since its essence is, the highest, the concept' (Hegel 1969, vol 2., p. 549). Nietzsche therefore sees in Hegel's philosophy the 'bridge of lies back to old ideals' and rejects Hegel's problematic practice of 'mediating' and 'fusing' (D Preface 4, KSA 3, p. 16). Hegel like Parmenides desired to know the absolute by means of reflection, '"to grasp the absolute within consciousness"'(PTAG 11, KSA 1, p. 847), and such attempts, Nietzsche is convinced, expose the tacit continuation of the staticist worldview (cf. KSA 15, p. 77).

Nietzsche is convinced that his philosophical predecessors, and also the natural sciences, had merely changed the appearance of the dominant staticist paradigm of being. Yet behind the macro-teleological idea of becoming, the idea of will as quasi thing-in-itself, and the positivism and objectivity of science, the belief in permanence as highest value remained unquestioned.

From whence this *chronophobia* and hysterical overvaluation of being? Nietzsche's writings are littered with attempts to provide ever new explanations of this phenomenon. Unfortunately, this part of Nietzsche's work is not often discussed in the exisiting literature. I believe, however, that it is of great importance because it allows us to see that Nietzsche holds a kind of error theory about staticism.

Against the Rejection of Time: Nietzsche's Error Theory

In note 9[60] of autumn 1887 Nietzsche presents a mini-genealogy of the idea of being that can be seen as paradigmatic for his belief in the primacy of becoming; at the same time this genealogy explains why humans cling so desperately to the idea of unchanging being. He wishes to subject to a genealogical critique both the concept of reality and the positive valuation of being. This genealogy *in nuce* starts with an instruction important for his overall idea of the genealogical method, that of *Selbstbesinnung*:

> 'Uncanny **self-reflection/auto-sensitization** [*Selbstbesinnung*]**:** not as individual but becoming conscious of oneself as human species. Let us come to our senses [*besinnen*], let us think backward: let us walk the short and the long paths' (Nachlaß Autumn 1887, KSA 12, 9[60])

Genealogy as a method or tool is not only a *(sich) besinnen* in the sense of 'contemplating' or 'reflecting upon', for example, the historicity of a value, but also always a *(sich) besinnen* in the literal sense, i.e., a 'returning to the senses', and thereby a returning and coming 'to one's senses'. It is worth quoting this genealogy of being in full:

> Man is searching for 'the truth': a world that does not contradict itself, does not deceive, does not change, a true world — a world, in which one does not suffer: contradiction, illusion, transistoriness — causes of suffering! He does not doubt, that such a world, as it ought to be, exists; he wants to find his way to it. …
> Whence does man take the concept of reality? —
> Why is it that man deduces suffering precisely from change, illusion, contradiction? And why not more so his happiness? ... —
> The contempt, the hatred of all that passes away, changes, transforms: — whence this valuation of the permanent?
> What is obvious here is the will to truth, just the desire for a world of permanence.
> The senses deceive, rationality corrects the errors: consequently, one inferred, that reason is the path to the permanent; the most non-sensory [*unsinnlichsten*] ideas must be closest to the 'true world'. — Most misfortunes come from the senses — they are fraudsters, beguilers, annihilators:
> Happiness is only warranted in what has being [*im Seienden*]: change and happiness are mutually exclusive. The greatest desire aims at a becoming one with being. This is the strange path to the highest form of happiness.
> In sum: the world, as it ought to be, exists; this world, the one we live in, is only error, — this our world ought not to exist.
> The belief in being turns out to be just <as> a consequence: the real primum mobile is the unbelief in becoming, the mistrust against becoming, the contempt for all becoming… (Nachlaß Autumn 1887, KSA 12, 9[60])

Nietzsche questions how human beings arrived at their belief in being and came to understand suffering as the consequence of 'change', 'illusion' and 'contradiction'. Why not equate change with happiness? At the core of this equation lies what I wish to call Nietzsche's error theory regarding staticism. It can be summarized as follows:

(i) ordinary human discourse is ineliminably committed to the staticist worldview (semantic thesis);
(ii) there are no relatively easily distinguishable, property-instantiating entities and objects (ontological thesis);
(iii) it follows that our ordinary natural attitude is false.

In addition to what can be called the semantic thesis (i) and the ontological thesis (ii) Nietzsche also offers an explanatory thesis:

(iv) human beings hold the staticist worldview because it allows them to reduce uncertainty, thereby alleviating suffering.

The explanatory step can be unpacked as follows: perceiving something *as* something involves a transformation that is error-prone. Since humans use their rational capacities to correct some of these initial errors, rationality seems the one and only remedy against the ills and contradictions of the senses. In our attempts to overcome the impractical unreliability of sense-impressions once and for all, 'non-sensory' ideas—for Nietzsche, entirely non-sensory (*unsinnlich*) amounts to nonsense (*Unsinn*)—are gradually regarded as closest to what is simple, true, and predictable, thereby creating a less painful environment. It is here that becoming and happiness can no longer coexist, so that they have become *einverleibt* or 'incorporated' (GS 1, KSA 3, p. 370) as mutually exclusive spheres, and static being comes to be the highest value (and Nietzsche really means incorporated: our species has adapted most successfully by organizing its world, thereby keeping uncertainty and pain at a minimum).

Note 9[60] shows that for Nietzsche this turn against the senses has two consequences that amount to two 'incorporated' commitments—one ontological, one ethical: the tacit *ontological commitment* entails that the world as it ought to exist, the staticist world, really exists ('die Welt, wie sie sein sollte, existiert'); and tacit *ethical commitment*, in turn, entails that the world of becoming therefore *ought not* to exist ('diese unsere Welt sollte nicht existieren').

This positive valuation of being leads to the strong belief in the existence of being and to the search for truth in a rational, abstract, measuring manner. Nietzsche concludes note 9[60] by restating this argument: the valuation of being arises as a consequence of the initial 'disdain for becoming' (ibid.).

Read as a genealogy *in nuce* this line of argument is therefore at the same time a reflection on the valuation of being, a rehabilitation of the senses, and thereby a sobering experience of 'coming to one's senses'. It is this argument that underpins Nietzsche's basic assumptions and accounts for both his own biased *ontological* commitment (that which really exists is better understood as *Werden*) and his concomitant *ethical* commitment: the world of any permanent *Sein* should therefore *not* exist—at least not within the same logical exclusive-disjunctive relation to becoming. But why is staticism so vicious?

Staticism and Nihilism

First, Nietzsche acknowledges that staticism and the Judeo-Christian morality it underpins has had real benefits as a successful defence against the earliest form of nihilism induced by fear and uncertainty. Nevertheless, he is adamant that 'the fear became less' (Nachlaß Spring–Autumn 1881, KSA 9, 11[26]) and that 'life is no longer so uncertain, accidental, chaotic in our Europe' (Nachlaß Summer 1886–Autumn 1887, KSA 12, 5[71]) and the level of strength which human beings have attained 'allows for a lowering of the means of taming' contingency. '"God"' or stable being as the ultimate guarantor of staticism 'is now a hypothesis much too extreme' (Nachlaß Summer 1886–Autumn 1887, KSA 12, 5[71]). And staticism, if not also 'lowered' and adapted to the current, lower level of uncertainty will now lead to a new type of nihilism. Why? It supports not just one peculiar valuation and meta-belief but rather an entire system of related valuations, lower-level beliefs—a two-world metaphysics (both a false ontology of what-there-is and a questionable epistemology of what we can know) within which permanence is valued highest. With such a web of beliefs in place, any value of the non-permanent is merely due to a kind of 'retension' of or 'protension' to some permanent state or realm, be that ontological or theoretic-epistemological, or ethical. The staticist viewpoint demands a revision (Nietzsche's early idea of a time-atom theory can be seen as an early attempt. Eternal recurrence is his late conception).

More correctly: the value of the non-static needs to be changed and for the first time taken seriously. Since logic and ontology in Nietzsche's view sprang from and subsequently confirmed and upheld the staticist error, traditional logic—and Nietzsche's attack on logic is always only an attack on traditional Aristotelian logic—can no longer be the tool to deliver reliable guidance. New frameworks and methodologies are needed within which philosophy can continue its interpretive-descriptive enterprise and avoid the trappings of the previous, nihilistic framework. If logic had been the science that derives certain and reliable truths from timeless laws, then a philosophy that wishes to undercut the staticist picture can no longer rely on it in the same way. It is here that history as genealogy becomes one of the new 'chronophile' investigative methods:

> Philosophy in the only way I still allow it to stand, as the most general form of history, as an attempt somehow to describe Heraclitean becoming and to abbreviate it into signs (so to speak, to *translate* and mummify it into a kind of illusory being) (Nachlaß June–July 1885, KSA 11, 36[27])

But history, too, can be practised as—and is in danger of being—an *ancilla metaphysica,* in thrall to staticism, describing the same hyperstable world as that projected in traditional metaphysics. Philosophy *as* history *proper* that takes the temporal disposition of the whole with its several simultaneous temporal-perspectival dimensions seriously must then create a very different, revised historical-philosophical approach, self-reflexively aware of the staticist fallacy. It, too, must incorporate an awareness of the latter. The contributions in *Nietzsche on Time and History* deal with the impact and importance of history for philosophy and the need gradually to unlearn the natural staticist standpoint.

However, and this is crucial and complicates matters considerably, Nietzsche's advice is not simply to do away with the staticist pictures.

The Staticist Picture: Nietzsche's Staticist Fictionalism?

In addition to Nietzsche's 'argument from anxiety' he repeatedly argues that the staticist picture stems from our Cartesian failure to conceive of ourselves as some kind of distinct, metaphysical, underlying substances or 'soul atoms' (Nachlaß June–July 1885, KSA 11, 37[4]) and that everything else, our world of subjects, objects, and causal relations, then simply follow:

> What separates me most deeply from metaphysicians is: I don't concede that the 'I' is what thinks. Instead, I take the I itself to be a construction of thinking, of the same rank as 'matter', 'thing', 'substance', 'individual', 'purpose', 'number': in other words to be only a regulative fiction with the help of which a kind of constancy and 'knowability' is inserted into, invented into, a world of becoming. (Nachlaß May–June 1885, KSA 11, 35[35])

Most of the so-called Continental interpretations of Nietzsche have focused—too much, in my view—on this critique of the self. Why too much? The well-known fragment contains an important qualification, a second premise if you will (provided we interpret Nietzsche's texts as arguments consisting of a number of explicit and implicit premises and assumptions). Nietzsche clearly insists at the end of 35[35] that the staticist picture and the self, though false, cannot be abandoned:

> However habituated and indispensable this fiction may now be, that in no way disproves its having been invented: something can be a condition of life and nevertheless be false. (Nachlaß Autumn 1884–Autumn 1885, KSA 11, 35[35])

It is here that we encounter a seemingly contradictory inconsistency that has been vexing commentators. Nietzsche's texts are littered with polemic reversals, on the one hand, and reversals of those reversals, on the other: the static picture is false (assuming that it contains belief in x, y, z,); and the static picture is an anthropological constant, a necessary condition of life. Any interpretation that is not in hermeneutic denial needs to account for both. If we, then, allow for the further assumption—and there is plenty of textual evidence that we should—that 'life' (in the above passage) is for Nietzsche a phenomenon of very high value, and if the static picture is a necessary condition of 'life'; and 'life' is of high, if not the highest, value, it follows that the staticist picture cannot simply be false per se.

One might argue then that Nietzsche is really a fictionalist about staticism: according to which staticism is false and yet human beings are committed to it for adaptive-pragmatic purposes. I do not believe that Nietzsche's analysis ends here; he demands something more than a quasi-staticism. Nietzsche is always aware of the dangers of such pragmatic-adaptive acceptance. The danger is that too much remains in place, too much remains acceptable, and that new variants of nihilism come in through the back door. Therefore, Nietzsche's critique of staticism as well as its rehabilitation (as necessary for life) needs to be re-situated rather than replaced. In European philosophy, there has been a tendency to negate logic too quickly without realizing that it is the very same logic of mutually exclusive alternatives which is still tacitly at work in its own abolition. There are good grounds for a different logical framework. For want of better terms I will call this Nietzsche's adualistic-dialetheic framework. Dialetheism, from Greek 'diplo-aletheia' or two-way truth, allows for true contradictions and can therefore cope better with 'transition states', borderline cases, and vague predicates.

The Dialetheic Status of Staticism

We are left in a state of tension: staticism is the case and is not the case. Immediately, most will argue that this is only superficially so: staticism might be, for example, false from a third-person, scientific point of view, and yet psychologically true from a first-person perspective. Think of *Human, All Too Human* where Nietzsche argues that although water has certain chemical properties, this is hardly what concerns the sailor in distress (HA I 9). Again others might say that staticism is indeed false *tout court*, that there are only fields composed of whatever 'ultimates' are assumed by our best scientific theories, and that the world as it is to us is merely epiphenomenal and in theory reducible to the best description our physics has to offer. But Nietzsche is clear that both worlds are (1) the same world, and (2) of equal importance; they are to be taken seriously *both* together *and* in opposition to each other. Neither ought to assume exclusively priority. A mutually exclusive opposition 'is after all only the contradiction [*Gegen-*

satz] typical for human beings' (Nachlaß Spring–Autumn 1881, KSA 9, 11[281]).

It is exactly this Heraclitean, deconstructive (if properly understood, see, e.g., Wheeler III 2000; Nancy 2000; Gemes 2001; Waldenfels 2002), adualistic-dialetheic tension that requires a rationality that can do more than simply either abandon staticism or subscribe to it fully in a quasi-staticist way. As is well-known, Graham Priest has, for example, argued for a logic opposed to the law of non-contradiction (see Priest 1995). In order to get into a different (logical) 'frame of mind', in order to arrange one's beliefs in a different, appropriate logical field (something which is *not always* appropriate), Nietzsche sometimes uses a certain dialetheic technique which allows him to express, I wish to argue, perspectival asymmetries as well as perspectival simultaneity of the above kind, between a first person 'experiential' perspective and third person 'descriptive' perspective. Both valid and necessary: any attempt to reduce the matter at hand to either the one or the other or to a mediating third term is conceived as unacceptably reductive. Brian Leiter's enlightening work on Nietzsche's critique of free will (Knobe/Leiter 2007; Leiter 2007) credits Nietzsche with the polemic reversal of the Cartesian error of granting absolute priority to the first-person perspective. Like most work that has been done on (not only Nietzsche's) philosophy of mind, Leiter shies away from trying to provide the more complex theoretical framework needed for an interpretation of consciousness and agency (not only in Nietzsche). Such a framework is already emerging within contemporary philosophy of mind though perhaps not yet in cognitive science and empirical psychology (see, e.g., McGinn 2004, Gray 2004, Abel 2004, Rockwell 2005, Freeman/Strawson 2006, Thompson 2007, and Cosmelli et al. 2007). Nietzsche's tensional asymmetries are by no means trivial. They cannot be done away with, Nietzsche held, precisely because reality is not simple but is perspectival, *con*stituted by these oppositions (see, e.g., Hatab 2008, p. 149; Reginster 2007).

Let us return to the explanatory step (4) one more time: it is certainly true that at times Nietzsche argues from an evolutionary point of view. The application of the law of non-contradiction, for example as an incorporated regulative rule, had a certain adaptive value. Existence demanded a complex matrix of choices within an environment that is itself not simply 'given' (as we know today also from Sellars and Quine) but also in part constituted and altered by practical and theoretical choices. Had the evolving organism failed to make a great number of either-or choices, it might simply have vanished. Whereas every regulative, seemingly constant 'fiction' or, better, 'practical belief' has its history (to be studied genealogically) and its time (when it is first selected), subject to changing conditions, some regulative fictions and 'habituated practices' might no longer stand the test of time. This 'test of time' will have to be properly examined and will amount to a test based on criteria such as practice or belief x is or is no longer necessary if it is either 'for' or 'against' life. And we might indeed be able to abandon some beliefs and replace them with ones better suited to our current form of life and its requirements.

I don't see any textual evidence that Nietzsche ever envisaged human beings as fit to abandon the staticist worldview. Again, as an analogy one might think again of the concept 'mental state' as it figures (i) within neuroscience and (ii) as a first-person, qualitative experience, i.e. the description of the C-fibres firing in the brain and the pain I am experiencing. We do not believe that all that is going on is the *quale*; rather we believe that there is a wealth of neurobiological, unconscious processes that we can even make visible. Despite all this, it will remain necessary to retain the qualitative, first-person state. We need, for example, a 'nondualist' (Rockwell 2005) or 'equal-status fundamental-duality monistic' framework (Strawson 2006, p. 241) that allows us to acknowledge that both descriptions somehow refer to, aim to describe the same 'mind-brain-world state' from a different perspective. Nietzsche believes, and I think he is right, that bringing such perspectives or interpretations or language games together within an adualistic-dialetheic framework does not leave things simply as they are (not in matters of the mind, knowledge, (meta)physics or politics). These perspectives enter into a relationship that will from now on change reality. We have good grounds to assume that consciousness as a qualitative, experiential state *is* also a neurobiological event, and yet the neurobiological event must account for much more than C-fibres in a state of electric excitation, namely their phenomenological 'experiential' features. The adualistic-dialetheic framework allows one to describe the necessary tension opened up in a field structured around combining the unity-asserting *both-and* and the difference-preserving *neither-nor*. Staticism might both 'be the case' (from and for a first person perspective) and 'not be the case' (from and for a third person perspective) and, yet, is reducible to neither one or the other. This has all been said before, in both anglo-analytic and continental traditions, but whenever something goes against beliefs held deeply or practices carried out mainly unconsciously, it is necessary to repeat it, rephrasing it continually until it finally sinks in.

This may perhaps be seen as a step too far. Staticism is false, as we saw following Nietzsche's argument, as it leads to nihilism. Nietzsche was much better at criticizing false views than at constructing theories. His focus on history and his rehabilitation of time is first and foremost concerned—and so are the fourteen essays of *Nietzsche on Time and History*—with the proof that staticism about persons, objects, entities such as nations, the law, truth, or the linear future of time and history itself is false. And yet, most contributions reach a point when a different conception is called for. Rather than summarizing the essays of *Nietzsche on Time and History* I will simply point towards these points of transition.

Nietzsche on Time and History

Nietzsche on Time and History falls in five parts: 'Time, History, Method'; 'Genealogy, Time, Becoming'; 'Eternal Recurrence, Meaning, Agency'; 'Nietzsche's Contemporaries'; and 'Tragic and Musical Time'.

Part one opens with an essay by Andrea Orsucci on 'Nietzsche's Cultural Criticism and his Historical Methodology'. Orsucci examines Nietzsche's treatments of ancient Greek civilization and primitive Christianity and traces Nietzsche's claims to his readings of, and critical engagements with, contemporary texts. It is the historical phenomena themselves that, according to Orsucci, Nietzsche's methods reveal as consisting of a complex simultaneity of temporal and historical layers, 'consistently concerned to identify and theorize the coexistence and mixing of very different traditions, cultures, and ways of thinking in any particular historical phenomenon' (Orsucci 2008, p. 12).

In 'Thucydides, Nietzsche, and Williams', Raymond Geuss analyses Nietzsche's preference for Thucydides over Plato. The reasons for Nietzsche's non-traditional preference are, first, that Thucydides portrayed human beings and their motivations in a non-moralizing way, and second, that he was opposed to the rationalistic, Platonic optimism symptomatic of two millennia of systematic philosophy. Between poetry on the one hand, and philosophy on the other, the bi-partite structure Nietzssche follows in the *Birth of Tragedy*, Thucydidean *Wissenschaftlichkeit* [scientific-mindedness]—'radically non-mythic, non-theological, and non-literary'—appears as a third possibility that informs Nietzsche's own interests not in the past per se but in dissecting 'those forms of collective human behaviour that are recurrent and thus comprehensible' (Geuss 2008, p. 43). In the late 1870s and early 1880s Nietzsche's notebooks indicate the importance he attributed to the 'strand of realist and empiricist thinking that Thucydides represents, and of seeing the demise of tragedy and of Thucydidean "enquiry" synoptically' (ibid., p. 46). It is the rejection of both optimism and pessimism, against the mutually exclusive alternatives 'to think either that these items [rationality, individual happiness, natural human development, socially desirable action] are set up so as to cohere, or that they are 'by nature' ineluctably fated to conflict in an unresolvable way—the refusal to be either an old-style philosophical optimist or a dogmatic pessimist'—that

Geuss finds at the heart of Nietzsche's fascination with Thucydides and Nietzsche's idea of the music-making Socrates: 'when Nietzsche wrote that "the Hellene was neither an optimist nor a pessimist" (Nachlaß Winter 1869/70–Spring 1870, KSA 7, 3[62]), this is what I assume he meant, and no Hellene could illustrate this more exactly than Thucydides' (ibid., p. 48).

Thomas Brobjer's 'The Late Nietzsche's Fundamental Critique of Historical Scholarship' focuses on the third essay of the *Genealogy* in which Nietzsche explicitly attacks the value of purely historical scholarship. Brobjer argues that the late Nietzsche's main objection to history as a science was not methodological but rather 'that history was placed above philosophy—that history and historical scholarship were seen as a goal or an end in itself rather than as a means' (Brobjer 2008, p. 52). Only once historical scholarship lives up to the demand for the philosophical creation of values does it find its proper justification.

Part twp of *Nietzsche on Time and History* opens with Tinneke Beeckman's essay on 'Nietzsche's Timely Genealogy: An Exercise in Anti-Reductionist Naturalism'. Beeckman revisits the link between Nietzsche's genealogical method and his Lamarck and Darwin inspired naturalism. The reactive, associated by Beeckman with Nietzsche's Darwin, and the active, associated with Lamarck, need to be considered side by side in order to appreciate Nietzsche's non-reductive naturalism: 'Adaptation is not active, but reactive. Nietzsche emphasizes Spencer's fatal mistake: to see life itself as an inner adaptation to external circumstances' (Beeckman 2008, p. 72).

According to Kevin Hill's 'From Kantian Temporality to Nietzschen Naturalism' it is central to any understanding of Nietzsche's view of time that Nietzsche struggled precisely with the idea that 'space and time … are mind-dependent in the sense that Kant and Schopenhauer intended, while also maintaining that the mind is itself something that occurs within nature, as Schopenhauer had maintained' (Hill 2008, p. 75). While the early Nietzsche had tried to resolve this matter by attributing time and space to a primordial intellect that 'produces space and time and by that produces the brain' (ibid., p. 76), the later Nietzsche arrives at a notion of naturalism that 'reinterprets things as complexes of power relations in which observers are always involved; he does not reduce things to sums of episodes within subjects' (ibid., p. 84).

John Richardson's examination of 'Nietzsche's Problem of the Past' sets itself the task of resolving the tension that lies in the fact that for Nietzsche the past is far too important to be ignored, but attention to it turns out to be harmful. The past is important for the simple reason that 'what one *is*' is what one has been selected (in an evolutionary sense) to be. The past 'has a kind of "presence" in us, constituting us now as who we are, determining the meaning of what we now do' (Richardson 2008, p. 91). Central to our understanding of the presence of the past in the present is Richardson's understanding of power wills that have been selected and structure us who 'express the aims of these wills, which carry their intentions ahead into us' (ibid., p. 91). Nietzsche's genealogical method is therefore a technique to become aware of the proto-intentional 'wills', to expose the social formation of values, and in a retrospective stance to bring into view 'the forces that really aimed the rules and values to which I commit myself' (ibid., p. 107). By 'cutting-into' our lives of desiring, willing, valuing etc. we are always in danger of falling into an alienating form of nihilism, and yet, Richardson argues that genealogy enables us to 'judge those designed-in purposes of our ways of thinking and acting—and decide whether we favour those purposes' (ibid., p. 108).

In the final paper of part two, 'Towards Adualism: Becoming and Nihilism in Nietzsche's Philosophy', I examine the relationship that holds between the concepts of 'being' and 'becoming' in Nietzsche's philosophy. I argue that Nietzsche's emphasis on 'becoming' is motivated by the anomaly of nihilism that is best explained as 'a function of the belief in being' (Dries 2008, p. 114). Nietzsche's philosophical agenda, his attempt to provide a 'counter-force' to nihilism, should be regarded as the reason for the initial, seemingly *radical* nature of his affirmation of becoming, which at first sight reintroduces a dualism between becoming and language, recapitulating the nihilism it had aimed to circumvent (ibid., p. 120). I argue that Nietzsche's ontology of becoming as will-to-power relations should be seen instead as a *less* radical presentation of becoming. Aiming at a non-reductive, adualistic practice of thought, he accounts for both the relative permanence of 'relations', 'entities' and 'objects' and their constantly changing, temporal complexity.

Part three of *Nietzsche on Time and History* is concerned with Nietzsche's attempt to describe the temporal disposition of the world as eternal recurrence and what this demands of the human being.

In 'Shocking Time: Reading Eternal Recurrence Literally' Lawrence Hatab argues that although Nietzsche did not present eternal recurrence as a cosmological theory or a scientific fact, it nevertheless must be taken literally, that 'a certain extra-psychological literality would better fit the world-disclosive and "revelatory" spirit of Nietzsche's accounts of eternal recurrence' (Hatab 2008, p. 148). In order to deal with the question of meaning that has hitherto blocked, in Nietzsche's view, the possibility of affirming life in its finite, temporal disposition as will to power, eternal recurrence emerges as 'Nietzsche's formula for "redemption" of time and becoming' (ibid., p. 150). Against nihilistic alternative models of time—

Hatab identifies six: *positivistic, salvational, teleological, cyclical, pessimistic,* and *novelistic*— 'eternal recurrence comes forth as the only conceivable temporal model that does not fall prey to a fugitive gaze *away* from life as lived' (ibid., p. 154).

Paul S. Loeb approaches Nietzsche's eternal recurrence through an examination of Camus's Sisyphus and suicide. According to Loeb, standard readings of eternal recurrence, Nietzsche's counter-ideal to the ascetic ideal, tend to emphasize the doctrine's supposed ability to bring about a reversal from the ascetic to the affirmative. Loeb argues that standard interpretations (such as Nehamas) overlook that on Nietzsche's own premises, affirmation, for example by reinterpreting one's past in view of one's present state, turns out to be impossible as it, in Loeb's view, leads to a falsification of the past. 'But the thought of eternal recurrence closes off all such escape and condemns the human animal to eternal meaninglessness' (Loeb 2008, p. 179). Nietzsche offers therefore a disconcerting counter ideal that will force 'the decadents *give in* to their dominant suicidal instincts' (ibid., p. 176), and only by 'overcoming' themselves come closest to affirming life. Eternal recurrence 'must *oppose* the ascetic ideal's ability to *block* the suicidal nihilism of degenerating life' (ibid.). It might be asked if Loeb relies on a notion of selfhood more static than is warranted. Based on an exclusive disjunction between affirmation and asceticism that demands, in Loeb's view, the voluntary suicide of the decadent, he argues that true life affirmation requires a cosmological understanding of eternal recurrence, a truly superhuman 'backward willing'. The latter ideas rely on Loeb's previous writings (see ibid., p. 182) to which I cannot do justice in this introduction.

The last contribution of part three, Herman W. Siemens' 'Nietzsche and the Temporality of (Self-)Legislation' deals with a fundamental problem: how does one reconcile the need for a stable legislation that stands 'in radical contradiction with the pluralism and dynamism of life-as-becoming' (Siemens 2008, p. 189). Siemens interprets first Nietzsche's conception of self-legislation in 'Schopenhauer as Educator' as a specific form of moral particularism coupled with an Emersonian notion of moral perfectionism. Schopenhauer's metaphysical solution is perceived as inadequate by Nietzsche as is Wagner's attempted artistic unification. Both Schopenhauer and Wagner, Siemens argues, fail 'the test of pluralism required for a life-affirming form of legislation' (ibid., p. 201). Siemens sets out to show that a different, agonistic and pluralistic and yet communal conception of self-legislation is to be found in the unpublished notes of the Zarathustra period. Here, the law should not only be understood as always only provisional and yet 'responsive to diversity, a law *for* many, not a law that subjects the many to One' (ibid., p. 202), it also requires us to combine an individual morality that 'cannot … be achieved in isolation' with a morality which is 'inseparable from the task of founding the kind of ethical community that makes it possible' (ibid., p. 207).

The essays of Anthony K. Jensen and Martin A. Ruehl provide a detailed account of Nietzsche's relationship with contemporary philology, on the one hand, and with Walter Burkhardt and the Renaissance on the other.

Jensen's '*Geschichte* or *Historie*? Nietzsche's Second *Untimely Meditation* in the Context of Nineteenth-Century Philological Studies' examines the polarization in classical studies or *Altertumswissenschaft* into '*Wort-Philologie*', 'approaching antiquity with the tools of textual emendation, codices, and literary criticism', and 'Sach-Philologie', often labelled as '"hermeneutical", "antiquarian", or "humanistic"' philology (Jensen 2008, p. 213). At first a follower of Ritchl who had tried to bridge both camps, Nietzsche would 'reject both traditions on the way to positing a third way of his own' (ibid., p. 216). Jensen's analysis enables him to throw new light on Wilamowitz's rejection of *The Birth of Tragedy*. Nietzsche aims to reveal the opposition between both 'scholastic factions' and life, depicts them 'as psychological types' rather than as scholars with different methodological preferences (ibid., p. 219). Nietzsche's own, monumental historical approach is modelled not on contemporary classicists but instead, among others, on Goethe whose character combines 'the healthiest aspects of antiquity for the sake of reinvigorating culture' (ibid., p. 224).

In 'An Uncanny Re-Awakening': Nietzsche's Renascence of the Renaissance out of the Spirit of Jacob Burckhardt' Martin A. Ruehl argues that with the exception of Greek antiquity no historical epoch fascinated Nietzsche more than the Italian Renaissance. In the 1870s, his study of the Renaissance, 'as a historical reference point and cultural ideal … allowed him to question a set of values and notions that had determined his early thought' and 'became a crystallization point, especially in the 1880s, for Nietzsche's most radically anti-humanist, anti-liberal ideas about tyranny and individuality, war and culture, violence and health' (Ruehl 2008, p. 229). Ruehl first discusses Burckhardt's portrayal of the *Civilization of the Renaissance* before tracing the various other sources of Nietzsche's 'Renaissancebild' and his selective appropriation of these sources against Wagner and Luther. It was the culture of the quattrocento, Ruehl argues, that furnished Nietzsche with an answer to Schopenhauer's pessimistic 'philosophical deconstruction of the *principium individuationis* and led him to rethink the significance of individual agency in history' (ibid., pp. 243–244) and ultimately led to Nietzsche's belief that only a few select superior human beings could bring about a cultural renewal. According to Ruehl's reading, in contrast to Burckhardt, Nietzsche focuses exclusively on the aristocratic elements of the Renaissance. Burckhardt had 'allowed for the growth of "individuality" and cultural productivity' (ibid., pp. 250–251), a republican alternative Nietzsche chose to ignore. Nietzsche's Renaissance-inspired individualism stands of course side by side with his ideal of an agonistic community.

The final two essays take a close look at Nietzsche's *Birth of Tragedy* and the importance of music for Nietzsche's views on time. Against influential interpretations of the *Birth of Tragedy* by among others Nehamas, de Man and Porter, Katherine Harloe argues for 'the positive character of its appropriation of Schopenhauer and Wagner', challenging the idea that it should be read primarily as a contribution to the major debate of post-Kantian German philosophy, namely 'that of the possibility of metaphysics' (Harloe 2008, p. 271). Harloe revisits the, in her view, simplified reading that Nietzsche's Dionysus–Apollo distinction mirrored Schopenhauer's own metaphysical distinctions and argues that it 'rests upon an oversimplification of what "Schopenhauer" could have represented for Nietzsche at the time' (ibid., p. 272). In fact, Nietzsche uses and relies heavily on Schopenhauer in his attack on Socratic optimism. Key passages often cited as a radical critique of Schopenhauer stem in fact from Schopenhauer himself. This, Harloe remarks, 'raises the possibility that *The Birth of Tragedy* deploys Schopenhauer not in parodic fashion ... to shatter all such illusions, but rather as a means of developing them in a new and superior form' (ibid., p. 281). Nietzsche can be shown to construct a historical narrative of the crisis of science and 'casts Schopenhauer in a leading role' (ibid., p. 282)

Finally, Jonathan R. Cohen analyses the importance of music for 'Nietzsche's Musical Conception of Time'. In a close reading of Nietzsche's critique of Wagnerian endless melody Cohen shows that Nietzsche promotes both loss of an essential notion of the self and yet maintains that 'structure is necessary for a flourishing and creative life' (Cohen 2008, p. 291). Hollingdale's translation obfuscates that Nietzsche's critique of Wagner is not based on Wagner's choice of irregular time measures but that Nietzsche asks about its *effect* and '*makes* endless melody be about rhythm, and thus by the same token about *time*' (ibid., p. 292). He criticizes that Wagner's melodies '"overflow" their measures' (ibid., p. 296) and that this leads to the loss of structure on the part of the listener and 'overrides the listener's own internal sense of structure' (ibid., p. 297). A larger issue emerges with regard to Nietzsche's conception of time: in the same way that Nietzsche rejects the idea of a thing-in-itself in favour of the world as it is experienced, he takes not the external metronome but rather 'takes the perspective of *the listener*' (ibid.) as the final measure of musical time. Cohen concludes that Nietzsche's emphasis on time as it is experienced corresponds to Nietzsche's insistence that each subject has its own internal rhythm and temporality, derived from, among other things, 'our internal physiological rhythms' (ibid., p. 299). The criterion for evaluating music then becomes its effect on us: 'it can help structure our internal rate of time—either directly or by providing a contrasting rhythm to serve as a beneficial tonic—or it can harm it ... And with no time-in-itself to fall back on, such undermining can be utterly destructive. It requires great strength to resist it and maintain one's own tempo' (ibid., p. 300).

References

Abel, Günter, 2004, Zeichen der Wirklichkeit, Frankfurt a. M.: Suhrkamp.

Beeckman, Tinneke, 2008, 'Nietzsche's Timely Genealogy: An Exercise in Anti-Reductionist Naturalism', in: Manuel Dries (ed.), Nietzsche on Time and History, Berlin: Walter de Gruyter, pp. 63–744.

Brobjer, Thomas, H., 'The Late Nietzsche's Fundamental Critique of Historical Scholarship', in: Manuel Dries (ed.), Nietzsche on Time and History, Berlin: Walter de Gruyter, pp. 51–60.

Cohen, Jonathan R., 2008, 'Nietzsche's Musical Conception of Time', in: Manuel Dries (ed.), Nietzsche on Time and History, Berlin: Walter de Gruyter, pp. 287–303.

Cosmelli, Diego / Jean-Philippe Lachaux / Evan Thompson, 2007, 'Neurodynamical Approaches to Consciousness', in: The Cambridge Handbook of Consciousness, ed. P.D. Zelazo, M. Moscovitch, E. Thompson, Cambridge: Cambridge University Press, pp. 731–772.

Dries, Manuel, 2008, 'Towards Adualism: Becoming and Nihilism in Nietzsche's Philosophy', in: Manuel Dries (ed.), Nietzsche on Time and History, Berlin: Walter de Gruyter, pp. 113–144.

Freeman, Anthony / Galen Strawson, et al., 2006, Consciousness and its Place in Nature. Does Physicalism Entail Panpsychism? Charlotteville, VA: Imprint Academic.

Gemes, Ken, 2001, 'Postmodernism's Use and Abuse of Nietzsche', in: Philosophy and Phenomenological Research, LXII/2, pp. 337–360.

Geuss, Raymond, 2008, 'Thucydides, Nietzsche, and Williams', in: Manuel Dries (ed.), Nietzsche on Time and History, Berlin: Walter de Gruyter, pp. 35–50.

Grey, Jeffrey, 2004, Consciousness. Creeping Up On the Hard Problem, Oxford: Oxford University Press.

Harloe, Katherine, 2008, 'Metaphysical and Historical Claims in *The Birth of Tragedy*', in: Manuel Dries (ed.), Nietzsche on Time and History, Berlin: Walter de Gruyter, pp. 271–286.

Hatab, Lawrence J., 2008, 'Shocking Time: Reading Eternal Recurrence Literally', in: Manuel Dries (ed.), Nietzsche on Time and History, Berlin: Walter de Gruyter, pp. 147–160.

Hegel, G. W. F., 1969, Wissenschaft der Logik, 2 vols., Frankfurt: Suhrkamp.

Hill, R. Kevin, 2008, 'From Kantian Temporality to Nietzschean Naturalism', in: Manuel Dries (ed.), Nietzsche on Time and History, Berlin: Walter de Gruyter, pp. 75–86.

Jensen, Anthony K., 2008, '*Geschichte* or *Historie*? Nietzsche's Second *Untimely Meditation* in the Context of Nineteenth-Century Philological Studies', in: Manuel Dries (ed.), Nietzsche on Time and History, Berlin: Walter de Gruyter, pp. 211–226.

Knobe, Joshua / Leiter, Brian, 2007, 'The Case for Nietzschean Moral Psychology', in: Leiter, Brian / Sinhababu, Neil (eds.), Nietzsche and Morality, Oxford: Oxford University Press, pp. 83–109.
Leiter, Brian, 2007a, 'Nietzsche's Theory of the Will', Philosopher's Imprint 7/7, pp. 1–15.
Loeb, Paul S., 'Suicide, Meaning, and Redemption', in: Manuel Dries (ed.), Nietzsche on Time and History, Berlin: Walter de Gruyter, pp. 161–188.
McGinn, Colin, 2004, Consciousnesss and Its Objects, Oxford: Clarendon Press.
Nancy, Jean-Luc, 2000, Being Singular Plural, Stanford: Stanford University Press.
Orsucci, Andrea, 2008, 'Nietzsche's Cultural Criticism and his Historical Methodology', in: Manuel Dries (ed.), Nietzsche on Time and History, Berlin: Walter de Gruyter, pp. 23–33.
Priest, Graham, 1995, Beyond the Limits of Thought, Cambridge: Cambridge University Press.
Reginster, Bernard, 2007, 'The Will to Power and the Ethics of Creativity', in: Leiter, Brian / Sinhababu, Neil (eds.), Nietzsche and Morality, Oxford: Oxford University Press, pp. 32–56.
Richardson, John, 2008, 'Nietzsche's Problem of the Past', in: Manuel Dries (ed.), Nietzsche on Time and History, Berlin: Walter de Gruyter, pp. 87–111.
Rockwell, W. Teed, 2005, Neither Brain Nor Ghost. A Nondualist Alternative to the Mind–Brain Identity Theory, Cambridge, MA: MIT Press.
Ruehl, Martin A., 2008, '"An Uncanny Re-Awakening": Nietzsche's Renascence of the Renaissance out of the Spirit of Jacob Burckhardt', in: Manuel Dries (ed.), Nietzsche on Time and History, Berlin: Walter de Gruyter, pp. 227–267.
Schopenhauer, Arthur, 1969, The World as Will and Representation, trans. E. F. J. Payne, 2 vols., New York: Dover.
Schopenhauer, Arthur, 1974, Parerga and Paralipomena: Short Philosophical Essays, trans. E. F. J. Payne, 2 vols., Oxford: Clarendon Press.
Siemens, Herman W., 2008, 'Nietzsche and the Temporality of (Self-)Legislation', in: Manuel Dries (ed.), Nietzsche on Time and History, Berlin: Walter de Gruyter, pp. 189–208.
Strawson, Galen, 2006, 'Panpsychism. Reply to Commentators with a Celebration of Descartes', in: Freeman, Anthony / Galen Strawson, et al., 2006, Consciousness and its Place in Nature. Does Physicalism Entail Panpsychism? Charlotteville, VA: Imprint Academic, pp. 184–280.
Thompson, Evan, 2007, Mind in Life: Biology, Phenomenology, and the Sciences of Mind, Cambridge, MA: Harvard University Press.
Wheeler III, Samuel C., 2000, Deconstruction as Analytic Philosophy, Stanford: Stanford University Press.
Waldenfels, Bernhard, 2002, Bruchlinien der Erfahrung, Frankfurt a. M.: Suhrkamp.

Part I

Time, History, Method

Nietzsche's Cultural Criticism and his Historical Methodology

Andrea Orsucci[*]

Friedrich Nietzsche adopts a notably 'untimely' approach to historical questions. That is, by considering history to reveal the heterogeneous and layered nature of any given historical phenomenon, its origins in exchanges, interweavings, and graftings between different cultural heritages, Nietzsche differs fundamentally from his contemporaries, and anticipates important trends in later historical thought.

The philologist Hermann Usener writes in his *Götternamen* of 1896 that apparently distinct civilizations are in fact separated by a 'boundary as thin as a knife-blade' (Usener 1929, p. 317). By thus insisting on the complex interweavings between different traditions and cultures, Usener reflects a trend in historical thinking which prevailed from the 1890s onwards among philologists, classicists, and theologians. My thesis, then, is that Nietzsche had already adopted a similar approach in the 1870s and 1880s.

Furthermore, Nietzsche's 'untimely' insistence on the heterogeneity of historical phenomena reflects his acquaintance and critical engagement with an enormously rich range of scholarly works in history, anthropology, and biology. Indeed, the complex origins and development of Nietzsche's own accounts of Western history and his general philosophical commitments regarding historical phenomena offer important insights into his philosophical practice in general. For they reveal that, rather than occupying himself merely with the timeless questions of traditional philosophy, Nietzsche is concerned both with concrete historical questions and with drawing substantial philosophical conclusions from his studies of them. Attending to the origins and development of these studies therefore serves to illuminate not only how Nietzsche arrived at such an 'untimely' approach to historical questions, but also its place in his broader philosophical practice, commitments, and conclusions.

---

* This essay has been translated from Italian by Tom Bailey.

In this essay, I will attempt to substantiate these claims with particular reference to certain revealing passages of Nietzsche's texts and to certain important examples of his readings. I will begin with his treatments of two particular historical phenomena, ancient Greek civilization and primitive Christianity, before proceeding to consider his broader pronouncements regarding the understanding of historical phenomena. In both cases, I will also attempt to trace Nietzsche's claims to his readings of, and critical engagements with, contemporary texts. Finally, I will consider Nietzsche's place in the historical thought of the second half of the nineteenth century and the first half of the twentieth century, in order to demonstrate his anticipation of important trends in later historical thought.

In many notes made between 1875 and 1878 and in various sections of *Human, All Too Human*, *Assorted Opinions and Maxims*, and *The Wanderer and his Shadow*, Nietzsche insists on the superiority of ancient Greek civilization on the grounds of its exceptional capacity to gather together heterogeneous elements appropriated from previous cultures. Indeed, in one section of *Assorted Opinions and Maxims* he even writes that, '[n]ot to create forms, but to borrow them from abroad and transform them into the fairest appearance of beauty — that is Greek' (AOM 221). In a typical note, he elaborates as follows.

> Natives of the Greek land: of Mongolian origin with [a] tree- and snake-cult. On the coast a marginalized Semitic strip. Here and there Thracians. The Greeks took into their blood all these elements, including all the Gods and myths ([there is] many a Mongolian in the fable of Odysseus). The Doric migration is a later event, after everything had already earlier been gradually submerged. (Nachlaß Spring–Summer 1875, KSA 8, 5[198])[1]

For Nietzsche, then, the ancient Greeks were 'joyous dilettantes' (Nachlaß Spring–Summer 1875, KSA 8, 5[65]), distinguished by their capacity to live with contradictions and dissonances inherited from previous cultures, rather than the examples of 'clarity, transparency, simplicity, and orderliness', or 'crystalline naturalness' (AOM 219), which were so often celebrated by the 'humanists'.[2]

In *Human, All Too Human* and his 1875 lecture course, 'Der Gottesdienst der Griechen', Nietzsche accounts for this exceptional combinatory capacity of the Greeks in terms of their intimacy with ancient beliefs, magic, and other forms of 'impure thinking', and their consequent ability

1 Regarding this note, see also Orsucci 1996, pp. 109–116. The claim regarding the Greeks' Mongolian origins derives from Nietzsche's reading of J. W. Draper, *Geschichte der geistigen Entwicklung Europas* (1871, p. 24).

2 See also Orsucci 1996, pp. 8ff.

to make concessions 'to the evil and suspicious, to the animal and backward, likewise to the barbarian, the pre-Greek and Asiatic, which still lived on in the foundations of the Hellenic nature' (AOM 220). In particular, in a lengthy section of *Human, All Too Human*, Nietzsche accounts for primitive religious rituals as superstitious attempts to 'impose a law on nature', to control a hostile and unpredictable natural environment. For, Nietzsche maintains, primitive man considered natural events to be the voluntary actions of embodied 'spirits', and therefore to be subject to influence by rituals directed towards these 'spirits' (HA I 111).[3] Nietzsche thus exploits his readings in the then-emerging field of ethnology, and, in particular, his close readings from 1875 onwards of works by contemporaries such as Edward B. Tylor, John Lubbock, and Wilhelm Mannhardt. For instance, Nietzsche found in Mannhardt's *Der Baumkultus der Germanen und ihrer Nachbarstämme* extensive accounts of the primitive belief that plants embody spirits, and of the rituals with which these spirits were encouraged to protect a community's crops—such as processions or burials of twigs or shrubs, and the conservation of the last sheaf of the harvest.[4]

Just as Nietzsche insists on the heterogeneous and combinatory nature of ancient Greek civilization in his earlier texts, in his later texts he makes corresponding claims about primitive Christianity and its relationship to the then-waning Greco-Roman civilization. In *Daybreak*, for instance, he writes that '[t]he Christian Church is an encyclopaedia of prehistoric cults and conceptions of the most diverse origin', with an exceptional 'power of causing the most various elements to coalesce' (D 70; see also AOM). As examples, he refers to primitive Christianity's appropriation of the pagan notion of punishment in the afterlife, and of the pagan proscription on suicide (D 72 and GS 131). Once again, these claims derive from Nietzsche's extensive readings—in this case, they can be traced to certain passages of the historian and moral philosopher W. H. Lecky's *History of European Morals from Augustus to Charlemagne*, which Nietzsche read in German translation (1879, vol.1, pp. 183–186, 336ff.).

Nietzsche returns to primitive Christianity's heterogeneous pagan origins in his late texts and notes, this time often reflecting his careful reading of the Ernest Renan's *Histoire des origines du christianisme*, and particu-

3 HA I 111 states particularly clearly the ideas elaborated by Nietzsche in his 1875 lecture course, 'Der Gottesdienst der Griechen' (KGW II.4, pp. 355–520).

4 Mannhardt writes, for instance, of the obscure idea of 'a spiritual being, a demon, whose life is bound to the life of the plant. The demon is born with the plant, and dies with it; in the plant he has his habitat, … his body' (Mannhardt 1875, p. 4; see also p. 609).

larly its seventh volume, *Marc-Aurèle et la fin du monde antique.*[5] In *The Antichrist*, for instance, Nietzsche describes how Christianity 'absorbed the doctrines and rites of every *subterranean* cult of the *Imperium Romanum*' (A 37). In a note written at the turn of 1888, Nietzsche similarly claims that 'Christian doctrine changed its emphasis continually', such that 'the "Christian" type gradually re-accepted everything which he had originally denied' and 'there ruled over Christianity: Judaism (Paul)[,] Platonism (Augustine)[,] the mystery cults (theory of redemption, symbol of the "cross")[,] asceticism (—hostility to 'nature', 'reason', 'the senses', — [the] Orient ...)' (Nachlaß November 1887–March 1888, KSA 13, 11[364]). In a passage of *On the Genealogy of Morality*, he even insists that primitive Christianity inherited a 'bucolic cloyingness' from the Greeks (GM III 22). It is unsurprising, then, that in *The Antichrist* Nietzsche insists that the complexity of pagan inheritances on primitive Christianity is such that it should not be considered a unitary phenomenon: 'the word "Christianity" is already a misunderstanding' (A 39), he writes. There and in contemporaneous notes he also attributes to Paul a crucial role in these appropriations from pagan culture by primitive Christianity— he writes, for instance, of 'Christianity as the formula for outbidding all the subterranean cults, those of Osiris, of the Great Mother, of Mithras for example— and for summing them up: it is in this insight that the genius of Paul consists' (A 58).[6]

As Nietzsche emphasizes in the third essay of the *Genealogy*, he also considers another particularly important root of primitive Christianity to lie in a certain sense of 'community' which then prevailed among excluded social groups. Thus he writes of the 'associations for mutual support, pauper-, invalid-, burial-associations, which sprung up on the undermost soil of the society of that time, and in which that principal medicine against depression, the small joy, that of mutual good deeds was consciously cultivated'. This particular '"will to mutuality", to herd-formation, to "community", to "cenacle"', Nietzsche insists, is the means to 'an arousal of the

5 Compare, e.g., Renan 1866, pp. 328, 338–339; 1869, pp. 202–206; 1877, p. 385; 1882, pp. 574–579.

6 See also GS 358, BGE 52, and Nachlaß Autumn 1887, KSA 12, 10[92 and 96], and, for a more extensive treatment, Orsucci 1996, pp. 281–317. Regarding Paul in particular, Nietzsche similarly claims in his notebooks that Paul translated Christ's teaching into 'the language of all the already existing subterranean religions', and thus made it able to express the new religious needs that emerged with the decline of Greco-Roman civilization, such as 'asceticism, world-denial, superstitious "purification"' (Nachlaß November 1887–March 1888, KSA 13, 11[294, 295]); see also Nachlaß November 1887–March 1888, KSA 13, 11[282]).

strongest, most life-affirming drive, even if in the most cautious of doses—the will to power' (GM III 18).[7] Thus, for Nietzsche, the rise of primitive Christianity owed much to this particular sense of 'community' and its arousing of a certain 'joy', or feeling of 'power', among excluded social groups. Furthermore, in this Nietzsche again appropriates from his readings: in particular, from Lecky on the suicidal tendencies common at the time of primitive Christianity's development, and from Renan on the anxieties which underpinned the pagan mysteries and on mutual support in small, excluded Jewish communities of the same period. Renan, for instance, writes in surprisingly 'Nietzschean' terms of the spread of new ideals of community in the Greco-Roman world, as ideals which give 'that impression of damp moulds, that murmur of prayers, ... a hot, limp atmosphere, which must have been so sweet for the [community] member' (1866, p. 361).[8]

7 See also GS 131 and 353, and GM I 10 and III 19 and 22. Nietzsche also elaborates on these claims in notes of Autumn 1887. He writes, for instance, of the Jews of the diaspora as a 'small people' who, excluded from society, cultivated a common identity by dedicating themselves to 'everything soothing, relieving, restoring, prayer, music, meals taken together and effusions of feeling, patience, indulgence, mutual support and service, above all the keeping quiet of the soul, so that the affects [of] anger, suspicion, hate, envy, revenge do not emerge ... Asceticism is not the essence of this life' (Nachlaß Autumn 1887, KSA 12, 10[92]; see also Nachlaß Autumn 1887, KSA 12, 10[92, 135, 157, 179, 181, 183, 188, and 191]).

8 Compare Nietzsche's terms in GM III 22 or A 21, for instance. Renan's thesis is that these notions of community were introduced into the Western, pagan world by groups of Syrians and Jews excluded from the prevailing Greco-Roman society. He writes, for instance, that 'The most characteristic aspect of the devoted Jewish life was always that of arousing much gaiety and cordiality. There was love in his small world: one loved a past, a common past; the religious ceremonies embraced life most sweetly' (Renan 1866, pp. 286ff.). For his accounts of these notions, and his use even of the term 'small people' and the metaphor of warmth adopted by Nietzsche, see Renan 1866, pp. 284ff. and 357; 1879, p. 319; 1882, pp. 547, 561–562, 570, and 590. With respect to attitudes towards suicide, on the other hand, Lecky writes, for example, of Ambrose's and Jerome's hesitation in condemning voluntary martyrdom, reflecting pagan attitudes that contrast radically with Augustine's later unqualified condemnation, and of the idea of a 'knowing suicide' to which many fifth-century ascetics and in particular the Circumcelliones aspired (see Lecky 1879, vol. 2, pp. 35–38; and also vol. 1, pp. 183–186, 336ff.). For a more extensive discussion of this, see Orsucci 1996, pp. 281–317. Franz Overbeck considered Nietzsche's identification of a 'will to mutuality' to be a substantial contribution to the understanding of the rise of Christianity, and, notably, to require that less importance be attributed to asceticism in this rise. In this, Overbeck emphasized, Nietzsche anticipated the claims of the theologian Adolph von Harnack, in his *Das Wesen des Christentums* (See Overbeck 1995a, vol. 4, pp. 165–

These treatments of ancient Greek civilization and primitive Christianity, two crucial elements of Western history for Nietzsche, reveal that from the mid-1870s onwards he practised the philosophical commitments regarding the treatment of historical phenomena which he made explicit only much later, in well-known passages of the second essay of the *Genealogy*. There Nietzsche insists on the distinction between the 'origin' and the 'goal' of any particular historical phenomenon, and, indeed, that 'the "development" of a thing, a practice, an organ is … least of all its progressus toward a goal, still less a logical and shortest progressus, reached with the smallest expenditure of energy and cost, —but rather the succession of more or less profound, more or less independent processes of overpowering which play themselves out in it' (GM II 12). Thus, he concludes, any developed historical phenomenon 'no longer represents a single meaning at all but rather an entire synthesis of "meanings" … [which] crystallizes into a kind of unity that is difficult to dissolve, difficult to analyse and … completely and utterly undefinable' (GM II 13). Or, as he puts it more pithily, 'only that which has no history is definable' (ibid.).[9] These well-known passages are therefore intended not simply to underline the differences between Nietzsche and the 'English' historians and philosophers to which he dismissively refers throughout the *Genealogy*. Rather, these passages make explicit the particular manner of treating historical phenomena which Nietzsche had practised since the 1870s, and especially in his studies of the nature and development of ancient Greek civilization and primitive Christianity.[10]

Furthermore, in thus making explicit his particular approach to historical phenomena, Nietzsche appropriates from certain conceptual developments in biological theory which gained ground in Germany after 1870. In particular, in 1886 Nietzsche read with great interest the botanist Karl Wilhelm Nägeli's 1884 text, *Mechanisch-physiologische Theorie der Abstammungslehre*. There Nägeli argues that evolution proceeds not by adapta-

---

167; 1995b vol. 5, p. 415; 1995c vol. 7.2, pp. 50–55, 198; also von Harnack 1985, pp. 74 and 104–106).

9 Notably, in a draft for this passage at GM II 12, Nietzsche refers to these philosophical commitments as 'the greatest triumph over the vis inertiae of the human intellect'.

10 I would suggest that these philosophical commitments are also demonstrated by Nietzsche's treatments of other historical phenomena. One example is his treatment of Buddhism as an appropriation of pre-existing, Brahmanistic notions of redemption in GM III 17, a treatment which reflects his readings of Paul Deussen's *Das System des Vedânta* (1883), and particularly Hermann Oldenberg's *Buddha. Sein Leben, seine Lehre, seine Gemeinde* (1881) (compare Oldenberg 1881, pp. 31, 33, 43, 47, 53–55, and 311).

tion, as Charles Darwin and Ernst Haeckel maintain, but by the cross-generational growth of 'idioplasmatic systems' of hereditary characteristics. Conflicts between the 'filaments' which constitute such complex systems, Nägeli claims, determine the continual evolution of organic forms of life, such that stronger 'filaments' eventually 'overwhelm' weaker ones. Moreover, although Nägeli considers this 'overwhelming' to occur very frequently, he nonetheless also maintains that, as he puts it, 'it takes some time before the tension generated by the formation of a device in the idioplasma becomes strong enough to overcome the resistances to it' (Nägeli 1884, pp. 184–185). When, in the *Genealogy*, Nietzsche chooses to make explicit his particular approach to historical phenomena, then, he exploits Nägeli's conception of biological evolution, so as to claim that, beneath the apparent homogeneity of any particular historical phenomenon, there exist different, inherited elements whose internal power conflicts explain the phenomenon's character and changes, and continue even beneath any surface stability which the phenomenon might display.

It is rarely noted that, by thus insisting on the heterogeneous nature of historical phenomena, Nietzsche anticipates important trends in later historical thought. For instance, in his *Der Historismus und seine Probleme* of 1922, Ernst Troeltsch maintains that Western civilization is distinguished from other civilizations by its 'complexity'—that is, by the heterogeneous and layered nature of its constitutive features. Troeltsch writes, for instance, that, even in the modern era, only European history is marked by the re-emergence of an extraordinary 'wealth ... of entirely different civilizations ... [which are] contained together as nowhere else' and which have 'branched into each other and ... grown together' over an extremely long period of time. European civilization is thus a conglomeration 'constituted by elements belonging to the most varied ... historical worlds', by heterogeneous contributions which 'continuously oppose each other and merge together again in different ways ... colliding with each other and mixing again with new forces and new ideas'. Troeltsch concludes that, in order to do justice to this mosaic, 'general historical-universal schemes' must be set aside. 'In this field,' he writes, 'preconceived formulas are of no use' (1922, pp. 716–719).[11]

11 Indeed, Troeltsch himself describes Nietzsche's historical claims as 'rather free ... indifferent to details', but nonetheless 'extraordinarily acute and penetrating' in their interpretations of 'historical movements through the antagonisms between impotence and herd spirit, on the one hand, and force and nobility, on the other'. Troeltsch further claims that, despite the efforts of 'third-rate editors', Nietzsche's works exercised 'a quite extraordinary influence on the general atmosphere of historical thought and feeling' in the early twentieth century, by effecting 'a devasta-

Also in the 1920s, Oswald Spengler similarly concerned himself with exchanges and interweavings between different cultural heritages. In particular, in his *Der Untergang des Abendlandes*, Spengler presents his research as a kind of 'comparative morphology' which pays attention to what he calls the 'chance and irregular reciprocal relations between civilizations'. Thus he treats such themes as the influence of Parsism on Judaism, the interactions between the pagan mysteries and Christianity, and the diffusion of Asian religions in the West (in a period in which, he writes characteristically, 'Rome had become a part of the East, a religious province of Syria'). Thus Spengler's philosophy of history is primarily concerned with what he terms 'pseudomorphoses', the ways in which the most characteristic features of a declining civilization have, to varying degrees, underground effects on succeeding cultures (Spengler 1995, pp. 524, 598, 784–785).[12]

Moreover, in interpreting Western history as a long series of 'pseudomorphoses' or graftings between different cultures, both Troeltsch and Spengler follow a trend which prevailed in the decades immediately following Nietzsche's collapse among classical philologists—examples include Hermann Usener and Richard Reitzenstein—and orientalists—such as Hermann Gunkel and Wilhelm Bousset. Crucially, however, in this they differ from Nietzsche's contemporaries, that is, from historians and philosophers of history active earlier in the second half of the nineteenth century, who generally paid little attention to interactions between civilizations. For instance, Nietzsche's colleague at the University of Basel, Jacob Burckhardt, interpreted the Renaissance as an extremely homogenous age, and, indeed, once wrote that '[t]he greatest difficulty in the history of civilization is the necessity of splitting up [*zerlegen*] a great spiritual continuum into single categories, which often appear arbitrary, in order to somehow produce a description of it' (Burckhardt 1922, p. 5). As Max Weber later noted, Burckhardt's 'ingenious formulations' soon appeared 'out of date' (1947, p. 95).

In the light of this crucial difference between philosophies and studies of history in the second half of the nineteenth century, on the one hand, and those at the turn of the century and early in the twentieth century, on the other, Nietzsche's own studies of historical phenomena and his broader

tion of value criteria and historical conventions … a growing distrust of the erudite specialisms of history, criticism and philology' (Troeltsch 1922, pp. 503, 506–507; see also pp. 4–5).

12 Spengler's notion of 'pseudomorphoses' was also adopted by Eduard Meyer in *Spenglers Untergang des Abendlandes* (see Meyer 1925, p. 15).

claims regarding historical studies are revealed to be decidedly 'untimely': they bear more similarity to those of Nietzsche's successors than to those of his contemporaries. That is, like Troeltsch and Spengler but unlike contemporaries such as Burckhardt, Nietzsche is consistently concerned to identify and theorize the coexistence and mixing of very different traditions, cultures, and ways of thinking in any particular historical phenomenon. In short, as his appreciative friend, the theologian Franz Overbeck, once put it, Nietzsche displays a 'sensitivity' to the historical 'mingling' of different civilizations which his contemporaries notably lacked (1995a, vol. 4, pp. 160–161; see also pp. 165–167, 536–537; 1995b, vol. 5, p. 415; 1995c, vol. 7.2, pp. 50–55).[13]

Unfortunately, however, the 'untimeliness' of Nietzsche's treatments of historical questions have been rarely appreciated by those who later engaged with or commentated on his philosophy. Consider, for instance, the highly influential interpretations of Nietzsche offered by Martin Heidegger and Michel Foucault. Heidegger simply ignores the wealth of detailed historical studies which Nietzsche's texts provide, and expressly excludes both the 'philosophy of culture' and the 'philosophy of history' from his study of Nietzsche, on the grounds that the former is irredeemably compromised by the Renaissance and by modernity and the latter is fatally undermined by its dependence on certain 'misleading constructions' (Heidegger 1989, p. 154; see also 1999a, pp. 113–116; 1999b, p. 118). Foucault arrives at much the same conclusions in a different way. By reducing Nietzsche's philosophy of history to a criticism of the 'superhistorical' history of the metaphysicians, on the grounds of an insistence on the 'singular randomness of events', Foucault fails to appreciate Nietzsche's own positive, and general, commitments regarding historical method (Foucault 1994, pp. 136–156). Thus, by marginalizing the substantial historical components of Nietzsche's philosophy, both Heidegger and Foucault not only profoundly misrepresent Nietzsche's concerns for concrete historical phenomena and questions of historical method, but also misunderstand his broader philosophical concerns and conclusions.[14]

13 Indeed, Overbeck himself practised this 'sensitivity' in his own studies of the origins of Christianity, and came to conclusions similar to Nietzsche's regarding the significance of excluded Jewish communities, conclusions which he found to correspond with those of von Harnack's 1900 work, *Das Wesen des Christentums*. See, for instance, Overbeck 1995a, pp. 157–161, 579, and von Harnack 1985, pp. 55–56, 74, and 104–106.

14 The influence of these readings perhaps explains why recent contributions to the study of Nietzsche and 'history' also marginalize these historical components. See,

In conclusion, I would suggest that such understandings of Nietzsche's philosophy also divest it of an insight which may be of contemporary importance. Indeed, Nietzsche's commitments regarding history were, at least in part, motivated by his own diagnosis of modern European culture. In a passage of *Beyond Good and Evil*, for instance, Nietzsche calls his time 'an age of disintegration' in which 'a person will have the legacy of multiple lineages in his body, which means conflicting (and often not merely conflicting) drives and value standards which fight with each other and rarely leave each other alone' (BGE 200; see also 242). In such an 'age of disintegration', a task of philosophers and historians is to reveal the multiplicity of Western history, the coexistence and mixing of different, pre-existing elements in our historical tradition. This is a perspective which might still merit our attention today.

References

Burckhardt, Jacob, 1922, Die Kultur der Renaissance in Italien, Leipzig: Kröner.

Deussen, Paul, 1883, Das System des Vedânta, Leipzig: Brockhaus.

Draper, J. W., 1871, Geschichte der geistigen Entwicklung Europas, Leipzig: Wigand.

Foucault, Michel, 1994, Nietzsche, la généalogie, l'histoire, in: Dits et écrits 1954–1988, vol. 2, Paris: Gallimard.

Harnack, Adolph von, 1985, Das Wesen des Christentums (1900), Gütersloh: Mohn.

Heidegger, Martin, 1989, Beiträge zur Philosophie (Vom Ereignis) (1936–37), in: Gesamtausgabe, vol. 65, Frankfurt a. M.: Klostermann.

Heidegger, Martin, 1999a, Die Grundbegriffe der Metaphysik. Welt-Endlichkeit-Einsamkeit (1929–30), in: Gesamtausgabe, vols. 29–30, Frankfurt a. M.: Klostermann.

Heidegger, Martin, 1999b, Die Überwindung der Metaphysik (1938–39), in: Gesamtausgabe, vol. 67, Metaphysik und Nihilismus, Frankfurt a. M.: Klostermann.

Lecky, William Hartpole, 1879, Sittengeschichte Europas von Augustus bis auf Karl den Großen, 2nd ed., Leipzig: Winter.

Mannhardt, Wilhelm, 1875, Der Baumkultus der Germanen und ihrer Nachbarstämme, Berlin: Bornträger.

Meyer, Eduard, 1925, Spenglers Untergang des Abendlandes, Berlin: Curtius.

Nägeli, Karl Wilhelm, 1884, Mechanisch-physiologische Theorie der Abstammungslehre, Leipzig: Oldenbourg.

Oldenberg, Hermann, 1881, Buddha. Sein Leben, seine Lehre, seine Gemeinde, Berlin: Hertz.

Orsucci, Andrea, 1996, Orient-Okzident: Nietzsches Versuch einer Loslösung vom europäischen Weltbild, Berlin: Walter de Gruyter.

Ottmann, Henning (ed.), 2000, Nietzsche-Handbuch, Stuttgart: Metzler.

Overbeck, Franz, 1995a, Werke und Nachlaß, v.4, Kirchenlexikon. Texte, Stuttgart: Metzler.

---

for example, the special issue of *Etudes germaniques* 55/2 on 'Nietzsche et l'histoire' (2000); and also the relevant entries in Ottmann 2000.

Overbeck, Franz, 1995b, Werke und Nachlaß, v.5, Kirchenlexikon. Texte, Stuttgart: Metzler.
Overbeck, Franz, 1995c, Werke und Nachlaß, v.7, pt. 2, Autobiographisches. 'Meine Freunde Treitschke, Nietzsche und Rohde', Stuttgart: Metzler.
Renan, Ernest, 1879, L'église chrétienne, Paris: Lévy.
Renan, Ernest, 1866, Histoire des origines du christianisme, v.2, Les apôtres, Paris: Lévy.
Renan, Ernest, 1869, Histoire des origines du christianisme, v.3, Saint Paul, Paris: Lévy.
Renan, Ernest, 1877, Histoire des origines du christianisme, v.5, Les Évangiles, Paris: Lévy.
Renan, Ernest, 1882, Histoire des origines du christianisme, v.7, Marc-Aurèle et la fin du monde antique, Paris: Lévy.
Spengler, Oswald, 1995, Der Untergang des Abendlandes (1919), Munich: Deutscher Taschenbuch Verlag.
Troeltsch, Ernst, 1922, Der Historismus und seine Probleme, Tübingen: Mohr.
Usener, Hermann, 1929, Götternamen. Versuch einer Lehre von der religiösen Begriffsbildung (1896), 2nd ed., Bonn: Cohen.
Weber, Max, 1947, Die protestantische Ethik und der Geist des Kapitalismus, in: Gesammelte Aufsätze zur Religionssoziologie, Tübingen: Mohr.

Translations

Beyond Good and Evil, ed. Rolf-Peter Horstmann, trans. Judith Norman, Cambridge: Cambridge University Press, 2002.
Daybreak, trans. R. J. Hollingdale, ed. Maudemarie Clark and Brian Leiter, Cambridge: Cambridge University Press, 1997.
The Gay Science, ed. Bernard Williams, trans. Josefine Nauckhoff, Cambridge: Cambridge University Press, 2001.
Human, All Too Human, trans. R. J. Hollingdale, Cambridge: Cambridge University Press, 1986.
On the Genealogy of Morality, trans. Maudemarie Clark and Alan J. Swensen, Indianapolis: Hackett, 1998.
Twilight of the Idols and The Antichrist, trans. R. J. Hollingdale, Harmondsworth: Penguin, 1968.

Thucydides, Nietzsche, and Williams

Raymond Geuss[*]

Who is a better guide to human life, Plato or Thucydides? Given this choice, virtually all European philosophers for the past two thousand years would have chosen Plato. Indeed one might almost say that to exhibit this preference defines what it is to be a philosophically minded person in the traditional mould. Plato has fascinating things to say about the human soul as an entity composed of parts that can conflict, about the nature of knowledge and the authority it should have in human life, and about how human excellence is related to the demands imposed on us by the necessity of living together. Even more significantly, Plato has presented all philosophers since his time with the model of what it is to 'have a philosophy' at all: it means having a systematically interconnected, abstract overview of and position on *all* the important features of human life which is argued for and justified in (purportedly) absolutely general terms. How could anyone think that the narration of a highly specific sequence of events that took place very long ago involving small groups of technologically rather primitive people squabbling in an obscure corner of the Balkans could conceivably compete with Plato's glorious project?

In the late nineteenth century Nietzsche broke radically with this founding assumption of Western philosophy. He did this not by developing one line or another of argument against Plato, but merely by raising the question about Plato's presumed self-evident superiority over Thucydides in a way that revealed that there was an issue of real philosophical substance and significance in the relation between the two on which it was possible to disagree. One way in which a philosopher can be original—

* Sir Bernard Williams, FBA (1929–2003) was Professor of Philosophy successively at University College/London, Cambridge, Oxford, and the University of California, Berkeley. From 1967 to 2003 he was a Fellow—and from 1979 to 1987 Provost—of King's College, Cambridge. In spring 1989 he gave the Sather Lectures at the University of California, Berkeley which were subsequently published as *Shame and Necessity* (Williams 1993). This essay is a slightly expanded version of a contribution to a memorial booklet printed privately by the Fellows of King's College. I am very grateful to Hilary Gaskin, Peter Garnsey, Jeremy Mynott, and Quentin Skinner for helpful comments on the original draft.

many would say, the most profound way in which a philosopher can be original—is not by giving an ingenious or particularly well-grounded or especially convincing answer to a pre-existing question, but rather by asking a novel question or finding an issue where no one before has seen one. Thus Nietzsche thought it was one of his strongest claims to originality that he for the first time explicitly and persistently asked questions like 'What is the value of our morality?' or 'Why do we assume that truth will always be of greater value than error?' and did not simply presuppose that the value of truth and morality was self-evident.

Nietzsche found Thucydides more illuminating about human life than Plato for two reasons. First, he held that Thucydides had an unprejudiced theoretical sympathy for, and hence understanding of, a much wider spectrum of possible human motivations than Plato had (D 168). *All* the characters in his history are allowed to exhibit the highest possible intelligence, clarity, and rationality in pursuing their respective enterprises, regardless of the judgements representatives of conventional morality would make on them (Nachlaß Spring 1884, KSA 11, 25[167]). Socrates, however, 'dragged moralizing into science', and Plato followed in his wake (Nachlaß June–July 1885, KSA 11, 36[11]). Such moralizing, Nietzsche thought, was a result of weakness, of a deep-seated inability to bear looking the facts of the world in the face,[1] it crippled Plato intellectually and prevented him from ever developing that most highly prized of Nietzschean traits: *Tatsachen-Sinn*,[2] a 'sense for the facts', that steely realism that is so abundantly evident on every page of Thucydides. Characters of whom Plato ethically disapproves, such as Thrasymachus or Callicles, are always shown in his dialogues to be confuted by Socrates. Vicious people, however, as we all know, do *not* always lose the argument. What Plato takes to be morally reprehensible behaviour *must*, he thinks, finally be a form of irrationality that is self-defeating, and this puts such narrow limits to his ability to understand humans that it renders him unfit to be a serious guide to the world in which we live.

---

1 See TI 'What I Owe to the Ancients' 2. I note that in this passage Nietzsche does a little rhetorical counter-moralizing himself, calling Plato not merely 'weak' and 'unable to face the facts', but 'a coward in the face of reality'. Needless to say, for reasons some of which emerge later in this essay, I do not think there is anything incoherent about this. At least one very important strand in Nietzsche is by no means opposed to any form of morality, but rather seems devoted to constructing a more realistic morality than that of Plato and the philosophical tradition (see A 59).

2 See A 59; also Nachlaß Summer 1883, KSA 10, 8[15] and GM Preface 7. See finally Williams 2002, pp. 12–19.

Another way of putting this might seem to be to claim that Plato could not have written such a characteristically clear-sighted, analytically rigorous, and uncompromising Thucydidean text as the Melian dialogue. Bernard Williams quite rightly corrects the implication this might naturally be taken to have when he points out that what is really at issue is not the empathetic, literary, hermeneutic, expository, or other human capacities of the individual Plato but what the explanatory motivational apparatus he recognizes and develops in his work would structurally require or admit: 'Thucydides' conception of an intelligible and typically human motivation is broader and less committed to a distinctive ethical outlook than Plato's; or rather—the distinction is important—it is broader than the conception acknowledged in Plato's psychological theories' (Williams 1993, pp. 161–162). Nietzsche is, of course, keen to connect these two—the man Plato and the Platonic philosophy—as closely as possible. Williams proposes a more subtle account of a kind with which we are familiar in other contexts. Many have thought that Freud the clinical practitioner exhibited a higher, deeper, or fuller 'understanding' of the human psyche than he was able to articulate in his theoretical constructs, so that the 'real' Freud is the Freud of the case histories, not the Freud of the meta-psychological writings. Hegel very clearly taught that any form of spirit (except his own) appealed to, used, and exhibited more complex structures than it could explicitly give an account of. So similarly, one might try to claim, Plato was, after all, an extraordinary literary and philosophical genius, who was capable even of the apparently deeply un-Platonic performance of depicting Alcibiades (in *Symposium*) as attractive; it might then well be the case that he exhibited in his dialogues—although he could not articulate—a much more subtle, flexible, and insightful practice of philosophy and understanding of human nature than his theories would have allowed. It was, of course, Plato's *theories* that were historically more influential than the practice, so in one sense it makes perfect sense to focus on them.

The situation here is further complicated by Nietzsche's claim that he had a low opinion of Plato's literary and stylistic gifts (TI 'What I Owe to the Ancients' 2). This, however, is such an extraordinarily obtuse or wilfully perverse judgement that one suspects that it must be a pose adopted for some strategic purpose or simply for effect, as when Nietzsche claims to prefer the music of Bizet to that of Wagner (CW Preface and 1–2). If Nietzsche really did find Plato 'boring', then perhaps there is simply nothing more to say about this particular lapse on his part, but there are clear ways, or at any rate the germs of ways, in which one could come to a very different judgement of Plato within a basically Nietzschean way of looking at the world. After all, in *Birth of Tragedy* Nietzsche emphasizes that tragedy in some sense killed itself ('sie starb durch Selbstmord', BT 11, KSA

1, p. 75); Euripides was the main executioner, with some help from Socrates. Tragedy, however, one could argue, was only really 'dead' when it was *replaced* by something else. In one sense what replaced it was Socratic rationalism and its extension, what came to be 'Western philosophy', but the process by which the replacement was effected requires a deeper account of how it took place than is often given. Excitable Hellenic youth did perhaps turns its back on the theatre because it had become boring,[3] but this does not yet explain why it chose to embrace the chaste and austere delights of linguistic analysis, logical argumentation, and (potentially) the Life of Reason instead. As Plato clearly realized (see the *Symposium*), the erotic fascination Socrates exercised during his life had something to do with this, but Nietzsche adds to this an observation about the important role the 'image' or 'picture' (*Bild*) of the dying Socrates played (BT 13, KSA 1, p. 91). Plato was 'enchanted' (HA I 261) by Socrates and 'threw himself down before this image' (BT 13, KSA 1, p. 91), and this quasi-erotic, quasi-religious bondage had significant historical consequences. As Nietzsche puts it in *Human, All Too Human*: 'It is by no means an idle question whether Plato, if he had remained free of enchantment by Socrates, might not have found an even higher type of philosophical man, which is lost to us forever. When one looks at the period before Plato one seems to be gazing into a workshop for forming such types [*Bildner-Werkstätte solcher Typen*]' (HA I 261). While the emphasis here is on Plato *finding* a type of philosopher, as if that were like a block of stone already roughed out for a statue in a mason's yard and needing simply to be discovered, this almost certainly underestimates the active shaping that would be required if the rough-hewn original were to be finished off, taken out of the shop, and set up so as to attract the appropriate continuing attention. There is a sense in which the image of Socrates is a fetish which Plato himself at least partly *created.* Socrates' impact, Nietzsche tells us, was in fact so overwhelming, that in order to tolerate him, Plato had to *transform* him (*umbilden*) (Nachlaß Summer 1883, KSA 10, 8[15], p. 337), to produce a very free portrait, a picture of Socrates that suited *Plato* ('Plato's freie Art ... sich Sokrates zurecht zu machen', ibid., p. 338). To present Socrates, whose life was essentially devoted to conducting private conversations with individuals,[4] as a figure who dies in some sense 'heroically' because of his commitment to the Life of Reason requires at least minimal artistic structuration and stylization of the material a real human life provides. Only when this image of the death of Socrates supplants those of the

3 As indicated by Dionysus' reaction at the start of Aristophanes' *The Frogs*.

4 See *Apology* 23b, 31c, *Gorgias* 484c–486d (admittedly by a hostile witness).

deaths of Patroclus, Ajax, and Hector can Reason really take over from myth, and Western philosophy succeed tragedy. Philosophy as we know it established itself as a continuing presence in Western culture partly because in his dialogues Plato was able to embody the erotic charge of Socrates in a striking image that transmitted it down the ages and hooked successive generations on the dialectic. This can be seen as a kind of artistic creation, a skilled production of highly and long-lastingly effective *Schein*, and it would, then, seem to be perfectly possible to find this achievement neither boring nor lacking in artistic merit.

This antecedent moralization of the basic categories in Plato's theory of human psychology vitiates his own positive ethical proposals. If he really has merely smuggled a set of tacit moral assumptions into his basic psychology, then it is not surprising that he can victoriously draw them out again as conclusions. To the extent to which Plato, and most philosophers after him, have done this while pretending to be engaged in some kind of disinterested enquiry, they are violating their own ostensible standards of good faith, truthfulness, and non-circularity of argumentation.

Nietzsche's second reason for preferring Thucydides concerns the issue of optimism or pessimism as the appropriate human attitude towards the world. Nietzsche correctly diagnosed the philosophical tradition as deeply optimistic.[5] This optimism had several related aspects. First of all, traditional philosophers assumed that the world could be made cognitively accessible to us without remainder: it was in principle possible to come to know any part of the world as it really was. Second, they assumed that when the world was correctly understood, it would make moral sense to us. Third, the kind of 'moral sense' which the world made to us would be one that would show it to have *some* orientation towards the satisfaction of some basic, rational human desires or interests, that is, the world was not sheerly indifferent to or perversely frustrating of human happiness.[6] Fourth, the world is set up so that for us to accumulate knowledge and use our reason as vigorously as possible will be good for us, and will contribute to making us happy. Finally, it was assumed that there was a natural fit between the exercise of reason, the conditions of healthy individual human development, the demands of individuals for satisfaction of their

5 Again Nietzsche thinks this is part of the legacy of Socrates, see BT 15, KSA 1, pp. 97–102.

6 Note that there are three distinct ideas here: (a) the world makes *some* kind of sense, (b) the world makes 'moral' sense, (c) the world makes a kind of moral sense in which human needs and at least some human aspirations have *some* standing.

needs, interests, and basic desires, and human sociability. Nature, reason, and all human goods, including human virtues, formed a potentially harmonious whole.[7] There was one human state and one course of human development which was 'correct' (or, as Aristotle would put it, 'natural') for us. 'Natural' human development would lead to a full development of human rational capacities. This is turn would make humans disposed towards socially desirable forms of conduct, and also individually and collectively happy. Over the two thousand years of history, there have been different accounts given of what 'correct' or 'natural' means, and there have been any number of minor reinterpretations of and deviations from the above scheme, but the basic structure of a philosophy centred around the claim of a harmonious fit between what is rational, what is good for us, and what is good for our society has been very widely retained. If one excludes a few Gnostics, the odd sceptic, and marginal figures like Schopenhauer, few philosophers or religious thinkers in the West have not been guided by it, at least as a tacit ideal.

In one respect the 'rationalism' of Socrates is, however, peculiar. Plato's Socrates may be wiser than others in that he does not think he knows what he does not know (*Apology* 21d), and he may strive constantly for greater knowledge and greater self-clarity, but his life is also fundamentally structured not around a form of well-grounded propositional knowledge, but rather around what he himself calls a 'great hope' (*πολλὴ ἐλπίς Apology* 40c4; also *Phaedo* 67b7–c3), the almost ludicrously optimistic belief that nothing bad can befall a good man. If Plato's account in *Apology*, *Crito*, and *Phaedo* is to be believed, in his last days Socrates refused to save himself by availing himself of existing possibilities of avoiding conviction and the death penalty, and then of escaping from prison, and succumbed to bouts of preachiness during which he exhorted his companions to be of 'good hope' (*ὑμᾶς ξϱή ... εὐέλπιδας εἶναι Apology* 41c8; compare also *Phaedo* 63b4–c7) with regard to death: it cannot be an inherently bad thing because it befalls both good and bad people alike. It is striking how heavy a weight this *ἐλπίς* is made to bear.

The contrast with Thucydides could hardly be starker. The power of *ἐλπίς* is a recurrent theme in his history, but *ἐλπίς* for him is almost invariably deluding and its power is overwhelmingly destructive.[8] 'Hope (*ἐλπίς*)

7 One of Williams' principal teachers, Isaiah Berlin, was a subtle analyst and highly outspoken critic of precisely this strand of traditional moral thinking in all its forms (Berlin 1969, LI, esp. pp. 8, 167–172).

8 For the best treatment of this aspect of Thucydides known to me, see Stahl 1966. See also discussion of *ἐλπίς* in Hesiod, *Works and Days* (West 1978, pp. 169–170) and further passages and modern works cited there.

and desire (ἔρος), the latter leading the way, cleverly hatching the enterprise, the former following, suggesting that chance will make the circumstances propitious for success (τὴν εὐπορίαν τῆς τύχης ὑποτιθεῖσα), between them cause the greatest destruction' (III.45.5; see also V.103). This view of ἐλπίς is also not simply an 'opinion' expressed by various speakers in the history, but Thucydides himself seems to delight in demonstrating its validity through the juxtaposition of speech and narrative. Thus, when Nicias in Sicily appeals to 'hope' (VII.77) in addressing his troops, the reader can hardly avoid feeling sure that he and they are about to suffer complete destruction, as in fact they do, and it is hard to believe Thucydides did not intend this sinister effect.

Thucydides seems largely immune to any of the forms of wishful thinking associated with Platonic optimism. He knows that good men suffer undeserved, irremediable, definitive catastrophic failure (Nicias); unworthy men reap the benefit of others' achievements (Cleon in Pylos); men exhibit pre-eminent virtue in some contexts and fall into decadence in others (Pausanias); there is no pre-existing 'meaning' in the world, only what we humans can construct by our weak powers and flawed efforts. Human rationality is real, but its motivational power is extremely weak, particularly in the face of human hopes, loves, desires, and fears, and the success of even the most well-founded and rational plan is at the mercy of external chance. Donald Rumsfeld, regardless of what one might think of the rest of his politics, is making a good Thucydidean point when he emphasizes the importance not just of 'known unknowns' in war and politics—factors for which some rational provision can be made, even if only on the basis of educated estimates—but also of 'unknown unknowns' which cannot be subjected even to crude rational approximation because they cannot be envisaged at all, and which thus lie strictly beyond the possibility of human ratiocination.

In what is in many ways his most impressive book, *Shame and Necessity*, Bernard Williams cites and endorses the above Nietzschean account: Thucydides should be seen to stand with Sophocles as the major representative of an attitude towards the world which is realistic, values truthfulness, and is lacking in the shallow 'optimism' of later philosophy (1993, pp. 163–164). This coupling of Thucydides and Sophocles might seem rather odd, and thus warrants some further attention.

We are used to believing that there was an 'old quarrel' between poetry and philosophy in pre-Socratic Greece (Plato, *Republic* 607b), although, as Andrew Ford in a recent work (Ford 2002) has persuasively shown, there is no evidence that this was the case and the claim is perhaps best understood as a bit of Platonic invention or disinformation. In the original published version of *The Birth of Tragedy* Nietzsche accepts this bipartite structure of the argument, although, of course, he wishes to reverse Plato's valuation and argue for the superiority of poetry over Socratic philosophy. Thucydides, who might have been thought to represent a third option, is not even mentioned in Nietzsche's text. After the publication of *The Birth of Tragedy*, however, Nietzsche does seem to have gradually become aware that reading the history of Greece simply through the lens of the struggle between Homer/Sophocles, on the one hand, and Socrates/Plato, on the other, leaves out something important. Thus in a note from 1885, he calls the attitude exhibited by the philosopher Democritus, the physician Hippocrates, and Thucydides the high point of scientific-mindedness (*Wissenschaftlichkeit*) that was attained in Greece, and speaks of the opposition between the ethical philosophy of someone like Socrates and 'science' (Nachlaß June–July 1885, KSA 11, 36[11]). *Wissenschaftlichkeit* means careful, methodical attention to the real facts of the situation being investigated. As Williams (2002, pp. 12, 152–153) points out, however, it does not necessarily imply commitment to the ideals of positivism to the extent to which these represent a code of restrictive practice. This negative canon would have it that a 'scientific' account must restrict itself to a purportedly value-free registering of observable facts, to the formulation of generalizations that have their full meaning by virtue of being connected to sets of observable facts, and to the use of conjunctions of fact and generalization for the purposes of causal explanation.

Thucydides' way of approaching his subject is like that advocated by the positivists in that his treatment is radically non-mythic (I.20–22), non-theological, and non-literary. Although he uses the Homeric poems as material to be studied in the interest of forming plausible hypotheses about certain aspects of earlier societies (I.3), he shows distain for the exaggerations, inventions, and factual insouciance of poets (I.10), and specifically disclaims any intention of trying to increase the appeal of his work by giving it a literary polish (I.22). His project is to exhibit what really moves people to act, and what then happens to them and to others as a consequence of how they act, not to write an edifying treatise or a partisan tract. Thucydides' account differs, however, from anything that positivists of the stricter observance would countenance in taking human beliefs, attitudes, emotions, valuations, even superstitions (II.54 plague and oracles; VII.50 about Nicias) very seriously indeed as things that need to be taken into account if one wishes to have a genuine understanding of what happens in the human world. His impartiality between the two warring sides in the conflict between the Athenians and Peloponnesians should also not be confused with a positivist commitment to an ideal of 'value-freedom'. He is in no way reluctant to express value judgements of his own when it suits him. These include not only low-level technical judgements such as praise of Pericles for his 'foresight'—in that the Athenians would have won the war if they had consistently followed his initial strategy (II.65)—but also overall moral evaluations such as that Nicias was a man who did not deserve the end he suffered (VII.86).

The work Thucydides wrote is not 'history' at all in the most usual sense in which we use the term, that is, a work that is centrally or specifically concerned with a study of the *past*.[9] Thucydides is specifically interested not in the *past*, but in understanding those forms of collective human behaviour that are recurrent and thus comprehensible (I.22). He pursues this aim by giving a narrative account of what was for him the present: current affairs, in some of which he was himself an actor. Part of this narrative account is an analysis of the motives and reasons of various individual and social agents. One could then say that he is trying to do something like what we might call 'social and political theory' or even 'behavioural science' (if the later term could be cleansed of all the associations it has acquired during the past century or so), but *only provided* one keeps clearly in mind that he does not think there are 'laws' of history or society which we can formulate abstractly and the mastery of which will allow us to control our fate.[10] One of the most important things one can learn from the study of 'human nature' is that this kind of control is an illusion.

Understanding human nature as exhibited in large-scale human action requires the correct sequencing of complex, spatially distant events, placing them in their proper order through time, and as Williams emphasizes in chapter 7 of *Truth and Truthfulness*, this requires having a general notion of a single, measurable historical time within which events in different places can be located. This is not a triviality because 'human beings can live without the idea of historical time' (2000, p. 169), and in fact they did so in Europe until the fifth century BC. Thucydides is extremely self-conscious and careful in introducing a single chronological scheme which will allow clear and unambiguous coordination of the diverse local calen-

9 Although we still use the term 'history' in a wider sense in expressions like 'natural history', the Greek word (*ἱστορία*) from which ours is derived has a very broad extension, meaning any kind of 'investigation', 'enquiry', or 'research'. Furthermore, we have no idea what, if anything, Thucydides himself would have called his work, had he finished it. In the text we have he never refers to it by using the word *ἱστορία*, and in any case the whole issue of the titles of works from antiquity is highly complex and obscure. In the late fifth century BC the giving of titles to works seems to have been a significantly more casual matter than it later became, and certainly not the object of sustained authorial concern.

10 On the non-instrumentalist nature of Thucydides' conception of understanding and explanation, see Stahl 1966.

dars used in the different Greek cities (see Gomme 1945, pp. 1–8). Williams seems to go so far as to attribute to Thucydides the 'invention' of the 'objective' conception of time (2000, pp. 154, 169–171). Again, this invention might be an essential precondition for doing history (rather than narrating stories of indeterminate historical location), but to invent a very conception of objective time is not, by itself, to do history, as we understand it.

It is a commonplace in the secondary literature on Thucydides that, in contrast, for instance, to Plato, he stands alone. There were plenty of followers of Plato, Platonists of one kind or another, in antiquity, but no Thucydideans.[11] In the strict sense Thucydides had no successors in doing his specific kind of 'investigation'. There were those who 'continued' his narrative, telling the story of the war between the Athenians and Peloponnesians from the point at which his (unfinished) account breaks off to the final destruction of Athens, but each of these 'continuators' had his own very different agenda, different aims, different literary styles and modes of proceeding from Thucydides. It is often claimed that the reason for this lack of direct influence was the extreme success of the discipline that came in some sense to be a competitor to Thucydidean 'enquiry': rhetoric. Rhetoric was in some ways the most immediately advantageous, practical skill a young man in a *πόλις* could learn, and thus came to be an increasingly powerful influence on the education of the young. For various reasons Thucydides' work was not especially useful for those wishing to learn to speak well in public. For one thing his style did not lend itself at all to emulation: it was too difficult and too obscure. Even Cicero, who was a fluent speaker of Greek and encountered it as a fully living language, calls the speeches in his work almost unintelligible. This was no model for clear, persuasive, public discourse (*Ipsae illae contiones ita multas habent obscuras abditasque sententias vix ut intellegantur; quod est in oratione civili vitium vel maxime*: *Orator* 30).

There is perhaps also a second reason that is connected with a more deep-seated incompatibility between the spirit of Thucydides' work and the demands of rhetorical training. In book III (82–84) Thucydides describes the long-lasting civil disorder in Corcyra. One result of this is that the accustomed meaning of words shifts. What used to be called 'senseless rashness' (*τόλμα ἀλόγιστος*) now comes to be called 'a manly spirit that looks out for its friends' (*ἀνδρεία φιλέταιρος*), and 'circumspection in every re-

11 Philistius is mentioned as 'Thucydidi imitator' (Quintilian X, 74; see also Cicero, *Brutus* 66), but so little of his work has survived we cannot know on what basis this judgement was made or whether it was well founded.

gard' (τὸ πρὸς ἅπαν ξυνετόν) comes to be considered to be, and is called, 'complete laziness' (ἐπὶ πᾶν ἀργόν). This shift in the application of customary evaluative terms was considered by Thucydides to be a clear sign of a seriously pathological state of society.[12] For the rhetorician, on the other hand, the fact that the same situation or character traits admit of a variety of different designations, each with a completely different moral and affective coloration—I am prudent; you are cautious; he is a coward—is a *precondition* of the exercise of his art, not a sign of degeneracy. Thucydides' final value judgements may be unconventional and hidden so deeply in his harsh and obscure prose as to require sustained attention and effort to comprehend them, but they are not, finally, slippery and ambiguous. He clearly did not think that by judicious redescription one could make the same course of action either good or bad, and surely one of the lessons one can hardly fail to learn from his work, if one studies it carefully and understands it correctly, is that it is, therefore, highly inadvisable in the long run to try to make actions *seem* good or bad *ad libitum*, even if one can succeed in producing an effective appearance. As long as rhetoric dominated political life and education there was no room for Thucydides' unique combination of superficial, analytic detachment from the demands of immediate political partisanship, compressed and convoluted literary style, and deep-seated, if idiosyncratic, moral realism (see Williams 2002, chs. 7 and 10). In the final analysis Nietzsche is closer to the mark when he connects Thucydides with incipient forms of *Wissenschaftlichkeit* such as one finds in Hippocrates (Nachlaß June–July 1885, KSA 11, 36[11]) than when he calls him a representative of the 'culture of the sophists' (TI 'What I Owe to the Ancients' 2).

Nietzsche's *Birth of Tragedy* is equally about the death of tragedy. One might think that a more complete account of that crucial period between the middle of the fifth and the middle of the fourth century BC when so many of the most characteristic European modes of thinking become visible would require, as a supplement and pendant to *The Birth of Tragedy*, a treatise on the murder by starvation of early Greek *Wissenschaftlichkeit*. One might call it 'Ugolino graecus, oder der Hungertod der frühgriechischen Wissenschaft'. Instead of Nietzsche's stark Aeschylean drama of two actors, tragedy and Socratic philosophy, there would have to be a

---

12 One might claim that modern political thought begins when Hobbes, who translated Thucydides, decides that the 'pathological' state Thucydides describes in Corcyra is the natural state from which the study of politics must begin. On these issues, see the seminal paper by Quentin Skinner, 'Hobbes on Rhetoric and the Construction of Morality', in Skinner 2002, pp. 87–141.

more Wagnerian drama with a fuller cast including two sets of infant-victims, the potential unborn children of Sophocles and those of Thucydides, and two murderers, Socratic philosophy and rhetoric. That Socratics and rhetoricians were also enemies is true, but irrelevant to the larger story; Plato's unrelenting guerrilla war against rhetoric (and the sophists) must not divert attention completely from the role he played in doing in and supplanting both tragedy and Thucydidean 'enquiry'. Thucydides' work instantiated and was clearly aimed at the cultivation of a kind of practical reasoning and political—and 'moral', if one wishes—judgement, which was supposed to have general scope; it was not positivist science. However, the nourishment his form of 'enquiry' absolutely needed was a keen interest in understanding clearly and exactly (*τὸ σαφές* and *τὸ ἀκριβές*) the *real*, causal details of human motivation, the contingencies of particular political situations, the historically and geographically specific structure of existing human institutions, topography (Sphakteria, Syracuse), dialectology (VII.44), and so on. In a society in which a very large number of the most active political agents devote themselves to trying to learn how to put together words pleasingly and convincingly so as to persuade their hearers, with little regard for truth, and the most reflective members are committed to the search for abstract definitions, general principles, dialectically sustainable hypotheses, and perhaps, in some cases, a 'vision of the idea of the good', Thucydidean political thinking informed by a study of the reality of what actually happens will be likely to wither away.

During the 1870s and early 1880s Nietzsche kept a series of notebooks in which one can find a large number sketches, drafts, and plans for a more comprehensive treatment of 'the Greeks' than that given in *The Birth of Tragedy*. None of it eventually materialized, but if one reads the notebooks, it seems clear that he became increasingly aware of the importance of the strand of realist and empiricist[13] thinking that Thucydides represents, and of seeing the demise of tragedy and of Thucydidean 'enquiry' synoptically.[14] Williams' later work can be seen as in some sense trying to do this.

Williams agrees with Nietzsche that there can be no simple return to ancient 'pre-Socratic' conceptions (Williams 1993, pp. 6–7, 9–11); he

13 These terms, like many of the others I am compelled to use, are philosophically loaded because of their later history. I must ask the reader to try to suspend as many of those later associations as possible.

14 The parallel between the fate of tragedy and that of Thucydidean investigation is not exact. For instance, tragedy was a long established institution with religious roots and an important civic aspect which was supported by public funds; 'enquiry', on the other hand, was a socially and politically much more fragile construct of uncertain standing.

knows as well as Nietzsche did that his own form of consciousness is possible only because of developments about which he has serious reservations (ibid., p. 9), and that this by itself would make simple return impossible, even if it were desirable on other grounds (which for many other reasons it is not). We have no alternative but to use the techniques of reflective analysis, formal argumentation, and modern, mathematically structured, empirical science that have been developed by representatives of post-Platonic philosophy, but we can try to use them to break through the bad faith on which traditional ethics rests.

To return to the two ways in which Nietzsche thinks Thucydides surpasses Plato: his more open-minded psychology, and his resistance to unfounded optimism (see above), each of them might contain a hint about how we could advance our own understanding. Although there can be no value-free psychology, not all values are 'moral' values, and not all moral values are of the kind originally recognized by Plato (and then bequeathed to the rest of the mainline of Western philosophy). We can try to become aware of the extent to which we presuppose certain values, and try to make our assumptions as realistic as possible. We can, that is, try to be as truthful and truth-loving as possible in developing an alternative to the deceitful, hypermoralized views of Plato, Aristotle, Kant, and the other major figures in the history of Western ethics. There is a story inspired by Christianity and sharpened by Kant which makes us look down on the ancients and especially ancient ethics for its deficient concept of will, or failure to put volition in the centre of human life. A complementary story, presented in a classic way by one of Williams' predecessors as Sather Lecturer, E. R. Dodds, in his book *The Greeks and the Irrational* (1951), denigrates ancient Greek culture as one based on 'shame'—the highly primitive reaction to loss of face vis-à-vis one's compeers—rather than on 'guilt', which is considered to be a more sophisticated and morally sensitive reaction. Contrary to this line of thought, Williams (1993) proposes that a psychology which is not based on notions of 'volition' (p. 36), the will (pp. 41–46), the distinction between the voluntary and the involuntary (pp. 66–68) or the idea of 'guilt' (pp. 75–102), but is centred on an expanded and reflectively clarified concept of 'shame', will actually contribute to a more realistic, substantial, and socially enlightened form of ethical thinking.

Williams, then, invites us to reflect on a possible historical path not taken, one from ancient shame, tragedy, and Thucydidean 'enquiry', rather than from Plato, Christianity, and guilt; not, of course, with the intention of inviting us to try to turn the clock back or embark now on the path not taken two thousand years ago—it was *not* taken and history cannot be turned back—but in order to inform our imagination for positive transformations of our own moral thinking.

On the question of optimism or pessimism, there are two different ways in which one could depart from the consensual optimism that characterizes most traditional philosophy. First, one could think that the world had a purposive structure, perhaps imposed on it by a malicious creator-god who delights in tormenting humans and who set it up in order to thwart essential human aspirations. The second view is that there is no inherent, discernible, purposive structure, sense, or meaning to the world at all, and no guarantee that the items in the optimists' package—rationality, individual happiness, natural human development, socially desirable action—are all compatible. To think either that these items are set up so as to cohere, or that they are 'by nature' ineluctably fated to conflict in an unresolvable way—to be either an old-style philosophical optimist or a dogmatic pessimist—is still to be prey to notions that are theological in their origins and implications. When Nietzsche wrote that 'the Hellene was neither an optimist nor a pessimist' (Nachlaß Winter 1869/70–Spring 1870, KSA 7, 3[62]), this is what I assume he meant, and no Hellene could illustrate this more exactly than Thucydides. As Williams puts it, the world is not 'intrinsically shaped to human interests' (1993, p. 163), either for good or for ill. This is a bleak view only relative to a set of exaggerated expectations about how the world *ought* to make sense that are themselves highly questionable (ibid., p. 68).

None of this at all suggests any serious form of 'Cartesian' scepticism about our knowledge of the world, the existence of truth, the importance of truthfulness, or our ability to make sense of our world and our lives (as opposed to discovering a pre-existing 'meaning' in the universe). This is the burden of Williams' final book, *Truth and Truthfulness*. Of course, we can know all sorts of things, and the very project of criticizing the Platonic-Aristotelean-Kantian tradition in ethics would be impossible without the apparatus of systematic enquiry and the evaluation of cognitive claims which philosophy, and latterly also science, have done so much to develop. There are some 'universal materials' out of which particular human ethical conceptions are constructed, and Williams (1993, p. 56) believes that there is no special problem in claiming that we can know this or what these materials are. However, he also holds that, contrary to what Plato and Kant thought, investigation of these universal materials *alone* will not throw adequate light on any particular concrete form of human ethical thought because there is no unique path from these materials to any particular historical conception. A more traditional philosopher would be inclined to suggest that this is merely a limitation, not an invalidation of the claim to pre-eminence of strictly philosophical analysis; abstract philosophical accounts, of course, give only an outline, which of course needs to be filled empirically, but this outline is a delineation of what is essential. Williams,

I think, would have rejected this suggestion. Thus, to take one case that he treats in some detail in *Shame and Necessity*, the concept of 'responsibility' has been an extremely prominent part of much ethical thinking during the past few hundred years, especially in the Kantian tradition. This concept puts together a number of different elements, which refer to universal features of human action: facts about causation, human intention, social needs for predictability, etc. There is, however, no unique way to put these elements together into an ethical or legal concept of responsibility. 'There is not, and there could never be, just one appropriate way of adjusting these elements to each other—as we might put it, just one correct concept of responsibility ... in different circumstances [we] need different conceptions' (Williams 1993, p. 55). The particular way in which the elements are connected will depend in a substantive, or—if one wishes to use this term—an 'essential' way on the particular social structure, political institutions, and vagaries of the human history of the society in which the concept has arisen and is used. The history, sociology, and politics of the case do not simply fill in the details of the picture: they are the picture. This is the most important thing we can learn from Thucydides, and we can perhaps learn it more easily from him than from Plato.

Given that our main source of knowledge about Socrates is through the Platonic dialogues, it is not surprising that there is a tendency to treat 'Socrates-Plato' as a single unitary philosophical personality, but, of course, in their better moments everyone knows that this is incorrect. We have seen above that Nietzsche accommodates a recognition of their duality through his account of the way in which Plato 'transformed' Socrates. In *Ethics and the Limits of Philosophy* (ch. 1) Williams claims that the genuinely philosophical impulse is the Socratic impulse of questioning, in particular asking the question how one should live; this is presumably intended to imply the possibility of distancing oneself philosophically not merely from Socratic optimism but also from the Socratic form of rationalism, especially as these are developed by Plato and the Western tradition. How then can this impulse be prevented from running away with itself and dissipating its energies in the sands of excessive abstraction, as it did in traditional forms of post-Platonic philosophy? At various points in *The Birth of Tragedy* Nietzsche appeals to the ideal of a 'Socrates who makes music'. In this essay I have been trying to claim that Williams' later work is similarly inspired by the ideal of what one might call a 'Thucydides who philosophizes'.

References

Berlin, Isaiah, 1969, Four Essays on Liberty, Oxford: Oxford University Press.

Dodds, E. R., 1951, The Greeks and the Irrational, Berkeley: University of California Press.

Ford, Andrew, 2002, The Origins of Criticism, Princeton, NJ: Princeton University Press.

Gomme, A. W., 1945, A Historical Commentary on Thucydides, vol. 1, Oxford: Oxford University Press.

Skinner, Quentin, 2002, Visions of Politics, vol. 3: Hobbes and Civil Science, Cambridge: Cambridge University Press.

Stahl, Hans-Peter, 1966, Thukydides: Die Stellung des Menschen im geschichtlichen Prozeß (Zetemata, Heft 40), Munich: Beck.

West, M. L. (ed.), 1978, Hesiod, Works and Days, Oxford: Oxford University Press.

Williams, Bernard, 1985, Ethics and the Limits of Philosophy, London: Fontana.

Williams, Bernard, 1993, Shame and Necessity, Berkeley: University of California Press.

Williams, Bernard, 2002, Truth and Truthfulness, Princeton, NJ: Princeton University Press.

The Late Nietzsche's Fundamental Critique of Historical Scholarship

Thomas H. Brobjer

There exist a number of misconceptions about Nietzsche's relation to historical scholarship and methods. Most discussions of this theme have been based on his early essay 'On the Uses and Disadvantages of History for Life' (1873), but I have recently shown that the views he presents there are not representative of the later Nietzsche. Already one or two years after the publication of that essay, he changed his view dramatically.[1] For example, in 1877 he writes: 'I want expressly to inform the readers of my earlier writings [i.e. *The Birth of Tragedy* and the *Untimely Meditations*] that I have abandoned the metaphysical-artistic views that fundamentally govern them: they are pleasant but untenable' (Nachlaß End of 1876–Summer 1877, KSA 8, 23[159]), and in 1883: 'Behind my first period grins the face of Jesuitism: I mean the deliberate holding on to illusion and the forcible annexation of illusion as the foundation of culture' (Nachlaß Autumn 1883, KSA 10, 16[23]). Furthermore, Nietzsche very rarely discusses or praises his essay on history after 1875, in stark contrast to all of his other books, and, after 1874, he never uses the several concepts and expressions such as *monumental*, *antiquarian*, and *critical* history, or *overhistorical*, which he coined and used in that book. I have also shown, contrary to most discussions, that Nietzsche was deeply influenced by and approved of the new historical views and methods which grew out of the works of Wolf, Niebuhr, Mommsen and Ranke (see Brobjer 2007). Nietzsche thus approved of historical methods and studies, and frequently used them in his philosophical investigations and critique. He was after all educated as a historian and classical scholar, and was a professor in the latter

1 In my article 'Nietzsche's View of the Value of Historical Studies and Methods' (Brobjer 2004) I show that not only does Nietzsche have a different view of history and historical scholarship after 1875/76, but also that he on many occasions explicitly rejected the view he proposed in 'On the Uses and Disadvantages of History for Life'. In this article and also in 'Nietzsche's Relation to Historical Methods and Nineteenth-Century German Historiography' (Brobjer 2007) I discuss many secondary resources on Nietzsche's relation to history.

field for ten years. Nonetheless, the late Nietzsche also expresses much hostility towards historical scholarship (and, in many ways similar to this, and related to it, towards natural science and scholarship generally). In this paper I will examine the reasons for this later explicit critique and what it entails. This later critique is primarily based on four factors or arguments: relating to objectivity, the idea of progress, the viewing of history as a means, not a goal, and on history as being reactive (like all sciences and scholarship). Thereafter I will discuss the only published text, *On the Genealogy of Morality* III 26, in which the late Nietzsche explicitly discusses modern historical writing.

What the late Nietzsche primarily objected to in regard to history was not the new methods introduced at the early part of the nineteenth century, but that history was placed above philosophy—that history and historical scholarship were seen as a goal or an end in itself rather than as a means. He saw this as a reflection of the nihilism which characterized modernity. More specifically, he objected to a number of aspects regarding history and historical scholarship, but most of this is the *consequence* of having accepted the historical revolution rather than standing in contradiction to it. He was a severe critic of the idea of progress, assumed by almost all major nineteenth-century historians (but not by Burckhardt). He regarded most historians as far too idealistic in their views (and still governed by religious faith) and accused them of lacking adequate knowledge of natural science and medicine. Perhaps even more pronounced is that he regarded most historians (and philosophers) as being much more anachronistic than they were aware of, especially in regard to moral and cultural values—he here thus regards them as bad historians according to their own and his criteria. Related to this, he regarded almost all historians as possessing egalitarian and anti-aristocratic values, and therefore only taking a stand for the suppressed and failed groups (relating to a revaluation of values). He objected to the almost exclusive concern with political history by the leading historians, and much preferred a broader cultural approach. He questioned both the possibility and the desirability of historical objectivity. He regarded history (and science) as by necessity reactive, and he felt that the historians were often indecent, in digging into private worlds or by questioning things greater than their own comprehension. But all these objections, with the possible exception of his view of objectivity, can be and were stated from within the historical turn which had occurred.

Throughout most of Nietzsche's so-called middle period, c. 1875–1880, he seems to have accepted the idea of progress, both generally and for the case of morality.[2] However, thereafter he strongly attacks it. Today, when we have a more complex and ambivalent view of the idea of progress, this may not seem to be a major point. However, during the nineteenth century, especially after the acceptance of Darwinism, this idea constituted a generally accepted dogma which penetrated into all fields and areas of intellectual activity. It constituted a premise and a major assumption of almost all historical writing. Nietzsche's critique and rejection of it was fundamental enough to suggest that he had a completely different view of history and historical scholarship than most of his contemporaries.

The late Nietzsche expresses his contempt for the idea of progress many times, but here it will be sufficient to quote one example: 'Mankind does not represent a development of the better or the stronger or the higher in the way that is believed today. "Progress" is merely a modern idea, that is to say a false idea. The European of today is of far less value than the European of the Renaissance; onward development is not by any means, by any necessity the same thing as elevation, advance, strengthening' (A 4).[3] The main reason for his rejection is also plainly stated—man is not, and has not become better and more valuable. Instead Nietzsche on several occasions says that we have become 'smaller'. Note that Nietzsche thus uses a different measure of progress than most commentators: Nietzsche asks if men have become, better and more elevated, J. S. Mill and others if they have become happier.

What does Nietzsche's critique of historical objectivity mean and entail? After all, objectivity is the keystone of the scientific and scholarly approach. We can begin by noting that his critique of objectivity in historical scholarship has a direct parallel in his critique of the natural sciences. It has two fundamental parts, one epistemological and one value-oriented.

---

2 For example, in HA I 236 he writes: 'To us, however, the very existence of the temperate zone of culture counts as progress.' Compare also: 'Wrath and punishment has had its time. —Wrath and punishment is a present to us from the animal world. Man will have come of age only when he returns this birthday gift to the animals. —Here there lies buried one of the greatest ideas mankind can have, the idea of progress to excel all progress. —Let us go forward a few thousand years together, my friends! There is a great deal of joy still reserved for mankind of which men of the present day have not had so much as a scent! And we may promise ourselves this joy, indeed testify that it must necessarily come to us, only provided that the evolution of human reason does not stand still!' (HA III 183). Further examples can be found in HA I 24 and 107; HA II 184 and 185.

3 Compare also GM II 12 and his discussions in TI 'Reconnaisance Raids' 37 and 48. See also Nachlaß June–July 1885, KSA 11, 36[48]; and Nachlaß End of 1886–Spring 1887, KSA 12, 7[8]; and Nachlaß Autumn 1887, KSA 12, 9[185].

This critique of objectivity in historical writing is present and important throughout his whole life, beginning already in his early critique of historical writing. He, for example, writes around the time of 'On the Uses and Disadvantages of History for Life' that 'The so-called objective writing of history is something unthinkable: the objective historians are crushed or smug characters' (Nachlaß Summer–Autumn 1873, KSA 7, 29[137]). When we return to the views of the late Nietzsche, we can observe that in a list of plans for books or lectures to write, from 1883, he states that 'there exists no "objective history"', for 'the appropriation of history is done under the guidance of stimuli and the drives' (Nachlaß Spring–Summer 1883, KSA 10, 7[268]). This is in essence an epistemological claim—that objectivity is impossible—which he repeats frequently. Both as individuals and as human beings, nothing is unaffected by our conscious and subconscious desires and values. When we pretend that to be the case, we falsify reality (and our relation to reality), and thereby fail to give a correct account of reality. Objectivity, he repeats, is merely a mask.

However, Nietzsche has a yet stronger critique of objectivity which is less epistemological and more value-oriented. A natural or 'healthy' relation to reality includes evaluations—we are the evaluative animal—and when we attempt to be objective we are denying ourselves as living organisms, we are being ascetic and nihilistic, and such a desire is an expression of self-rejection and self-contempt.

> The 'desire-to-be-objective', e.g. in Flaubert, is a modern misconception ... It is self-contempt ... 'denial' of self. But there exists no 'thing-in-itself', gentlemen! What you attain is science or photography, i.e., description without perspectives ... In fact, there is very much displeasure [*Unlust*] in the modern historical and natural historical mania—one flees from oneself and also from the making of ideals, the making better. (Nachlaß Spring 1884, KSA 11, 25[164])

Shortly thereafter he repeats these two claims, perhaps even more clearly: 'To mock the school of the "objective" and the "positivists". They want to get around the setting of values, and discover and present only the facts' (Nachlaß Summer–Autumn 1884, KSA 11, 26[348]).[4] And he exemplifies with a historian, one that he generally values, to show how dishonest and incorrect this is: 'But one can see, e.g., in Taine: in the background he has preferences: for the strong expressive types, e.g., also for those who enjoy more than for the puritans' (ibid.). But is Nietzsche able to solve the problem of how to combine history with the setting of values, without it merely

4 Compare Nachlaß End of 1886–Spring 1887, KSA 12, 7[60].

appearing as prejudice? To answer that we need to examine his discussion of different types of historical writing.

Nietzsche's Explicit Discussion of Modern Historical Scholarship

Only once does the late Nietzsche explicitly discuss at any length modern writing of history, in the third essay of *On the Genealogy of Morality*. He divides it into two main forms, positivistic history and idealistic (or judging and artistic) history. Unfortunately, he does not explicitly tell us what sort of historical scholarship and historians we, in his view, ought to become, but I will argue for an interpretation which is consistent with what he states.

He begins by posing the question of the value of modern historical work, in a manner which resembles that of 'On the Uses and Disadvantages of History for Life' (but now it is less idealistically founded than in 1873, but still closely entwined with values).

> Or does modern historiography perhaps display an attitude more assured of life and ideals? (GM III 26, translated by Kaufmann)[5]

In his response to this question, he describes and discusses the two fundamental ways in which modern history is pursued. He begins with what can be called positivistic history (which attempts to be scientific and objective):

> Its noblest claim nowadays is that it is a mirror; it rejects all teleology; it no longer wishes to 'prove' anything; it disdains to play the judge and considers this a sign of good taste—it affirms as little as it denies; it ascertains, it 'describes'... All this is to a high degree ascetic; but at the same time it is to an even higher degree nihilistic, let us not deceive ourselves about that! One observes a sad, stern, but resolute glance ... here nothing will grow or prosper any longer, or at most Petersburg metaphysics and Tolstoian 'pity'. (GM III 26)

This is contrasted by modern idealistic history, where historical scholarship is combined with values and attempts to judge history:

> As for the other type of historian, an even more 'modern' type perhaps, a hedonist and voluptuary who flirts both with life and with the ascetic ideal, who employs the word 'artist' as a glove and has today taken sole lease of the praise of contemplation ... I know of nothing that excites such disgust as this kind of 'objective' armchair scholar, this kind of scented voluptuary of history,

5 Nietzsche is here asking for a 'more assured' than that of the natural sciences, which he had discussed in the previous section.

half parson, half satyr, perfumed by Renan, who betrays immediately with the high falsetto of his applause what he lacks, where he lacks it. (ibid.)

When Nietzsche compares these two types, there is no doubt that he prefers the scientific or positivistic to the idealistic sort:

Oh how these sweetish and clever fellows [the idealistic historians] make one long even for ascetics and winter landscapes! No! the devil take this type of 'contemplative'! I would even prefer to wander through the gloomy grey, cold fog with those historical nihilists! ... The 'contemplatives' are a hundred times worse. (ibid.)[6]

He even goes so far as to praise the former sort of history:

All honour to the ascetic ideal insofar as it is honest! so long as it believes in itself and does not play tricks on us! (ibid.)

In contrast, he goes on to vehemently criticize idealistic history at length. However, at the end of the section he points out that there are advantages with this overproduction of false ideals:

Europe is rich and inventive today above all in means of excitation; it seems to need nothing as much as it needs stimulants and brandy: hence also the tremendous amount of forgery in ideals [and he explicitly mentions 'the Christian-moral ideal'] ... With this overproduction there is obviously a new opening for trade here ... don't let this opportunity slip! Who has the courage for it?—we have in our hands the means to 'idealize' the whole earth! ... But why am I speaking of courage: only one thing is needed here, the hand, an uninhibited, a very uninhibited hand. (ibid.)

The section ends here, and the last two sections of the third essay do not continue this argument. However, it is clear what Nietzsche is referring to at the end of this discussion, namely the idea of eternal recurrence, which, according to Nietzsche, would crush those who have no ideals or false ideals (thus the need for courage), and the idea of the revaluation of all values, which would allow us to '"idealize" the whole earth' (GM III 26).

That Nietzsche is referring to those topics is not only suggested by the fact that he was at this time intensively working on writing a magnum opus in which these two topics were going to be foundation stones,[7] but also by the fact that he, at the beginning of the next section, explicitly states that he

6 Nietzsche, while making this comparison, even hints that he would prefer yet another type of 'historian' or scholar: 'Indeed, if I had to choose I might even opt for some completely unhistorical, anti-historical person' (GM III 26), and exemplifies this by the philosopher Dühring, of whom he normally is highly critical. In the early draft to this section, given in the Kommentar (KSA 14, p. 382), he explains why he would prefer this alternative: because Dühring attacks the whole history, we are persuaded to become its 'historian' and last judges.

7 See my 'Nietzsche's *Magnum Opus*' (Brobjer 2006).

will continue this discussion in that work: 'I shall probe these things more thoroughly and severely in another connection (under the title "On the History of European Nihilism"; it will be contained in a work in progress: **The Will to Power**, Attempt at a Revaluation of All Values)' (GM III 27).[8] This work was never completed, but among Nietzsche's notes we have the long, so-called Lenzer Heide-fragment, entitled 'European Nihilism', written on 10 June 1887 (KSA 12, 5[71]),[9] i.e., shortly before he wrote *On the Genealogy of Morality*, in which the discussion of nihilism is closely related to the idea of eternal recurrence, which itself is referred to as 'a hammer' and a most dangerous idea.[10] When Nietzsche reviews the *Genealogy* in *Ecce homo* he states that the essays constitute three preliminary studies for a revaluation of all values. He further claims, while summarizing the content of the third essay, that 'a counter-ideal was lacking — *until Zarathustra*' (EH III GM), meaning his own mature philosophy (in which the revaluation of values constituted a central tenet), as expressed metaphorically in *Thus Spoke Zarathustra* and which he attempted to develop and express more philosophically in the planned magnum opus.

What can we make of this account, and what does it say about Nietzsche's view of historical writing and research? We are given two or three ways of writing history, but Nietzsche seems to reject them all. It would be the wrong approach to ask which of these ways Nietzsche approves of, or with which one he shows most sympathy. The question of which *methods* to use for one (for any person) to become a good composer, painter, or historian is never sufficient for Nietzsche. Experience, appropriate values, a healthy view of the world (for example, not being a pessimist or being full of resentment), and personal characteristics are always required. Nietzsche's general approach is therefore not to approve of specific methods, but to argue that one should copy and learn from great predecessors. Compare, for example, how he recommends his students to become good classical philologists: 'How does one become a philologist?', where he answers not by methods but by exemplars: 'Start from the conception of the great philologists' (KGW II.3, p. 366).

---

8 In an earlier version of this section, Nietzsche had written 'my main work [*Hauptwerk*] which is in progress [*mein in Vorbereitung befindliches Hauptwerk*]' instead of 'a work in progress' (Kommentar, KSA 14, p. 382).

9 KSA 12, 5[71]. In the note immediately before this one, 5[70], Nietzsche also speaks of 'a history of values'.

10 Compare my discussion in 'Götzen-Hammer: The Meaning of the Expression "To Philosophize with a Hammer"' (Brobjer 1999).

What we see is that Nietzsche approves of positivistic or scholarly historical research as a means to understand the world and to avoid illusions and misconceptions. He is, for example, continually critical of those who, like the priests, 'lack all the preconditions of comprehension: who lack the neutrality of the historian' (Nachlaß November 1887–March 1888, KSA 13, 11[302]),[11] and all bad historians (such as most philosophers and those who discuss the history of morals).[12] However, we should realize that this sort of history can never achieve total objectivity, that values and personal preferences can never be fully excluded. More important is that this sort of history—when regarded not as a means but as an end—is and leads to nihilism.

Is a higher sort of historical research and writing possible? It seems as if Nietzsche believed that to be the case—that is what he suggests at the end of this section; a history which is intertwined with values—not the old corrupt and decadent Christian, moral, and modern values, but with revalued values. He says little about what sort of values these are—since he is still working on that and hopes to present that in his forthcoming magnum opus, entitled precisely 'Revaluation of All Values'—and thus he does not say anything about how that sort of constructive or evaluative history could appear.

However, a hint can be gained from the fact that the late Nietzsche emphasizes the importance of cultural history, and he praises the works of Ritschl, Burckhardt, and Taine, who all were skilful masters of modern historical methodology. Equally important is that his own book, *On the Genealogy of Morality*, shows us a way to be historians of values. Furthermore, Nietzsche explicitly states what sort of history and historian he hopes for, and offers the name of Thucydides, his favourite ancient historian. He calls him the perfect expression of 'realist culture', and continues:

> Thucydides: the great sum, the last revelation of that strong, severe, hard factuality which was instinctive with the older Hellenes. In the end it is courage in the face of reality that distinguishes a man like Thucydides from Plato: Plato is a coward before reality—consequently he flees into the ideal; Thucydides has control of himself —consequently he also maintains control of things. (TI 'What I Owe to the Ancients' 2)

11 This note was written shortly after the *Genealogy*.

12 See my longer discussion of this in 'Nietzsche's View of the Value of Historical Studies and Methods' (Brobjer 2004).

Nietzsche's longest and most detailed discussion of Thucydides occurs in his lectures on Greek literature, which he held twice in 1874/75.[13] He there refers to Thucydides as 'the greatest researcher and thinker among the historians' (KGW II.5, p. 225). Most of this discussion was made by the scholar and philologist Nietzsche, but later he added a long footnote with a more personal touch, in which he describes what sort of historical research and writing he dreams of encountering in the distant future: 'A history developed in the spirit of Thucydides and suffused with a still more profound philosophy than his was, remains my hope' (KGW II.5, p. 258).

According to Nietzsche, life, as well as history and science, are meaningless, unless we create a meaning for them.[14] Once the belief in God gave us and nature meaning, and subsequently truth and morality, but we now risk having only nihilistic approaches left. What we need is to create purpose and meaning, through a revaluation of all values, and that was the final task which occupied Nietzsche during the last years before his mental collapse. He never finished that work, and thus the question of how a corresponding non-nihilistic, value-affirming writing of history could be constructed also remained unanswered. We are left with the late Nietzsche having affirmed historical methods and approaches for the purpose of understanding the world and of disclosing illusions, as long as these methods and approaches are not made into fundamental goals, but remain merely means and an aid in dealing with and supporting our comprehension of the world.

References

Brobjer, Thomas H., 1999, 'Götzen-Hammer: The Meaning of the Expression "To Philosophize with a Hammer"', in: Nietzsche-Studien, 28, pp. 38–41.

Brobjer, Thomas H., 2004, 'Nietzsche's View of the Value of Historical Studies and Methods', in: Journal of the History of Ideas, 65, pp. 301–322.

Brobjer, Thomas H., 2006, 'Nietzsche's *Magnum Opus*', in: History of European Ideas, 32, pp. 278–294.

Brobjer, Thomas H., 2007, 'Nietzsche's Relation to Historical Methods and Nineteenth-Century German Historiography', in: History and Theory, 2/46, pp. 155–179.

13 KGW II.5. Historical literature is discussed on pp. 224–269, Thucydides is mentioned throughout, but specifically on pp. 235–246.

14 Compare Nietzsche's view of nature and cosmos, e.g., in Nachlaß June–July 1885, KSA 11, 38[12] (the final section, 1067, of *The Will to Power*) and Nachlaß Spring 1888, KSA 13, 14[188].

Translations

Human, All Too Human, trans. R. J. Hollingdale, Cambridge: Cambridge University Press, 1986.

On the Genealogy of Morality, trans. Walter Kaufmann and R. J. Hollingdale, New York: Vintage Books, 1966.

The Antichrist, in: *The Portable Nietzsche*, ed. and trans. Walter Kaufmann, New York: Viking Press, 1954.

Twilight of the Idols, in: *The Portable Nietzsche*, ed. and trans. Walter Kaufmann, New York: Viking Press, 1954.

Part II

Genealogy, Time, Becoming

Nietzsche's Timely Genealogy: An Exercise in Anti-Reductionist Naturalism

Tinneke Beeckman

Introduction

Nietzsche was the first thinker to develop a specifically philosophical notion of 'genealogy', thereby granting the word a sense quite different from the usual meaning, namely 'family descent' or 'pedigree'. Genealogy represents a new historico-philosophical method, which is history practised correctly. This 'correctness' means a genealogy is an active, critical, philosophical investigation, which expresses above all an urge to enquire rather than to formulate a definite solution to a problem. A genealogy also means a 'meticulous' interpretation of history (Foucault 1971, p. 145).

Nietzsche's concept of a genealogy is original vis-à-vis two streams of thought. Firstly, he subverts the usual intention of tracing a pedigree. This philosophical effort is widely discussed in academic writings, without it necessarily containing a naturalist viewpoint.[1] Secondly, Nietzsche offers revealing insights for those interested in the question of cultural evolution and thus for contemporary attempts to develop a naturalist theory of culture. Although the naturalist aspect is not always present in discussions of Nietzsche's work, I believe that Nietzsche's genealogy is a striking example of philosophical naturalism. Striking, because it offers something quite different from a reductive naturalist analysis of culture.

In this article, I shall concentrate on this second aspect and discuss how his genealogy continues to give tools for thinking about cultural phenomena in a contemporary and challenging way.

---

1 See Foucault 1971 and Geuss 1994. Geuss discusses five different aspects of the traditional interpretation of a pedigree: in the interest of a positive valorization (genealogy wants to legitimize), the pedigree assumes a singular origin, which is a source of that value and traces an unbroken line of succession from the origin to that item, and during the series of steps, the value in question is preserved. Nietzsche, according to Geuss, breaks with all these aspects.

I shall begin by briefly outlining what I mean by 'naturalism'. Then I will give two Nietzschean 'criticisms' of a reductionistic view on naturalism that are crucial to the idea of a correct genealogical method. The first criticism is related to the explanatory capacity of naturalism and to the problem of functionality and utility. The second aspect of naturalism is the question of the driving force for cultural evolution and change. For Nietzsche, this force is the 'will to power'. In this context, I shall refer to Nietzsche's antagonism towards Darwin as contrasted with Lamarck.

With regard to both comments, I will indicate how Nietzsche's approach offers very relevant and original questionings of reductive naturalism (e.g., evolutionary psychology and memetics). Today, the 'social sciences' are described as bulwarks of resistance against a biological, naturalist interpretation of the human being, that makes it seem that the true choice is between either accepting natural selection as the main explanatory mechanism for what is specifically human or having a hopelessly old-fashioned and narrow-minded viewpoint.[2] I believe this not only to be a false opposition, but also one that regrettably prevents a more stimulating dialogue between scientific research and philosophical reflection.[3] Seen in this light, a genealogy is a narrative that tries to explain a cultural phenomenon by describing a possible way of its coming about. This implies that culture evolves, although not necessarily solely according to the principles of natural selection.

One factor complicates the analysis of Nietzsche's naturalism, namely that he often seems rather confused about what Darwin actually said, for instance concerning teleology and the place of man in nature. When Nietzsche criticizes Darwin for supposing evolution means progress, he aims at Lamarck for the simple reason that Darwin never supposed humanity was on its way to perfection. This matter, unfortunately, occupies quite some space in Nietzsche's argumentation.

Another element is more relevant: Nietzsche felt a closer affinity to Lamarck than to Darwin. In the Nachlaß of 1885, he exclaims, for example, that without Hegel and Lamarck there was no Darwin—Darwin is just 'eine Nachwirkung', an after-effect (Nachlaß April–June 1885, KSA 11, 34[73]). The basic reason for this preference is clear: Lamarck offers another interpretation of the forces of evolution, hereby anticipating the 'will to power'.

---

2 E.g., Barkow et al. 1992. For a criticism of their view, see, e.g., Kitcher 2003.

3 See also Williams 2002.

What is Naturalism?

Nietzsche's genealogy offers a naturalistic account of history. Firstly, Nietzsche wants to understand the origins of morality without reference to supernatural entities. This idea is very prominent throughout his work. He resumes the essence of the *Genealogy* as such in *Ecce Homo*. In the Preface to the *Genealogy*, he stresses the continuity with *Human, All Too Human*. In this work, he strives to explain the highest manifestations of cultural activity in terms of the 'lowest'. 'Perhaps,' he asks himself, 'the whole of humanity is no more than a stage in the evolution of a certain species of animal of limited duration: so that man has emerged from the ape and will return to the ape' (HA I 247). In *Beyond Good and Evil*, he wants 'to translate man back into nature' (BGE 230). In the Preface to the *Genealogy*, Nietzsche recalls his attachment to a more theological comprehension of morality in his youth. But very soon, he stopped seeking the origin of good and evil beyond this world and rephrased the problem as: under which circumstances did man invent good and evil as judgements of values, and what is the value of these values in itself? The first impulse to effectively try to understand morality as a human creation, Nietzsche continues, came to him through Paul Rée's work. So, concerning the genealogy of morals, Nietzsche wants to understand the value of value, considering man as a part of nature. This is the first meaning of 'naturalism'.[4]

Secondly, naturalism refers to the natural sciences. Indeed, it is not a coincidence that the very idea of a genealogy coincided with major scientific developments (e.g., Lamarck, Darwin). As such, Nietzsche forces us to position ourselves with regard to scientific methods or results, and I believe this to be one of the lasting challenges of his work. How can we develop a philosophical method that is not in contradiction to science, but that allows at the same time for the best explanation of phenomena with a broader than 'scientific' scope? This means that we should not stick to the scientifically 'provable' solutions for human phenomena: reductionism is not an option. But the balance between recognizing the relevance of scientific results and a truly philosophical approach requires the skill—flexibility and strength—of a ballerina dancing *en pointe*.

In the context of naturalism, culture poses the following problem. Human beings are part of nature. They are genealogically related to other animals, but they are also different, for instance in the fact that they have language. Explaining the differences with other animals is the main concern of a naturalist approach. Claiming that culture is somehow different

4 For a further elaboration of this theme, see also Hatab 2005.

from mechanisms at work in 'the rest of nature' calls into question whether cultural evolution does not occur according to entirely different principles. Perhaps we are in need of different terms from the ones used to understand other species. It is exactly in this context that Nietzsche's genealogy has some suggestions to offer.[5]

The Question of Functionality (Anticipating a Criticism of Evolutionary Psychology)

The explanatory power of a naturalist theory seems to be generated by the analysis of functionality. But immediately, this means entering dangerous territory, because Nietzsche cautions precisely against a too narrow interpretation of functionality.

Nietzsche begins the *Genealogy* with a remarkable statement concerning his methodology. He opposes his genealogical project to another form of reading history by distinguishing between a 'correct' way of genealogical thinking, namely one that results in a real history of morality, and 'the back-to-front and perverse kind of genealogical hypotheses' (GM Preface 4). 'English psychologists'[6] (and Paul Rée) are at least partly responsible for our misunderstanding of moral history because they produce misleading histories. The first psychologist Nietzsche discusses is his former friend Paul Rée. According to Nietzsche, Rée is wrong to follow the English tendency by reducing the history and genesis of morality exclusively to a concern for utility. Nietzsche assumes this criticism is applicable to Darwinism in general, since he links Rée's thoughts directly to Darwinian theory. For instance, he contends: 'But he [Rée] had read Darwin' (GM Preface 7). Although Nietzsche mistakenly compares Rée and Darwin, his attitude is probably inspired by Rée's own quotations of Darwin and Lamarck in the introduction to his work.

What is the confusion? Nietzsche seems to consider Darwinism as a theory that supposes progress, in the sense of moral progress. The thought that humanity is on a path to perfection irritates Nietzsche considerably (Nachlaß Spring 1888, KSA 13, 14[123]). He therefore continues to explain that he doubts this idea of progress, because he fails to see in what

5 The central question for Nietzsche concerned ethical life, but this question posed in a naturalistic way, as Williams (2002) shows, means we ask about human ethical life in relation to the rest of human nature, thus in relation to motivations and drives of human psychology.

6 Nietzsche probably has Lecky's *History of European Morals* (1869) in mind when he refers to the 'English'.

sense the strong have won, on the contrary.[7] He thus applies Darwinism on a moral level, which certainly was not Darwin's intention. In fact, nothing about natural selection supports the idea of a progression towards perfection, only the idea of an increasing complexity. It is necessary to distinguish between scientific interpretation and the interpretation of value. Darwin was well aware that his theory could lead to such misinterpretation, which is why he reminded himself in his notebook 'never to say "higher" or "lower"' (Laland/Brown 2002, p. 65). This misreading of Darwin was unfortunately quite frequent: throughout the nineteenth century, Darwin's name had been unfairly linked to theories about social life, which were derived from Lamarck and Spencer. I shall return to a comparison between Lamarck and Darwin. Here, it is rather unfortunate that Nietzsche did not understand Darwin well enough to realize that his criticism was unjustified.

Nietzsche is actually criticizing social or German Darwinism, inspired by Lamarck, whose views were linear and progressive. Each species has its own, first ancestor (contrary to Darwin's tree of life). Species have an inherent striving to evolve greater complexity. The human being is the pinnacle of creation. Lamarck's central hypothesis to explain evolution was the inheritance of acquired characteristics. Offspring could inherit characteristics that the parents acquired during their lifetime. Unfortunately for Lamarck, biology proved him wrong. The fact that his conception of biology is untenable, however, does not mean Lamarck's importance has diminished in the context of cultural change.

In a famous passage in the *Genealogy*, Nietzsche highlights the core of his methodology. His most important objection is that naive genealogists of law and morality, like Rée, confuse 'origin' and 'purpose'. This inevitably leads to a false interpretation of history.

> But 'purpose in law' is the last thing we should apply to the history of the emergence of law: on the contrary, there is no more important proposition for all kinds of historical research ... —namely that the origin of the emergence of a thing and its ultimate usefulness, its practical application and incorporation into a system of ends, are *toto coelo* separate; that anything in existence, having somehow come about, is continually interpreted anew, requisitioned anew, transformed and redirected to a new purpose by a power superior to it. (GM II 12).

In other words, understanding something's utility does not shed light on its origin. Nietzsche rejects what he sees as a typical confusion of utilitarians. I believe Nietzsche's critique is equally relevant in the context of natu-

---

7 Another example see Nachlaß Spring 1888, KSA 13, 14[123].

ralism, although he did not realize to what extent his own incisive intuitions in this passage were thoroughly Darwinian—he was probably unable to recognize them as such. Darwin states: 'Nothing can be more hopeless than to attempt to explain the similarity of pattern in members of the same class, by utility or by the doctrine of final causes' (Gould 2002, p. 100). He himself cautioned 'against the mistake of inferring current function or meaning from ancestral function or meaning' (Dennett 1995, p. 465). Daniel Dennett, for example, admits that this fallacy is easily and frequently committed, particularly by evolutionary psychologists. What Nietzsche is effectively describing here comes close to Stephen Gould's idea of pre-adaptation or 'exaptation'. Gould's conception of 'exaptation' is methodologically important because it alters the way in which the genesis of function can be historically reconstructed (Gould/Vrba 1982). The possibility of 'exaptation' is certainly not a revolutionary invention with regard to adaptation since it continues to invoke some notion of function. It is significant, however, because it implies that a non-adaptive trait can exist without being selected against, so long as it is not maladaptive, before assuming a function different from the one it had previously served. This has enormous implications for what it means to practice 'decent' genealogical thinking, because it suggests that there is no direct line between what traits exists today and possible functions in the past. Thus we cannot reconstruct the coming into being of certain traits by reasoning in terms of 'best' or 'most adequate' solutions to problems of adaptation. A solution just has to be adequate enough, or present enough without disturbing or diminishing the functioning of the organism, in order to be transmitted.

In his criticism, it is as if Nietzsche is hinting not so much at Darwinism, but rather at sociobiology, recently resurrected under the name 'evolutionary psychology'. Spencer and Rée believed that they could detect the simple, straight path to altruism. Nietzsche's genealogy, by way of contrast, reveals that there is more than one point of origin, so a genealogy cannot imply a direct retracing over time. The history of morality (or any kind of cultural phenomenon) cannot be described in terms of a linear development. The idea that a value or a phenomenon has many different origins is consistent with Darwin and is an essential element of historical thinking.

Nietzsche gives a methodological warning, when he insists that a functional account can be given of items that were previously not seen as having that same function. Especially the link between utility and adaptation is a possible trap. This leads us to the next question, namely what driving force causes cultural changes and thus the changes of function?

The Driving Force of Evolution and Change: Time and Abundance

In 'On Uses and Disadvantages of History for Life' (UM II) Nietzsche opposes the so-called objective historical practice to 'life'. Later, when he develops his genealogical method more fully, he sees the 'will to power' as the driving force of change.

For Foucault, Nietzsche is *the* philosopher of power, and herein lies the contribution of his genealogical method. It is generally accepted that will to power does not denote any conscious, individual desire for power.[8] But how can will to power be understood in a naturalist reading? A good starting point may be found in *Twilight of the Idols*. There Nietzsche describes the will to power in opposition to the so-called 'struggle for life':

> Anti-Darwin. — As far as the famous 'struggle for existence' is concerned, this seems to me to be more of an opinion than a proven fact at the moment. It takes place, but as an exception; the overall condition of life is not a state of need, a state of hunger, but rather abundance, opulence, even absurd squandering. —Where there is a struggle, there is a struggle for power ... You should not confuse Malthus with nature. (TI 'Reconnaisance Raids' 14)[9]

In this passage, Nietzsche pointedly unites three thoughts. One, what Darwin says, is untenable. Two, this is because he focuses on the wrong struggle, namely the struggle caused by a lack, a shortage (hunger and distress), whereas life is abundance (wealth and luxury). Three, the correct representation of the struggle is a struggle for power.[10]

In order to understand how the will to power is an alternative to a Darwinian solution, I shall turn to Nietzsche's preference for Lamarck over Darwin. I believe Nietzsche saw in Lamarck an opportunity to rethink forces of cultural evolution. This assumption has consequences for a naturalist genealogy in a contemporary context.

The basic idea of Darwinism is natural selection. Darwin was struck by Malthus' analysis that population growth would always reach a point where food was not sufficiently available for everyone. The resulting

8 In this sense, my interpretation is thoroughly 'Deleuzian' (Deleuze 1962).

9 In *The Gay Science*, Nietzsche makes a similar point: 'But a natural scientist should come out of his human nook; and in nature it is not conditions of distress that are dominant but overflow and squandering, even to the point of absurdity. The struggle for existence is only an exception, a temporary restriction of the will to life. The great and the small struggle always revolves around superiority, around growth and expansion, around power—in accordance with the will to power which is will to life' (GS 349).

10 He repeats this 'Darwinian mistake' several times in the Nachlaß, especially in the piece entitled Anti-Darwin (Nachlaß Spring 1888, KSA 13, 14[123]).

shortage is a basic feature of demographic evolution. Darwin elaborated this idea: within a group or species, the existing variation between the members implies that those whose anatomical, physiological, and behavioural characteristics best fit the challenges of the environment also have the greatest chances of surviving and reproducing. Natural selection is a process of variation, selection, and inheritance. Repeated over and over again, the descendants of the original species will have new and distinctly different features. Complex organisms are the result of the accumulation of innumerable slight variations. One might therefore argue that the sole evolutionary 'force' is the *passing of time*: only time is abundantly present in the process of evolution. Put more precisely: only the limitless working of adaptation over time creates change, unfortunately at the expense of a huge number of 'victims'.

As an abstracted idea, some thought natural selection simply too good an explanation not to be applied to culture itself. Dawkins introduced the notion 'meme' for a cultural replicator. Memes possess variation, heredity, and differential fitness. Thus meme evolution is cultural evolution according to natural selection.[11] The world is full of brains—homes for memes (Blackmore 1999, p. 37). There are more memes than can a find a home. Understanding cultural evolution means finding the strategies and struggles of memes to inhabit brains. The continuous and severe struggle for existence is the basic model of evolution. This struggle is the logical consequence of an irresolvable lack or shortage: suffering and change are intertwined.

Nietzsche offers a fundamentally different model. He disagrees with a Darwinian explanation of cultural chance, because he cannot accept Darwin's emphasis on the environment as a generator of change. Instead, he prefers a concept of inner power or transformation. This is where Lamarck comes into the picture. He has two basic ideas: the first is the linearity and progress of evolution, whereas the second is the 'inheritance of acquired characteristics' as the driving mechanism of evolution. Nietzsche did not believe in the progress predicted by evolutionary theory and, as we saw earlier, (mistakenly) attributed this idea above all to Darwin. Concerning the specificity of cultural evolution, Lamarck's second idea is relevant. It is important to note that Lamarck, the zoologist, did not apply his ideas to culture. Nietzsche, however, aims precisely at understanding human productions.

11 Besides natural selection, modern Darwinism emphasizes chance events such as genetic drift and mutation, but these are for an evident reason not relevant in the attempt to explain cultural evolution.

What is the connection between Lamarck's ideas and Nietzsche's will to power? The environment, Lamarck supposes, alters the form of organisms. So organisms have to adapt to changes in the environment, by changing their habits. These, in turn, affect the form, through what he called 'inheritance of acquired characteristics': alterations in the habits of organisms can be transmitted directly to the offspring. The classic and simple example is: how come a giraffe has a long neck? Because the parents spot the fresh leaves high up the tree. Subsequently, their will to have a longer neck directly influences the next generation.

Unfortunately for Lamarck, this theory was refuted by Weismann in 1930: Lamarckism is genetically untenable. Evolution just does not work this way. Since cultural evolution implies a different perception of time, Lamarck's hypothesis is more attractive than Darwin's. During a certain period Lamarck's hypothesis seemed biologically appealing, because it allows for evolutionary change to occur in the direction of an adaptation within a very short period of time. Darwinian evolution implies that information is acquired solely by way of a genetic system and can only be transmitted at a specific point in time, namely conception. Consequently, an organism cannot evolve faster than it reproduces. An organism or a being develops abilities to react to short-term changes phenotypically, but long-term changes require at least a generation (Callebaut 1993, p. 388). This is why Darwinian evolution takes a very long time and makes a great number of 'victims'.

Culture, however, involves non-genetic learning. Therefore, it is useful to distinguish between Lamarckism at the genetic level and Lamarckism at the cultural level. Daniel Dennett suggests that the refutation of Lamarckism at the cultural level (as phenotypic transmission) is less obvious (Dennett 1995): the idea of the inheritance of acquired characteristics cannot be refuted for culture the way it can be for biology. Thus, a preference for Lamarck over Darwin in the context of cultural evolution does not mean leaving the naturalist or scientific perspective.

But the question of refutation is only part of the story. At first sight, Lamarck's theory is not altogether different from Darwinian evolution, apart from the genetic impossibility. After all, Lamarck maintains the idea that changes occur in function of adaptations. The crucial difference, however, especially with regard to Nietzsche's genealogy, is his introduction of a transformative and purely active power that intervenes prior to the moment of adaptation. Lamarck has a specific word for the driving force of evolution: 'la volonté', the will.

Time is no longer the force that enables evolution to happen. With his preference for Lamarck, Nietzsche seems to conclude that evolution is indeed, abundance, but not of time. Furthermore, this notion of 'power' makes the Lamarckian view on culture irreconcilable with the idea of natural selection. Returning to the contemporary debate, this aspect has important implications, although these are not recognized as such. According to Dennett, for instance, assuming cultural evolution is Lamarckian comes down to saying culture is about learning. But the application of Lamarck's biological theory to cultural change is more fundamental than the rather tautological idea that culture involves learning. Or put differently, something more fundamental opposes the application of natural selection to cultural phenomena. According to Aunger, memetics can become a science if it can answer the following questions: whether selection is directional rather than neutral and whether it allows for the identification of the selective agent (Aunger 2000, p. 14). But there can be no 'agent' in the analogy with natural selection, because there is no concept of active force. Brains, it seems to me, are just passive receptors of invading memes. Surprisingly enough, some debate whether memetics is a Lamarckian theory, weary and cautious not to come up with an 'unscientific idea'.[12] The answer sounds more like: unfortunately not.

For Nietzsche, Lamarckism applied to culture leads to an essential difference: action versus reaction. The problem with Darwinism is that it only captures reaction. Physiology and biology thus lack an elementary conception of activity. Adaptation is not active, but reactive. Nietzsche emphasizes Spencer's fatal mistake: to see life itself as an inner adaptation to external circumstances. This interpretation, however, denies life an essential activity: the priority of the spontaneous, assertive, affirming, and transformative forces which Nietzsche calls the will to power.

Nietzsche's turn against the reactive forces displayed by the natural sciences is, however, only the beginning of his genealogical journey. He subsequently elaborates a theory of how the conflicting forces influence consciousness and the inner 'eye' of morality—and because they affect the interpretation of history they require a genealogical correction.

Conclusion

I hope to have shown that the Lamarck–Darwin debate in Nietzsche's thought is more than a simple recapitulation of nineteenth-century science. To this day, one of the truly puzzling questions concerns the driving forces

12 Whether this is the case would depend on the description of meme genotype and meme-phenotype analogies (Blackmore 1999).

of evolution that are called upon to explain phenomena.[13] Struggle, lack, and shortage are central ideas in many domains of the humanities: in economy, social theory, psychoanalysis, and cultural studies, to name but a few. There always seems to be more desire than can be fulfilled, more individuals present than can survive, more units that want to multiply than can be transmitted. And it is only 'natural', and thus inevitable, that some are left out. But what if, Nietzsche seems to suggest, a different principle guides human interaction, namely the struggle between active and reactive forces? Such a principle would also give better insights into creativity and other 'human' characteristics. It is not naturally in contradiction either with science or with naturalism (after all, memetics is not a science and evolutionary psychology has its difficulties). It might avoid the trap of reductionism and reintroduce plurality and multiplicity that are so typical for all human creations, including morality.

References

Aunger Robert, 2000, Conclusion, in: Darwinizing Culture: The Status of Memetics as a Science, Oxford: Oxford University Press, pp. 205–232.

Barkow, Jerome / Cosmides, Lena / Tooby, John, 1995, The Adapted Mind: Evolutionary Psychology and the Generation of Culture, New York: Oxford University Press.

Blackmore, Susan, 1999, The Meme Machine, Oxford: Oxford University Press.

Callebaut, Werner, 1993, Taking the Naturalistic Turn, or How Real Philosophy of Science is Done, Chicago: University of Chicago Press.

Deleuze, Gilles, 1962, Nietzsche et la philosophie, Paris: Presses Universitaires de France.

Dennett, Daniel, 1995, Darwin's Dangerous Idea: Evolution and the Meaning of Life, London: Penguin.

Foucault, Michel, 1971, Nietzsche, Genealogy, History, in: Bachelard, Suzanne / Canguilhem, Georges / Dagognet, François / Hyppolite, Jean, Hommage à Jean Hyppolite, Paris: Presses Universitaire France.

Geuss, Raymond, 1994, 'Nietzsche and Genealogy', in: European Journal of Philosophy, 2, pp. 275–292.

Gould, Stephen, 2002, The Structure of Evolutionary Theory, Cambridge, MA: Harvard University Press.

Gould, Stephen / Vrba, Elisabeth, 1982, 'Exaptation—A Missing Term in the Science of Form', in: Paleobiology, 8, pp. 4–15.

Hatab, Lawrence J., 2005, Nietzsche's Life Sentence: Coming to terms with Eternal Recurrence, New York: Routledge.

13 For a recent update on the interaction between culture and environment, see Richerson/Boyd 2005.

Kitcher, Philip, 2003, 'Giving Darwin his Due', in: Hodge, Jonathan / Radick, Gregory (eds.), The Cambridge Companion to Darwin, Cambridge: Cambridge University Press.
Laland, Kevin / Brown, Gillian, 2002, Sense and Nonsense. Evolutionary Perspectives on Human Behaviour, Oxford: Oxford University Press.
Lecky, W. E. H., 1869, History of European Morals from Augustus to Charlemagne, 2 vols., London: Longman Green.
Richerson, Peter / Boyd, Robert, 2005, Not by Genes Alone. How Culture Transformed Human Evolution, Chicago: University of Chicago Press.
Williams, Bernard, 2002, Truth and Truthfulness. An Essay in Genealogy, Princeton: Princeton University Press.

Translations

Human, All Too Human, trans. R. J. Hollingdale, Cambridge: Cambridge University Press, 1986.
On the Genealogy of Morality, ed. Keith Ansell-Pearson, trans. Carol Diethe, Cambridge: Cambridge University Press, 1994.
Twilight of the Idols, in: The Anti-Christ, Ecce Homo, Twilight of the Idols, ed. Aaron Ridley, trans. Judith Norman, Cambridge: Cambridge University Press, 2005, pp. 153–229.
The Gay Science, ed. Bernard Williams, trans. Josefine Nauckhoff and Adrian del Caro, Cambridge: Cambridge University Press, 2001.

From Kantian Temporality to Nietzschean Naturalism

R. Kevin Hill

Interpretation of Nietzsche's mature thought often focuses on his method of genealogical critique, his perspectivism, his ontology of becoming, and his naturalism. I want to suggest that Nietzsche's naturalism is best understood as a thesis about the nature of space and time, a problem he inherited from Kant and Schopenhauer. Briefly, Nietzsche could not accept both the idea that space and time, and hence those real things, events or processes that fill them, are mind-dependent in the sense that Kant and Schopenhauer intended, while also maintaining that the mind is itself something that occurs within nature, as Schopenhauer had maintained. And though Nietzsche was concerned about the status of both space and time, he was especially concerned with the status of time.

In the fragment 'On Schopenhauer', Nietzsche early argued that Schopenhauer's use of Kant's theory of space and time was unacceptable. For Schopenhauer, the mind which constructs empirical phenomena is the human brain, itself an empirical phenomenon. Nietzsche is quick to point out that this position leads to incoherence.

> [The] intellect is supposed to appear, i.e., the flower of knowledge is alleged to have suddenly and abruptly blossomed forth from a world that does not exist. Moreover, all of this is supposed to have taken place in a realm devoid of time and space, without the mediation of causality: yet that which has its origin in such an unworldly world must—according to Schopenhauer's tenets—be a thing-in-itself: either the intellect is now a new attribute, forever combined with the thing-in-itself; or there can be no intellect, because an intellect could never have come into being. But there does exist one [an intellect]: consequently, it could not be an instrument of the world of appearances, as Schopenhauer claims, but a thing-in-itself, i.e., a will. Thus Schopenhauer's thing-in-itself would be both *principium individuationis* and the ground of necessity: in other words: the world as it exists. ('On Schopenhauer' 4, Nachlaß Autumn 1867–Spring 1868, KGW I.4, 57[55])

The collapse of Schopenhauer's view leaves us with only two possibilities. We can abandon Schopenhauer's identification of the Kantian intellect with the brain and return to a purer Kantian position which locates the human intellect outside nature. Alternately, we can preserve the identification of the human intellect with the Kantian intellect, but reject the claim that the intellect produces the space and time of nature.

Because Nietzsche was 'reluctant to deduce space, time and causality from our pitiful human consciousness' (Nachlaß September 1870–January 1871, KSA 7, 5[81]), he initially responded to this conundrum with the daring and extreme expedient of considering space and time as the mind-dependent projections of an intellect other than our own: the primordial unity, the primordial intellect, the Dionysian world-artist. This way Nietzsche can say we all share a common space and time, while evading the difficulties raised by identifying the brain with the human intellect. For now it is not the human intellect which, in producing space and time, produces the brain and therefore itself. Rather the primordial intellect produces space and time and by that produces the brain.

However, by 1873, in 'On Truth and Lying', Nietzsche appears to be asserting the Kantian view but with a twist:

> But everything which is wonderful and which elicits our astonishment at precisely these laws of nature, everything which demands explanation of us and could seduce us into being suspicious of idealism, is attributable precisely and exclusively to the rigour and universal validity of the representations of time and space. But these we produce within ourselves and from ourselves with the same necessity as a spider spins. (OTL 1, KSA 1, p. 885)

However, Nietzsche also seems to be asserting the paradoxical Schopenhauerian view that identifies the Kantian intellect with the brain.

> Someone could invent a fable like this and yet they would still not have given a satisfactory illustration of just how pitiful, how insubstantial and transitory, how purposeless and arbitrary the human intellect looks within nature. (OTL 1, KSA 1, p. 875)

I see no way of reconciling these strands other than by assuming that there are two different natures: the nature which is the object of natural science and the nature which contains the brain, nature as it seems to be and nature as it is. Such a view is compatible with the notion of the primordial unity that projects its own space and time. Nietzsche had expressed this notion in public one year earlier, in *Birth of Tragedy*, and in notes dating back even earlier. Such a view merely suggests that we cannot know through geometrical intuition what the geometry of the primordial unity's space is. For all we know, it may be the same as the geometry of our own phenomenal space. Early Nietzsche, following Kant, links natural science (i.e., Newtonian physics) to our forms of intuition. Since our forms of intuition generate the space and time referred to in Newtonian theory, it follows that Newtonian theory does not give us access to nature as it really is, the nature within which we, and our brains, are embedded. Though real nature may be Newtonian, it may very well not be. We do not know if phenomenal space resembles nature's space. The result is scepticism.

This contrast between the two spaces and times, one known but imaginary, the other real but unknown, persists long after Nietzsche had disowned the 'artist's metaphysics' of *Birth of Tragedy*: '[O]ur space is valid for an imaginary world. Of the space, which belongs to the eternal river of the things, we know nothing' (Nachlaß Spring–Autumn 1881, KSA 9, 11[155]). And even later: 'By my showing the subjective genesis, e.g., of space etc. the thing in itself is neither refuted nor proven. Against Kant—' (Nachlaß Summer–Autumn 1884, KSA 11, 27[68]). The mature view differs from the view in *Birth of Tragedy* and 'On Truth and Lying'. First, Nietzsche rejects the notion of the projection of a common space and time by the primordial intellect. Second, though the notion of the Kantian intellect developing over time is one Nietzsche carried over from Schopenhauer, Nietzsche 'Darwinizes' it. The specific character our forms of intuition possess is due to their selective advantage over equally possible competitors. Had rival intellects appeared, they would have perished, leaving the field to nature's favoured Euclideans.

> The categories are 'truths' only in the sense that they are conditions of life for us: as Euclidean space is such a conditional 'truth'. (Between ourselves: since no one would maintain that there is any necessity for men to exist, reason, as well as Euclidean space, is a mere idiosyncrasy of a certain species of animal, and one among many …) (Nachlaß Spring 1888, KSA 13, 14[152])

Now as the original formulation of the 'neglected alternative' thesis had it, the relevant contrast was between transcendentally ideal space and time and transcendentally real space and time, the space and time of things-in-themselves. But Nietzsche rejects the intelligibility of a contrast between appearances and things-in-themselves. If the very idea of the thing-in-itself is unacceptable, then one would think that Nietzsche would be forced back into a strict Kantianism, for the very idea of a contrasting domain that might differ from the domain in which we find phenomenal space has lost all purchase. Matters are further confused by Nietzsche's repeated characterization of phenomenal space as a falsification. This not only suggests the existence of a contrasting domain, but even of some sort of access to it. We see assertions seeming to confirm this, as when he says, 'I believe in absolute space as the substratum of force: the latter limits and forms. Time eternal. But in themselves there is neither space nor time' (Nachlaß June–July 1885, KSA 11, 36[25]). In this last passage in particular, he seems to give with one hand what he takes away with the other: how can space and time be 'absolute' and 'not exist in themselves'?

We begin to get a clue to what Nietzsche has done with his Kantian inheritance in the following: 'Denial of empty space and the reduction of mechanics to the tyranny of the eye and the touch. Denial of actio in distans' (Nachlaß Autumn 1885, KSA 11, 43[2]). There is one possible explanation for these Kantian themes in the late Nietzsche's view of space. The contrast is not between the space human beings construct versus the space of the thing-in-itself. The contrast is between the space we directly experience by virtue of our evolved, innate, space-producing psychological mechanisms versus the space our best empirical theory posits. However, this latter contrast does not commit Nietzsche to the claim that the space of natural science is mind-independent. Possibly, nature as the late Nietzsche understands it is mind-dependent (i.e., is not to be identified with the Kantian thing-in-itself). Meanwhile we can still distinguish between the world as it appears in naive, uncorrected experience, and the world as it appears in scientifically corrected experience.

The heart of Nietzsche's *naturalism* is his commitment to the existence of two spaces and times. One pair is produced by the human intellect. The other pair is that space and time within which the human intellect is embedded (see Richardson 2006, pp. 214–216). The central claim of Nietzsche's naturalism is that everything we are as individuals takes place in nature. Now it will be familiar to contemporary readers of analytic philosophy to identify nature with whatever it is that our best empirical theories posit as existing. For this sort of naturalism, a posit is acceptable if we can reduce the theory in which the posit appears to our best empirical theory (typically, physics).

In 'On Truth and Lying', Nietzsche embeds human beings and their cognitive capacities within nature, while being a sceptic about natural science's ability to represent nature. Apparently for early Nietzsche in 'On Truth and Lying', nature cannot be identified with the posits of natural science. This suggests that early Nietzsche's naturalism is actually a form of transcendental realism about the space and time we are embedded in, coupled with the thesis, inherited from Kant, that natural science only represents the phenomenal world. But that cannot be Nietzsche's last word, because by the 1880s, Nietzsche is claiming that there are no things-in-themselves, and yet he continues to endorse some sort of naturalism, not only in the sense that we are embedded in a space and time not of our making, but also in the sense of a much higher epistemic value associated with natural science itself. This suggests that the late Nietzsche parts company with Kant by refusing to identify the space of physics with phenomenal space, the space produced by our minds in organizing our sensory data. If that is correct, the common denominator between early and late Nietzsche is a rejection of the idea that there can only be a space and time whose origin is the human mind; the difference between the early and late Nietzsche is that the early Nietzsche associates the space and time which the human mind produces with the space and time with which physics is concerned, whereas the late Nietzsche is willing to say that the space and time which is not the product of the human mind is the one with which physics is concerned. Whether Nietzsche can coherently maintain this while denying the existence of things-in-themselves remains to be seen.

Yet thus far, we have seen no underlying *reason* motivating Nietzsche's departure from transcendental idealism in favour of naturalism. All the twists and turns of his thoughts on space and time go back to the original antithesis we mentioned above: the impossibility of reconciling a Kantian theory with a Schopenhauerian identification of the Kantian intellect with the brain. A reasonable conclusion would be to return to the original Kantian position: distinguish the Kantian intellect from the brain and allow the former to produce the latter without bootstraps. To grasp why Nietzsche believed that path was unavailable, we must turn to his views on time.

Nietzsche also began his thoughts about time as a Kantian, given his Schopenhauerian starting point. While he came to regard Schopenhauer's naturalized Kantianism as unstable, he also concluded that Kant's theory of time fails on its own terms. In *Philosophy in the Tragic Age of the Greeks*, in the discussion of Parmenides, Nietzsche identifies Parmenides' view of time with Kant's, and then endorses Afrikan Spir's objection:

> 'Now first, it is plain that I can know nothing of a succession as such, if I do not have the consecutive links together at once in my awareness. The representation of a succession is therefore not at all successive, consequently it is also altogether different from the succession of our representations. Second, Kant's assumption implies such apparent absurdities, that it takes a miracle to see how he could let them go unheeded. Caesar and Socrates are on this assumption not really dead, they live still even as well as they did two thousand years before and only seem dead to me because of the equipment of my "inner sense". Future persons now live already, and if they have still not stepped forward as living, for this too the equipment of the "inner sense" is responsible. Here one must ask above all: How can the beginning and end of conscious life, together with all its inner and outer senses, exist only in the apprehension of the inner sense? The fact is, one cannot deny the pervasive actuality of change. Throw it out the window, it slips back in through the keyhole. One says "it seems to me clear that conditions and representations change"—however, this appearance itself is something objectively existing and in it succession has undoubted objective actuality, inside it something really follows one after the other. —Furthermore one must notice that the whole criticism of reason could have rationale and legitimacy only on condition that our representations appear as they are. Then if representations had appeared differently than they really are, one would be able to make no valid claims about them, and therefore one could erect no theory of knowledge and no "transcendental" investigation of objective validity. However it is beyond doubt that our representations appear to us as successive.' (PTAG 15)

Spir's weak objection is that transcendental idealism about time is simply incredible if it means that 'Caesar and Socrates are on this assumption not really dead'. The stronger objection is that 'our representations appear to us as successive'. Even if we do construct appearances from our sensory data and impose spatial form on them in the process, sensory data themselves are presented to us in temporal succession. Therefore, there is at least one domain in which temporal succession is real: the domain of our own representations. What is more, any operations on our representations can only be conceived of as successive, and Kant's transcendental psychology is shot through with references to constructive *processes*; yet the very idea of an atemporal process makes little sense on the face of it.

We have, then, a path from within Kantian thought away from transcendental idealism, at least about time. For even if we do not speculate about the character of the world outside the mind, transcendental idealism about time is undermined even within the mind itself. We have no choice but to embed the mind within a time extending before the existence of the mind and continuing long after its destruction. Having been forced to make this move, there seems little point in not embedding the mind within space as well, at which point there are no further problems of *Kantian* origin to prevent the identification of the human intellect with the brain.

There is an apparent problem with the interpretive claims we have made thus far. Recall that we said that Nietzsche rejects Kant's claim that the space and time of nature are the space and time our cognitive apparatus imposes on our sensations. Now for Kant, there are really only two possible positions: that space and time are transcendentally ideal or that they are transcendentally real, where the former means that they are mind-dependent and the latter means that they are features of things-in-themselves. For Kant, the realm of nature is precisely the spatiotemporal realm. So this question about the status of space and time, for Kant, just is the question of whether nature is mind-dependent or whether nature is real 'in itself'.

Nietzsche agrees with Kant that our minds impose a spatial form on our sensations such that space appears to be Euclidean and in some regions empty. The same holds true for time. Though the 'geometry' of time is simpler by two dimensions, it does have a linear structure by which one can infer, e.g., that if t_2 is after t_1 and t_3 is after t_2, then t_3 is after t_1, etc. Also, just as regions of space can be empty, so too can moments of time. Nietzsche diverges from Kant in his refusal to identify this phenomenal space and time with the space and time of nature. He is willing to posit a space and time of nature, which may differ from our intuitional space and time as to geometry, emptiness, etc. Consequently, in rejecting Kant's identification of these two spaces (and times), Nietzsche must also reject

Kant's thesis that the geometrical properties of physical space are knowable a priori.

However, on Kant's terms, this would be to reject the transcendental ideality of physical space and time, and to affirm their transcendental reality. For Kant, this is just to say that physical space and time (and whatever fills them) are things-in-themselves or transcendentally real relations between things-in-themselves. Notoriously, Nietzsche in his latest writings denies that the very idea of a thing-in-itself is coherent, and thus denies that there are any such things. Thus, one would think, if space and time cannot be transcendentally real (because nothing is transcendentally real, the very concept being incoherent and empty) they must be transcendentally ideal. This would collapse Nietzsche's position into something not unlike Kant's again.

Here, however, appearances can be misleading. Kant himself allows that natural science can posit entities, events, processes, etc. which are 'empirical' but unobservable for reasons relating to contingent limitations on our cognitive capacities. Even outside science, in the domain of ordinary life, there are countless cases of such 'empirical' speculation. We may never know the precise number of people who shot at President Kennedy; if there was more than one person, these other parties went (presumably) unobserved, left insufficient traces behind to confirm their existence, etc. Nonetheless, it is an empirical fact, not a noumenal fact, that the number of assassins was one (or two, or three). Historians and conspiracy theorists may posit the existence or nonexistence of these additional assassins, based on available empirical evidence. In natural science, there are theoretical posits which may not be directly observable, which are posited based on available empirical evidence. Kant is not an instrumentalist about such posits. They fall within the domain of the phenomenal. They are no less a part of the world of empirical fact for being hidden, small, or far from observers. Whether this is the stance that Kant ought to have taken is a question I leave aside; the point is that for Kant, quarks would be 'empirically real and transcendentally ideal' despite being unobservable.

What I want to suggest, then, is that for Nietzsche, the space (and time) of nature are empirically real but unobservable posits not unlike atoms. The fact that we cannot directly perceive the world in anything but Euclidean terms is analogous to the fact that we cannot perceive objects smaller than a certain size. However, based on the empirical evidence we possess, we can suppose that there are imperceptible quarks; similarly, based on the empirical evidence we possess, we can suppose that the space of nature is non-Euclidean. Whether our predisposition to perceive the world as Euclidean helps or hinders us in getting the geometry of physical space right is a question about nature. It is independent of whether nature as a whole is mind-dependent.

Nietzsche's understanding of what a thing-in-itself is *is* simply the notion of a completely mind-independent object; in denying that there are things-in-themselves, Nietzsche can be understood to be asserting a claim analogous to Berkeley's claim that to be is to be perceived. We must expand the scope of the perceived to encompass both the perceivable but not presently perceived. We should also include theoretical posits of the same general character as medium-sized objects, which for peculiar reasons are unobservable (the extremely small and the extremely remote). The sum of such objects, events, processes, etc. is what Nietzsche conceives nature to be. However, there is no obstacle to Nietzsche regarding the entire natural world as mind-dependent. Nietzsche's persistent claims to the effect that we falsify the world because our cognitive capacities have evolved under conditions which select for expedient falsification, are completely independent of his metaphysical claims that the world is mind-dependent, or that the very idea of a mind-independent world is incoherent. The distinction between the falsified world and the not-falsified world should be made within the sphere of the mind-dependent. The sphere of the mind-independent is empty.

One worry concerning the interpretation I have offered so far concerns the status of unperceived posits of our empirical theory. If we assimilate Nietzsche's view to Kant's, it seems we would have to regard Nietzsche as some sort of phenomenalist. We have already seen that when Schopenhauer defended his own idealism from incoherence, he fell back on quasi-phenomenalist construals of statements concerning unobserved objects. Yet Nietzsche rejected these construals in 1868 in 'On Schopenhauer'. We have no further discussion of phenomenalism in his texts, so we must assume that this was a commitment he stood by. Without phenomenalism, how can Nietzsche distinguish a space and time produced by our sensibility from a space and time of nature? Such a nature is both mind-dependent in Berkeley's sense, and yet contains objects not currently perceived by us.

Nietzsche has available to him only the three options Berkeley had available to him: either some mind perceives all of nature (God, and therefore an option late Nietzsche would reject), or the unobserved objects really do not exist (which is incredible), or there are enough finite minds in existence to guarantee that nothing we want to posit in our empirical theories goes unobserved. I believe that Nietzsche was committed to the third, peculiar view, though not in a completely arbitrary or ad hoc fashion.

According to Nietzsche, the cosmos is essentially as Roger G. Boscovich characterized it: fields of force are the fundamental reality, and these fields permeate all of existence—where they are not, there is literally nothing, not even empty space and time. However, Nietzsche also appears to have endorsed a form of panpsychism regarding these fields of force. His reasoning for this is not our primary concern; he thought the idea of force makes no sense unless we understand forcing and being forced to be something undergone, felt, something (in our sense of the word) mental. Thus every field of force will have its corresponding 'feel' as it presses on other fields and is pressed upon in turn. If force possesses rudimentary awareness of its milieu, and force is pervasive, then the Berkeleian problem of unobserved items is solved. Everything is a perceiver because it undergoes influence from other fields of force. Also, everything is being perceived because there are fields of force upon which it expresses itself. Panpsychism thus allows Nietzsche to escape from the most untoward consequences of the *esse est percipi* principle. It allows Nietzsche to continue to affirm the existence of a nature within which we are embedded. Though it is permeated with mind, Nietzsche's nature *transcends* us. Our knowledge of it may well be imperfect, thus affirming a distinction between how things seem and what is so.

Some readers understand Nietzsche's perspectivism to be more fundamental than the thesis that the world is will to power. For them, it may seem that I have not done justice to Nietzsche's thesis that there is no thing-in-itself, that *nature* brings the thing-in-itself in through the back door. Conversely, one might think that calling Nietzsche's nature 'mind-dependent' stretches the meaning of that expression to its breaking point. I think the resolution of this difficulty is to see Nietzsche's concept of the thing-in-itself as meaning less than it is typically taken to mean.

A 'thing-in-itself' is an object whose characteristics obtain independent of all observers. To deny that there is such a thing does not mean that things become reduced entirely to episodes within the experiences of observers without residue. There is a difference between regarding things as collections of mental events, and regarding things as *powers* (*Mächte* or *Kräfte*). An alternative to eliminating the thing altogether, or reducing it to the sum of observers' experiences, is to regard it as a locus *affecting* observers. The issue is not whether the thing has mind-independent qualities, but whether its mind-dependent qualities inhere in it or in its perceivers. It would be natural to ask the question, 'What, apart from the affections it produces in observers, is the object really like?' If one replies, 'Well, the object surely has characteristics, but we will never know what they are,' then one is committed to the object being a thing-in-itself. If, conversely, one says that the very idea of what the object is like apart from how it affects observers makes no sense, then one is denying that there is a thing-in-itself in Nietzsche's sense. However, clearly there is something out there, from which flows the sum of the effects; this something is nothing other than the locus. Nietzsche reinterprets things as complexes of power relations in which observers are always involved; he does not reduce things to sums of episodes within subjects. This view, I believe, is remarkably like Berkeley's in certain respects, as seen in the following passages.

> The properties of a thing are effects on other 'things': if one removes other 'things', then a thing has no properties, i.e., there is no thing without other things, i.e., there is no 'thing-in-itself'. (Nachlaß Autumn 1885–Autumn 1886, KSA 12, 2[85])

> That things possess a constitution in themselves quite apart from interpretation and subjectivity, is a quite idle hypothesis: it presupposes that interpretation and subjectivity are not essential, that a thing freed from all relationships would still be a thing. Conversely, the apparent objective character of things: could it not be merely a difference of degree within the subjective? (Nachlaß Autumn 1887, KSA 12, 9[40])

Notice the strikingly Berkeleian character of these remarks. One could almost say that Nietzsche is reliving an earlier philosophical drama, and has cast Kant in the role of Locke and himself in the role of Berkeley. It is true that for Nietzsche, as for Kant, the mind is not a genuine *substance*. But it would be a mistake to think that because Nietzsche cannot believe that only mental *substance* exists, Berkeleian patterns of argument are not afoot. The rejection of a strong 'veil of perception' doctrine, a critique of the idea that a substratum of properties can be conceived independent of the properties it instantiates, the identification of sensible properties with sensory states, and the doctrine that the only causality possible is between minds, all bear the marks of deep Berkeleian commitments.

If we take this view, it is still easy to see how Nietzsche, like Berkeley, could think that his view mitigated sceptical difficulties considerably. It is not that we cannot be barred from access to reality because there is no reality to be barred from. It is that we cannot be barred from access to reality because there are no characteristics of reality apart from those which produce experiences.

This view, as I understand it, is reminiscent of Berkeley's in yet another way. For on Berkeley's view, it is not just that there are only experiences and experiencers. Rather, experiences are caused by something external to them that places them in us: God. God's intentions, then, occupy the same role that mind-independent things do in a realist view: they are not the experiences, nor the sum of experiences, but a locus that produces them. Nietzsche himself seems to see this similarity, when he makes the very Berkeleian remark that '"Will," of course, can affect only "will"—and not "matter" (not "nerves," for example—)' (BGE 36). The difference from Berkeley, then, is not really at the level of fundamental metaphysics, but at that of theology. Where Berkeley has one God, Nietzsche has many gods, for that is what each 'thing' is: a *petit* Berkeleian deity producing experiences in the minds it affects.

Nietzsche's metaphysical commitment to the reality of time is also central to his 'genealogical' critique of Kant. First, rejection of a beginning to time is of a piece with Nietzsche's rejection of the Judeo-Christian tradition, with its conception of a creator God. Second, Nietzsche's concern to vindicate the reality of time is not only of a piece with (though not strictly required by) the historicist aspect of genealogical critique; Nietzsche sees Kant's rejection of the reality of time as symptomatic of a more general failing of philosophers since Parmenides. By making time transcendentally ideal, Kant can distinguish between an apparent world, in which there is destruction, and a real world, in which there is no destruction. This contrast, in turn, allows Kant to associate that which he values (noumenal agency) with a domain distinct from the world of experience and immune to decay. Such a line of thought is symptomatic, Nietzsche thinks, of an inability to cope with, and subsequently a hatred for, life itself.

> You ask me which of the philosophers' traits are really idiosyncrasies? ... For example, their lack of historical sense, their hatred of the very idea of becoming, their Egypticism ... Death, change, old age, as well as procreation and growth, are to their minds objections, —even refutations. Whatever has being does not become; whatever becomes does not have being ... But since they never grasp it, they seek for reasons why it is kept from them. 'There must be mere appearance, there must be some deception which prevents us from perceiving that which has being: where is the deceiver?'—'We have found him,' they cry ecstatically, 'it is the senses! These senses, which are so immoral in other ways too, deceive us concerning the true world ...' (TI '"Reason" in Philosophy' 1)

The above described line of thinking is hardly the official basis for Kant's doctrine of the phenomenality of time. However it is surely fair to say that Kant is happy to exploit the phenomenal/noumenal contrast. It is crucial to his arguments licensing faith in such transcendent goods as free will, immortality, and God, as we see in the second *Critique*. Nietzsche's assessment of such a metaphysics is that

> Any distinction between a 'true' and an 'apparent' world, whether in the Christian manner or in the manner of Kant (in the end, an underhanded Christian)—is only a suggestion of decadence, a symptom of the decline of life. (TI '"Reason" in Philosophy' 6).

References

Richardson, John, 2006, 'Nietzsche on Time and Becoming', in: Keith Ansell-Pearson (ed.), A Companion to Nietzsche, Oxford: Blackwell, pp. 208–229.

Nietzsche's Problem of the Past

John Richardson

Nietzsche has a problem with the past. He thinks we all have a problem with it, indeed several interlocking problems, whose chief root he tries to identify. His repeated attention to this topic, coming at key points in his texts, amounts almost to a fixation.

My aims are to point out this repeating theme, which I think has been under-recognized, but more importantly to suggest the underlying reasons Nietzsche has for making the past a problem. And I'll sketch how he ultimately handles this problem—a solution that draws jointly upon his genealogical method, his ideal of freedom, and his thought of eternal return. For this purpose I will use and extend certain lines of interpretation and argument that I presented in *Nietzsche's New Darwinism* (Richardson 2004).

Let me start by reminding of a few places in Nietzsche's corpus that treat the past explicitly and with emphasis.[1] In each of them our—humans'—relation to the past is a *problem* we have difficulty addressing—and Nietzsche offers to help us with it.

a. First an early passage. 'On the Uses and Disadvantages of History for Life' focuses on certain kinds of misuse of *history* that Nietzsche thinks are symptomatic of the present age: 'we are all suffering from a consuming fever of history and ought at least to recognize that we are suffering from it' (UM II Foreword). And then in section 1: 'there is a degree of sleeplessness, of rumination, of the historical sense, which is harmful and ultimately fatal to the living thing' (UM II 1, KSA 1, p. 250). This criticism of history and of a certain 'historical attitude' which characterizes our age is probably the most emphasized point in this essay.

Nietzsche claims that this modern misuse of history is connected to (rooted in) something broadly and in fact essentially human: what distinguishes us from animals is that we remember the past, but this memory is also our great burden: 'Man ... braces himself against the great and ever

1 For the sake of (a kind of) economy, I'll largely confine myself to these four texts: *Untimely Meditations* II, *Human, All Too Human*, *Thus Spoke Zarathustra*, and *On the Genealogy of Morality*.

greater pressure of what is past: it pushes him down or bends him sideways'; 'it was' is 'that password which gives conflict, suffering and satiety access to man so as to remind him what his existence fundamentally is—an imperfect tense that can never become a perfect one' (UM II 1, KSA 1, p. 249). Another set of early passages I'll use but won't quote now are the opening sections of *Human, All Too Human*, which detail how a primitive past is still present—still works—in all of us.

b. Next a passage from Nietzsche's 'midday', in *Zarathustra*. His famous explication there of the eternal return—the climax of the book—presents this idea as responding to a deep human worry over the past. The past disturbs us, because it is utterly beyond the will's reach: the will 'can't will backwards': 'Willing liberates; but what is it called that puts even the liberator in fetters? / "It was": that is the will's gnashing of teeth and loneliest sorrow. Powerless with respect to what has been done—it is an angry spectator of all that is past. / Backwards the will is unable to will; that it cannot break time and time's desire—that is the will's loneliest sorrow' (Z II 'On Redemption'). And soon after: 'This, yes this alone, is what revenge itself is: the will's ill-will toward time and its "It was"' (ibid.).

Later in this section Zarathustra has an inkling how eternal return can solve the problem, but he is only able to embrace that thought in part three. By willing eternal return, the will is able to redeem the past and to say (truly) 'Thus I will it, thus shall I will it', thus meeting the challenge that had been set in 'On Redemption'. And this is the dramatic turning point of the book, which therefore hinges on the problem of the past. Eternal return is needed above all to meet this challenge; its chief function is to change our relation to the past, and solve that problem with it.

c. Finally a late passage, in the *Genealogy*. The second essay's opening genealogy is of *memory*: this capacity didn't come to us from our animal past, but had to be trained into humans by ages of brutal punishments. This memory was imposed against the grain of our natural 'forgetfulness', which is an active repression of the past, requisite for healthy and effective functioning: 'there could be no happiness, no cheerfulness, no hope, no pride, no present, without forgetfulness' (GM II 1).

Originally, what humans were trained to remember were their past promises, including especially their promise to obey the social rules. People were trained to 'remember' them not just as past facts, but in the strong practical sense of *keeping allegiance* to them, taking themselves to be bound by them: 'an active desire not to rid oneself, a desire for the continuance of something desired once, a real memory of the will' (GM II 1). This memory for the rules was a necessary condition for increasingly close, large-scale, and efficient social life.

So society needed to 'burn this memory into us'. Consider this complex passage on the past, and memory: 'perhaps there was nothing more fearful and uncanny in the whole prehistory of man than his mnemotechnics ... Something of the terror that formerly attended all promises, pledges, and vows on earth is still effective: the past, the longest, deepest and sternest past, breathes upon us and rises up in us whenever we become "serious"' (GM II 3).

This training in memory is the key first step in the 'taming' and 'civilizing' of humans, their socialization. *But*, Nietzsche says, both this new capacity (memory) and that long training for it by means of terror, pain, and punishment have had the overall, pervading effect of *sickening* and depressing us, even today. It's this training in memory, above all, that has rendered us 'the sick animal'.

Now seeing Nietzsche returning so often and so critically to the past and memory raises the simple question: *why* is the past so important to him? And we can mean this question in at least two ways: (a) What psychological factors induced Nietzsche to worry over the past (and his relation to the past) in this way? (b) What does Nietzsche *avow* as the past's importance—i.e., what reasons or grounds does he offer?

I'll say just a few things about (a), on the personal-psychological weight of the past for Nietzsche, and what might be at the root of this. Certainly there are reasons to think that his philosophical attention to the past reflects a 'fixation' running down at the level of his own psychological character.

His professional field, classical philology, is itself a major statement of his special fascination with the past. We should take seriously the oddity in this, that a philosopher who later prided himself on being so far ahead and futural should have originally occupied himself with not just history but *ancient* history. Famously, he later regretted this decision—and that he hadn't read in the sciences instead. So he expresses a retrospective regret at his own retrospectiveness.[2]

I think we find another expression of Nietzsche's personal problem with the past in the hostility he so characteristically shows to his predecessors—and especially to those who have clearly influenced him. His hyperbolic rejections of Socrates, Plato, Kant, Schopenhauer, and Darwin can be read to show him bothered by their influence—by the debt he may owe to

2 Notice, in the 1886 Preface to HA II, how Nietzsche says that all of his works except one (surely *Zarathustra*) 'are to be dated back—they always speak of something "behind me"'. Each of them describes a viewpoint Nietzsche had lived through at varying distances in his past. So most of his writing is retrospective.

others' ideas, and the threat to his independence and originality. This is mixed with a sense of the flaws and failures in these sources. He expresses some of this in *Human, All Too Human*: 'He who has come to a clear understanding of the problem of culture suffers from a feeling similar to that suffered by one who has inherited a fortune dishonestly acquired … He thinks with sorrow of his origins and is often ashamed, often sensitive about them. The whole sum of the energy, will and joy he expends on his property is often balanced by a profound weariness: he cannot forget his origins' (HA I 249).[3]

I don't know, and do ask, whether Nietzsche's relation to his personal past was troubled—whether he struggled against unpleasant memories. Are there things in his past that he regrets, and feels guilty or ashamed of? Perhaps the Lou Salomé episode—his lack of success with her, his own resentful reaction to her and Paul Rée? Perhaps his father—as a 'priest', as having gone mad? Perhaps his rather embarrassing mother and sister, surely deflating to his own grand ambitions and self-conception?

If this could be filled in, it would license a psychologistic suggestion: that Nietzsche found himself obsessed with or fixated on the past in ways he found troubling and self-undermining—and that he was both expressing and working this personal issue through in his basic philosophical thinking.

He purports to have eventually succeeded in this. For in *Ecce Homo* we find an utterly positive account of his past: he claims to be completely content with it, seeing it all as the path by which he 'becomes who he is'. (We can read his glad emphasis here as a sign, perhaps, of how much discontent he overcame.) In the introductory paragraph: 'How could I not be grateful to my whole life?' He depicts himself, perhaps, as exemplifying the kind of gratitude towards the past, and will to have it just as it was, that is involved in willing eternal return.[4] Biographically, again, we may wonder whether he really did achieve the reconciliation with the past he so long wanted.

However, more important than these biographical speculations is the task of clarifying the main structure of Nietzsche's philosophical views about the past and memory—to see how his various reflections on memory might fit together into a coherent theory. So I'll go on now to try to formu-

---

3 In this light HA II 110 may be read as a confession.

4 In the same paragraph he says, 'I looked backwards [*rückwärts*], I looked out [*hinaus*], I have never seen so many and such good things at once' (EH 'Why I Am so Wise', KSA 6, p. 263). These terms for the retrospective and prospective stances echo those in a key passage on eternal return in *Zarathustra*, which we'll look at in section 3 below. Nietzsche here claims the simultaneous satisfaction of both stances that I'll argue eternal return is supposed to represent.

late this 'problem of the past' more fully and exactly, and to settle his answer or response to it. Here I mean to sketch the gist of the problem, as Nietzsche came to see it in his maturity.

I'll present the problem as lying in the tension or apparent contradiction between two large points:

1. that the past is far too important to ignore (being important in ways and for reasons we don't suspect),

2. but that attention to the past seems to be harmful to us.

So, it seems, we're damned whether we do or don't pay attention to it.

1. The Past's Importance

The past is important because we don't and can't 'leave it behind': it is the secret meaning of who we are (and what we do). I think this is a point on which Nietzsche disagrees with both common sense and science. He thinks the past 'gives our meaning' not just in the (scientific and commonsensical) sense that it *did* make us, and hence explains us as a cause (something externally determining)—the relevance the past usually seems to us to have. Rather, the past 'gives our meaning' in the stronger sense that it has a kind of 'presence' in us, constituting us now as who we are, determining the meaning of what we now do. This is because, first, there are structures or mechanisms in us that were made long ago in very different conditions—that are 'remnants' of those past times. And it is because, second, these structures were made by *wills*, and hence express the aims of these wills, which carry their intentions ahead into us. Together these points make the past constitutive of the present: of who I am, of the meaning of what I do.

The first point is more to the fore in *Human, All Too Human*. Nietzsche states it most broadly at the opening of section 223 'Wither we have to travel': 'Direct self-observation is not nearly sufficient for us to know ourselves: we require history, for the past continues to flow within us in a hundred waves; we ourselves are, indeed, nothing but that which at every moment we experience of this continued flowing [*Fortströmen*]' (HA II 223). This section builds to the lesson that one might study the sedimented layers of the past *within oneself*, discovering all of our cultural and even organic past deposited there. Such a one 'will rediscover the adventurous travels of this becoming and changing ego [*dieses werdenden und verwandelten ego*] in Egypt and Greece, Byzantium and Rome, France and Germany, in the age of the nomadic or of the settled nations, in the Renaissance and the Reformation, at home and abroad, indeed in the sea, the forests, in the plants and in the mountains. —Thus self-knowledge will become universal knowledge [*All-Erkenntnis*] with regard to all that is past' (ibid.).

This passage from the second volume of *Human, All Too Human* states the lesson behind much of the discussion that opens volume one. Notice, for example, the description of cruel people as 'stages of earlier cultures that have remained behind: the mountain ranges of humanity here openly display the deeper formations that otherwise lie concealed ... In our brains there must also be furrows and whorls corresponding to that state of mind, just as reminders of our existence as fishes should be ascertainable in the form of individual human organs. But these furrows and whorls are no longer the bed along which the stream of our sensation now rolls' (HA I 43).[5] Note again the reference to past evolution as *layering* structures into us.

By the time of the *Genealogy* the second point—the intentionality of those past 'causes'—is clear. Those past shapings of us were for meanings and purposes that our parts and processes continue to bear. So the book elaborates (a) how our practices of punishment, bad conscience, and religion have been designed for certain functions, and (b) how the genealogy of this design shows the 'meanings' these practices now have. The practices *carry along* the aims their selection-design has given them, so that Christian values, for example, still carry the slavish and resentful purposes with which they were first formulated.[6]

Here let me step back to give my current reading of Nietzsche's notions of *will* and *drive*, as the bearers of that intentionality.[7] These are the basic *explanatory* terms in Nietzsche's account of human beings—and indeed of 'life' more generally. Wills and drives are identified by what they are 'to' or 'for'. They therefore involve a *kind* of teleology. Explaining by wills or drives involves explaining by what they are 'to' or 'for': you cite the *outcome* as explainer, as why the drive did what it did. So for example with 'will to power': this explains action/doing as 'done as a means to power'.

However, it is a persisting point for Nietzsche that we *cannot* explain this 'to' (or this 'done as a means to') by the model usual in philosophy and in common sense: the model of a personal and conscious aiming,

5 See also HA I 250: 'the past is still too powerful in their muscles'.

6 This is why a critique of moral values depends on 'a knowledge of the conditions and circumstances out of which they have grown, under which they have developed and shifted' (GM Preface 6).

7 Here I summarize an interpretation developed in *Nietzsche's New Darwinism* (Richardson 2004).

which represents and thereby 'looks ahead' towards a goal. Nietzsche thinks this model is deeply false. We need to think our way out of it, despite the tendency of our language's very structure to pull us back in. Wills and drives can't be 'mini-subjects' or micro-consciousnesses. They can't have their 'towardness' by representing their goals in thought or consciousness or anything mental (not even an unconscious mentality). They don't 'look ahead' or 'anticipate' in any straightforward sense.

On the other hand, wills can't be mere causal tendencies to bring about certain outcomes. Nietzsche's term is richer than this. Power isn't just the tended-outcome of a will. If it were, 'power' would drop out of the explanation. Nietzsche insists on the need to supplement mechanism with a 'will'. So we need a kind of 'intentionality' and 'directedness', but again not in a cognitive or mental way. Nietzsche expresses his need for such a non-mental directedness in those clearly metaphorical attributions to life of a foresighting purposiveness, e.g., when he speaks of 'life's healing-artist instinct' (GM III 16). He forces us to look for a way that aiming could be something non-mental.

So how do wills/drives get their aims? The suggestion I present in *Nietzsche's New Darwinism* is that wills have their ends precisely in their having been selected for certain outcomes. So what makes a will '*to* power' is not that it holds power in view, or 'looks ahead' to it, but that a selective process has 'made' this will 'so that' it tends to bring about power. The causal tendency to bring about power would not itself be enough to make it 'to' power. It must also have been selected to bring about power, such that power explains why the tendency is here. This 'etiological' account of wills (owing ultimately to Larry Wright and Ruth Millikan) makes that selective process determinative of drives' aims and meanings.

What's the 'selective process' that determines a drive's aim?

a. Originally *natural selection*, whereby drives serve 'life'. I've tried to show that Nietzsche takes over, though little mentioning it, the basic Darwinian insight: that selection for replicative success gives functions to organs and goals to organic processes. Even here he thinks he has a correction to Darwin, on the consequence (or tendency) of natural selection: what gets selected are especially drives and instincts for *power*, and not, as Nietzsche thinks Darwin thought, instincts for survival.

b. Nietzsche has a second correction to Darwin: he thinks there's a new kind of selection that works in the human case, by virtue of our *social* character. I call this 'social selection'. Humans become capable of not just 'drives' but 'habits', i.e., behavioural dispositions transmitted not genetically, but by social copying. By this, a disposition for some behaviour is 'selected' by its spreading through the population. It is incrementally and cumulatively modified ('designed') to facilitate this spread.

Nietzsche thinks this new kind of selection designs these habits for a different ultimate end. Whereas natural selection designs structures that further my 'reproductive fitness', social selection designs habits that favour the success of my society—and especially by increasing society's cohesiveness and homogeneity. So social selection's main tendency is to design practices that make us want to be similar to one another. Its tendency, in other words, is to render us 'herd animals', by developing a 'herd instinct' in us. So: natural selection selects *drives to power* because they serve the *organism's replicative success*, but social selection selects *habits of herding* because they serve the *society's cohesion and strength.*

It's this social selection that mainly explains why I have the habits of acting and thinking as I do. I acquire these by copying from the social nexus, and they have their meaning and point in that wider context. The 'meaning' of the habit or practice is then *not* anything I believe myself doing it for, as I do it, but the functions designed into it by the social processes that spread it to me. These functions are 'what the drive is doing in me', though often unbeknownst to me.[8]

Consider an objection: don't I choose the habit because it appeals to some psychological need or taste or desire? And isn't the latter the explanation why I have it, and what it's doing in me? But this misses, first, how the habit is rather, in Nietzsche's view, a kind of virus that uses this desire as a point of entry, for purposes of its own. This is how Nietzsche thinks Christianity has worked: it appeals to certain weaknesses and sicknesses for entry into persons, but then treats-aggravates that weakness for purposes of its own (see, e.g., GM III 16). And it also misses how many of our desires are themselves inserted into us by the same socializing process—and above all the desire to 'do as others do'.

So all of these wills/drives are (as it were) a great many 'machines' designed for various purposes and built into us beneath our notice. It's these machines, and the functions designed into them, that explain most of what we do. We need to realize that there are these many mechanisms in place in us (as parts of us), which are unconsciously plastic towards certain outcomes—which are 'for' these outcomes in the sense of having been designed to accomplish them. It's precisely because we *don't* see how there

8 On the efficacy of these long early stages see *Human, All Too Human* I 2: 'everything essential in human development occurred during primeval times, long before those four thousand years with which we are more or less acquainted; during these years, humanity may well not have changed much more.' And later *On the Genealogy* III 9 speaks of 'those enormous stretches of time characterized by the "ethic of custom", which lie before "world history" as the real and decisive principal history that established the character of humankind'.

can be such directednesses in us, except for the overt and cognitive sort (i.e., consciously, deliberately willing an outcome) that we fail to see all the meanings our behaviour has, 'beneath', 'before', even 'instead of' our conscious meanings.

Ultimately, perhaps, Nietzsche is attacking what might be called a 'psychology of presence'. He opposes our commonsense confidence that we determine what we want and mean by our present acts of intending. 'A thought comes when "it" wants, not when "I" want' (BGE 17). And what the thought wants is determined by the processes that selected-designed the thought. We might call this a 'temporal externalism' about meaning.

This social design of our habits and practices builds into us our *values*, and in doing so gives these values their *meanings*. So the *Genealogy* uncovers the meaning of Christian values: they've been designed as a 'slave morality', i.e., have been structured to appeal to and further the interests of the reactive, sick, and suffering. This design builds into us meanings we don't understand. It makes us 'intend' things we're unaware of.

Human, All Too Human I 18 already states the point:

> [W]hen the sensate individual observes itself, it takes every sensation, every change for something isolated, that is, unconditioned, without connection: it rises up from within us without any tie to earlier or later things. We are hungry, yet do not originally think that the organism wishes to be sustained; instead, that feeling seems to assert itself without any ground and purpose, it isolates itself and takes itself as arbitrary. Therefore: the belief in the freedom of will is an original error of everything organic.

So the 'will' of the organism to be sustained is not available in the sensation (or experience) of hunger. And the point extends much more widely: all of our desires and values are doing further things, serving further purposes, than what shows up in them.

Recall also the famous genealogy of punishment in GM II. In investigating 'why' we punish, Nietzsche turns not to our present and conscious intentions. Instead he considers the practice as a complex set of procedures, and asks how different parts of this complex have been designed at different times *in the past* for different functions now. These many functions have been *layered into* the practice, which *now* has *all* these meanings. And it has also built into us a host of metaphysical errors: 'What we now call the world is the result of a host of errors and fantasies, which emerged gradually during the overall development of organic beings, merged together as they grew, and are now passed on to us as the accumulated treasure of the entire past' (HA I 16).

For all these reasons, we are *in thrall to our past*. This is the first threat in the past: that it controls us. It deprives me of a freedom I have always supposed myself to have. It means that I lack the responsibility I've claimed for myself in acting and valuing. Of course we've always believed that the past was the threat to our freedom, but because we might be *causally determined* by the past. Instead, it's because the meanings of what we do are *logically constituted* by past selection.

Now given these ways the past controls us, it seems attention to the past is requisite, both for the sake of understanding ourselves, and for the sake of realizing the autonomy or agency we suppose ourselves already to have.

And this expectation is confirmed by Nietzsche's frequent insistence that philosophy must become 'historical': 'A lack of historical sensibility is the original failing of all philosophers ... They do not want to learn that humanity has come to be, that even the faculty of cognition has also come to be' (HA I 2). Later: 'From now on therefore, historical philosophizing will be necessary, and along with it the virtue of modesty.' And: 'The steady and laborious process of science, which will someday finally celebrate its highest triumph in a genetic history of thought' (HA I 16).

2. Problems with Retrospection

And yet, when we do retrospect or remember or study the past, it seems this tends to *hurt* or *damage* us.[9] Or at least, Nietzsche often remarks on such negative effects. We looked at three sets of passages (from *Untimely Meditations*, *Zarathustra*, and the *Genealogy*) that presented this harm in three different ways. Now we need a more comprehensive account of Nietzsche's considered judgement. How, given what we've seen, does he think we should proceed to study and assess retrospection?

Clearly, we need to study retrospection's genealogy: to see how this stance arose and developed by the same kinds of historical processes we've been treating. That is, we need a retrospective study of our retrospection, in order to understand what it's doing in us now. This is why it is—paradoxically?—Nietzsche's (retrospective) genealogy of retrospection that best shows how the latter is harmful. Its damage lies in the 'functions' that have been designed into these retrospective abilities and practices.

So how did our capacity and propensity to remember and (in general) 'regard' the past arise and develop—and what functions has it thus been selected for? I take this to be a or the central topic of the *Genealogy*'s sec-

9 This is one more place we find Nietzsche in dispute with Plato—and the value he places on 'recollection' of the Forms.

ond treatise, which let's examine. This essay, I'll try to show, presents memory as the decisive ability acquired by the earliest humans, upon and around which our striking higher capacities were then built.

First a background point. The second essay opens with a certain picture of the causality of this evolution: it presents 'nature' as 'setting itself the task' to 'breed an animal that may promise' (GM II 1). Nietzsche speaks as if nature foresightingly aims this process towards the 'sovereign individual' who, he soon says, emerges eventually from it (GM II 2). Later in the *Genealogy* he offers this picture again, to explain the social role of priests; he develops 'what I think life's healing-artist instinct has at least attempted through the ascetic priest' (GM III 16). Such passages challenge the reader: Nietzsche clearly posits some kind of overall 'design', and yet we can't believe that he means literally this designer he mentions—'nature', treated as an agent working towards represented goals. This agentive nature must be a stand-in for some other causal process working at the level of the society or species as a whole, and somehow 'in the interest' of such aggregates. As I've said, I think Nietzsche must mean a *selective* process, working by aggregate effects over populations.

Memory evolves by selection *in social groups*. As animal, we were capable only of the *projective, forward-turning stance* that Nietzsche associates with *will*. This is what we share with all the rest of 'life'. Memory—our capacity for a second, *retrospective and back-turning stance*—arose within society. Or better: memory arose at the very time that society formed, and *so that* society could be the more possible and successful.[10] Nietzsche gives the name *Sittlichkeit der Sitte* (ethic of custom) (GM II 2) to this long early phase of our history, in which memory made social custom and vice versa. Humans' retrospectivity, which distinguishes us from all the rest of nature and life, was originally selected-designed to facilitate our 'socialization'—to make us creatures of habit and custom.

Indeed it might not be too strong to say that originally memory *just was* the ability to *acquire (social) habits or practices*, distinct from the innate drives. Memory was first and foremost the ability to *remember the rules*, even when one's drives pushed hard the other way. It was the ability 'to keep a few primitive requirements of social co-existence present for these slaves of momentary affect and desire' (GM II 3). One remembers

10 Nietzsche later suggests that the beginnings of memory precede society, in the more primitive relation between 'buyer and seller, creditor [*Gläubiger*] and debtor [*Schuldner*]', which he says is 'older than even the beginnings of any societal associations and organizational forms' (GM II 8). But clearly the main work developing memory is done socially—and indeed Nietzsche goes on at once to focus on the community [*Gemeinwesen*] as the most important 'creditor' (GM II 9).

not to steal the fruit in the market-place, even when one's hunger drive impels. So memory's original work is to 'give one pause', restraining one from acting on the immediate excited drive, by inserting a glance 'back' at one's commitment to the social rules.

Early humans were trained to remember this commitment by 'burning' into their bodies and senses certain vivid and powerful experiences of the horrific punishments inflicted on those who break the rules. After many generations of this training, 'one finally retains in memory five, six "I will nots", in connection with which one has given one's promise in order to live within the advantages of society' (GM II 3).[11] These dramatic punishments train into us the ability to interpose between drive and action that memory of the rule. And the capacity to remember all of the rules is the ability to impose, on top of one's drives, that new layer of social practices—to which one is committed to subdue one's drives.[12] All of this shows how the first function of memory was to 'socialize' us—to make us abide by the rules necessary for social existence.

This new power and propensity to remember gets inserted into an existing context of drives. It struggles to control the latter, but of course they also struggle to control it. They try—by Nietzsche's drive-psychology—not only to enact themselves despite its restraining efforts, but also to infiltrate, modify, and use that new capacity for their own purposes. We need to bear in mind in what follows this counter-action by the drives upon retrospection.

Now of course this is only the very beginning of a genealogy of our retrospective powers, which have obviously evolved very richly from this start. In fact, Nietzsche treats these powers as embryonic for major further developments in our human cultural history, which we should look at. As the retrospective stance is broadened and enriched into further powers, the latter largely take up the original function of memory—to socialize us, or in Nietzsche's terms to 'tame' and 'herd' us. We can distinguish two sets of such powers developed from that root.

a. First there are ways that memory, our backwards view, founds both *religion* and *morality*. Nietzsche stresses how belief in gods develops out of a retrospective view—the feeling of indebtedness to ancestors. The social group reveres its ancestors as the founders of the customs and laws

11 Punishment effects 'a lengthening of memory', 'a sharpening of prudence, mastery of the appetites' (GM II 15).

12 Among the drives subdued in this way is that to *revenge* oneself for injuries received—one learns to treat these as offences against 'the law', to be punished by it. Thus 'the eye is trained for an ever more impersonal appraisal of deeds, even the eye of the injured one himself' (GM II 11).

that have made the group strong (and the good life within it possible): 'all customs, as works of ancestors, are also their statutes and commands' (GM II 19). This retrospective feeling of debt binds each member more tightly to those customs. And as the group grows stronger, those founders are magnified into gods—and members' debt to them is magnified as well.[13]

More than religion, morality is Nietzsche's target, and this too is an offshoot of that primordial power to 'remember the rules'. Morality is a next phase of social values, evolved from the ethic of custom. Though a very complex phenomenon, its key ingredient I think is what Nietzsche calls 'bad conscience' or 'guilt', and which he also genealogizes in the second essay. Our bad conscience, which so poisons the moral stance for Nietzsche, develops from memory in the following way.

Learning to remember the rules is learning to constrain or suppress the aggressive drives that would threaten social life. Yet these drives crave expression, and can't be utterly stifled. So a way of making them subserve our socializing-taming is found:[14] these aggressive drives are turned back against themselves, against the 'entire animal old self' (GM II 18). Members are trained to feel guilty about their instincts, and this feeling is a way of venting some of those instincts—especially those to inflict pain—venting them on oneself. So here that retrospective stance which draws back from the drives' engagement re-aims those drives against themselves, as a means to its own fuller control. Our memory for the social norms is reinforced by the habit of paining ourselves with regrets at the drives that tend to violate those norms. It's this co-opting of members to punish themselves with a retrospective guilt that distinguishes morality from custom.

b. Besides its roles in religion and morality, that retrospective stance is at the root of another main human achievement, which Nietzsche likewise views with a famous suspicion. This is our *reason*, our *cognition*—the attitude or stance in which we understand and know. Theory and science are at root retrospective.[15] We should notice how often Nietzsche takes memory to *represent and epitomize* the theoretical attitude—an attitude we take to be different and broader than memory. So his critique of 'history' becomes a critique of all science, and his critique of 'memory' becomes a

13 Nietzsche struggles against a related feeling of indebtedness when he fights (as we saw above) to distinguish himself from the philosophical predecessors who most influenced him.

14 It is 'found' by selection at the social level—and not necessarily by the conscious discovery and design of determining individuals.

15 Notice how UM II 10 says that science 'sees everywhere things that have been, things historical' (KSA 1, p. 330).

critique of all self-reflection (all looking at one's aims or values from outside them).

Nietzsche treats this theoretical stance, the 'will to truth', in not the second but the third essay of the *Genealogy*, where he identifies it as a (surprising) manifestation of the ascetic ideal, indeed as its core (GM III 27). We can better see why he thinks so by seeing how this stance develops out of that primordial memory.

Before and without this memory, the stance of our original willing was and is dominantly *futural*: our drives make us lean ahead towards their ends, and see and assess present conditions as they bear on those ends—as opportunities or obstacles. In this willing we understand ourselves in what we, in our drives, are trying to be (do, have).

Nietzsche thinks of memory as the root of a *second stance* humans become able to take. By that primeval training to remember the rules, humans learn to 'step back' from that immediate willing in their drives—to 'insert a pause' in enacting those drives. Instead of focusing ahead on what our impulses can achieve, our attention 'turns back' to some content independent of them.[16] Moreover, one then binds and constrains those futural aims, in line with this content.

Centrally, this independent content was and is the social rules. The pause in enacting the drives is the ability to remember these rules—what one must not do. Although I'm calling this 'retrospective', it is perhaps less a matter of what 'the mind's eye' sees than of what the mind's ear hears—verbal formulations of those rules. As well, one remembers all the more particular commitments one has made, besides, to these norms: the promises made to others and oneself.

Now these memories are not mere reveries, but effective and practical: I constrain my forward-pressing impulses and aims in view of past commitments which I keep present. And the point and purpose of the content I notice—my promises and commitments—is precisely to change what I do, how I act. But, I think Nietzsche thinks, this capacity to 'turn back' from the aiming in my drives into a space or attitude apart from them, resistant to them, is the germ of humans' capacity for theory—for the project of knowledge or truth.

16 This way that (what I have been calling) memory turns us away from the drives, and towards something separate from them, shows how it goes beyond the kind of memory that subserves the drives, and that surely *did* occur in our animal past. The latter is the power by which animals remember where the water source was, for example; this doesn't interrupt their thirst drive, but is entirely steered by it. So it doesn't count as the memory or retrospection we're examining.

Our theoretical attitude, in which we try to know things 'as they are', objectively, depends on that ability to pull back from our usual engagement—to put a pause in our effort. But it uses this pause not to recall practical aims distinct from the drives (those social norms and promises), but instead to 'just look' at things, in that space apart from drives' effort.

To be sure, we must recognize that this 'space apart' from my drives isn't wholly apart, because of that counter-action of the drives upon retrospection, noted before. That new power is in competition with the drives, and is affected by them as they are by it. So Nietzsche frequently stresses how our theory expresses our drives and does not achieve the separation it aspires to.

Moreover, in that 'space apart' I don't find myself alone—not even alone with my drives. Like that practical memory, theory too depends on remembering rules: the vocabulary and methods of whatever practices of knowing I have learned. So theory is not an individual and solitary stance, but fully as social and intergenerational as our memory of the social norms. I must remember those rules and guide my observing and describing by them.

This begins to explain why Nietzsche thinks that this theoretical attitude, our 'will to truth', belongs to the ascetic ideal. It is ascetic, ultimately, precisely because it takes this stance contrary to the willing in our body and drives. It turns away from our aims and ends, and binds us to something independent of them. Indeed, in this regard our will to truth is an ultimate form of that contrary stance—'that ideal itself in its strictest, most spiritual formulation' (GM III 27).

This new stance, in which we front the world differently than in willing, becomes most *contrary* to willing when it turns to study this willing itself. Knowing, just as retrospective, already absents itself from our effort to enact our drives. But when it turns back to look at those drives themselves—at the aims by which we really set our behaviour—it works actively against them. When we turn this retrospective eye upon our values and aims, we chill and kill them. So the historical or genealogical study of our values is the most ascetic of all.

Already in 'On the Uses and Disadvantages of History for Life' Nietzsche writes:

> A historical phenomenon, known clearly and completely and resolved into a phenomenon of knowledge, is, for him who has perceived it, dead: for he has recognized the delusion, the injustice, the blind passion, and in general the whole earthly and darkening horizon of this phenomenon, and has thereby also understood its power in history. This power has now lost its hold over him insofar as he is a man of knowledge: but perhaps it has not done so insofar as he is a man involved in life (UM II 1, KSA 1, p. 257)

Moreover, because our will to truth is a development out of that prior ability to 'remember the rules', it is entangled in some of the latter's long and harsh design; it has been partly aimed by that design. For example, it participates in the ideology against the drives, and in guilt over failures to abide by what the retrospective stance reveals. In acquiring this will we inevitably receive with it elements with these remnant functions.

Overall, this genealogy of memory or retrospection shows it in a very unfavourable light. It reveals how this back-turning stance has been designed for a group of functions we were unaware of, and which are often immediately unappealing—not functions we would *want* our attitude or stance to perform. I'll come back to give Nietzsche's assessment of them in a moment.

But first we should notice that there's another, more positive way in which this retrospective stance changes us. Besides these ways it lies at the root of religion, morality, and theory, it also changes the kind of *self* we can have or be. Before memory was acquired, a human was simply a collection of drives—drives of differing strengths, which thereby naturally formed a hierarchy or power-structure of impulses.[17] Such a human—or proto-human—identifies with the ends of these drives, or perhaps with the ends of whichever drives happen to be active or dominant at the time. He is who these drives are currently trying to be.

But once memory is developed, it functions in a new way of identifying oneself. I now 'find myself' not just in these ends my drives aim me at, but in the past commitments and promises I have made. I find myself retrospectively, in ways I have bound myself to be. And this changes as well my prospective or projective stance. I now aim, not just at the end of whatever drive is now strongest in me, but at fulfilling my promises—at being the person I have committed myself to be. This gets fixed in me as a new drive—but a most anomalous one. By this binding I am someone somewhat independent of those (other) drives: I can act quite against them, to keep my commitments. So I can have and feel a kind of *power over* my drives. My 'agency' lies in the continuity I give my life by this retrospecting committedness.

All of us feel, I think, this gratifying sense of control, in our identity as deciding subjects. But although Nietzsche does think there is a major achievement here, he believes it is less and different than we suppose. This self-identifying involves a great error or illusion. I take my 'moments of

17 This was, perhaps, too much the only way I treated our drive-structure in *Nietzsche's System* (Richardson 1996). And Nietzsche himself sometimes operates with this simpler picture.

decision'—the choices in which I commit myself to be a person *like this*—as the complete and sufficient explainers of what I then am and do. And in these moments of choice I feel my reasons as the only reasons that select and explain how I choose. But in fact, my own reasons are selected by those aggregate processes that 'designed' values for social functions. My reasons are subordinate to those functions—I have been aimed by them, and don't really aim myself as I naively suppose.

The crucial such function is of course to socialize me—to induce me to live by the rules, and more generally to fit my behaviour into the general practice. What I bind myself to is—for the most part—these social rules, so that my commitment to them serves, ultimately, not my own interests but those of society. These seemingly free subjects are really a kind of automata, each of them committed to that same morality which holds them to their social tasks.

So, like our theoretical attitude, our 'sovereignty', as a development from that retrospective stance, retains deep design features that rather spoil it, Nietzsche thinks. That positive sense of power we have in this stance proves to be in important ways an illusion and even a deception. These herding values disguise their own work by purveying this sense of personal responsibility and control. Nevertheless, it is this image of a personal sovereignty that inspires our immediate aversion to these herding forces, once our genealogy reveals them. I discover that I'm very much *not* what I've learned to pride myself as being.

3. Solving the Problem of the Past

Together, the points in sections 1 and 2 confront us with a dilemma: the past is more important than we suppose, but attending to it seems to hurt rather than to help. We're in thrall to our past, but working on the problem by studying the past seems only to make it worse, by diverting us into a stance that undermines our natural drive-effort, co-opting it for social ends.

Early on, I suggest, Nietzsche has a simple (and somewhat mechanical) way of answering this problem. He insists on the need to *limit* this retrospective stance. We must learn to subordinate our knowing to our willing. Happiness (*Glück*) can't occur in that stance, since it depends on forgetting (UM II 1, KSA 1, p. 250). So science and knowing must be controlled: 'science requires supervision; a hygiene of life [*Gesundheitslehre des Lebens*] belongs close beside science' (UM II 10, KSA 1, p. 331).

Nietzsche does not make clear just how this limiting or constraining of our retrospection would work. He says that 'the unhistorical' (forgetting) and 'the suprahistorical' (art and religion) will be used as 'antidotes' to our historical stance (ibid.). There may also be some kind of restriction on the topics that get studied and known. A healthy culture, and a healthy individual, will not turn this retrospecting scrutiny on the values most crucial to it. This is a most important theme in the notes from the early 1870s.

When Nietzsche enters his positivist phase this naturally changes his position on the worth of this retrospective, knowing stance. Perhaps we should say that he now enters into and occupies it, and becomes preoccupied with exposing as much about us as he can to the cold eye of study. So he no longer fully recognizes those problems in the attitude—what it misses, what it injures. This (temporary) loss of the sense of what's best for him was the means by which he cured himself of Romanticism (HA II Preface 2).[18]

We find this less troubled stance expressed in *Human, All Too Human* I 292, which counsels contentment with the way we're still liable to religious, artistic, and other such 'unclear [*unreinen*] thinking'—since we can use this susceptibility for truth. This presence in us of past irrationalities had seemed a threat to our knowing, but really it helps us 'forward on the path to wisdom'. For our rootedness gives us insight into the past, useful for our future: 'Turn back and trace the footsteps of mankind as it made its great sorrowful way through the desert of the past: thus you will learn in the surest way whither all mankind can and may not go again.'[19] So the way we are 'thrown' (*geworfen*) into this mixed condition turns out to be useful for our effort to understand and learn from the past.

But—Nietzsche later thinks—this positivist embrace of knowing lost sight of what knowing costs us. In his step 'back to health', in his maturity, he regains 'the perspective of life', and his sense of how that knowing stance is dangerous and undermining to it. This stance is responsible for the death of God, and threatens to destroy all of our values as well and to land us in nihilism. But now Nietzsche's response is not to restrict or oppose our knowing, but to use it in a new project that realizes all of its po-

---

18 As usual in the retrospective 1886 Prefaces and 1888 *Ecce Homo*, Nietzsche claims that an implicit purposiveness was at work in him. His 'still healthy instinct' was *using* this cold and sceptical objectivity in a spiritual self-surgery. And in *Human, All Too Human* II Preface 5: 'so I, as physician and patient in one person, compelled myself to an opposite and unexplored clime of the soul.' Similarly *Ecce Homo* says that *Human, All Too Human* 'is the monument to a crisis … I used it to liberate myself from things that did not belong to my nature'—such as idealism (EH III HA 1).

19 See also HA I 616 on how absorption in past world-views gives us a valuable perspective on the present as a whole. Also HA II 179 and HA II 223, quoted above.

tential. He sees a way to *reform and intensify* our relation to the past by putting it to work in a new *healthy* project, a project that betters our 'life' and 'power'.

I will try to show how Nietzsche's new solution to the problem has as its two main components *genealogy*, as an epistemic ideal, and *freedom*, as an ideal for willing; they are the new versions he offers for our two basic stances, projective and retrospective, and which he thinks will allow a reconciliation of them. And I'll present *eternal return* as Nietzsche's emblematic image for this reconciliation.

Let me start by suggesting another way to read a very familiar passage of *Zarathustra* 'On the Vision and Riddle':

> 'Behold this gateway, dwarf! I continued. It has two faces. Two ways come together here: nobody has ever taken them to the end. / 'This long lane back here: it goes on for an eternity. And that long lane out there—that is another eternity. / 'They contradict themselves, these ways; they confront one another head on, and here, at this gateway, is where they come together.' (Z III 'On the Vision and Riddle' 2)

I want to propose a different way to interpret these two paths: in terms of the two 'stances' we humans are capable of. When Nietzsche speaks of the paths 'out' (*hinaus*) and 'back' (*zurück*), he may mean not just (and even not mainly) the future and the past, but these two different stances or modes of comportment (of intentionality).[20] The path ahead refers to our projective thrust towards ends—to our willing. And the path behind refers to that 'retrospective' pause or interruption in this willing, which distinguishes humans and reaches its fullest form in our knowing.

So when Nietzsche says that these stances contradict one another, he is (partly, I suggest) referring to the incompatibility we've seen between these attitudes: how the retrospective stance undermines and negates our effort to will ahead. This reading lets us connect this crucial passage in *Zarathustra* to the diagnosis we've seen Nietzsche makes of our human condition as deeply disturbed by its special power of memory—by the way we're now 'conflicted' between our willing and our knowing. And this passage, Z III ' On the Vision and the Riddle' 2, is crucial because it sets up the problem that eternal return is meant to solve.

This reading of the passage is reinforced, I think, by recalling Z II 'On Redemption', which gave an earlier statement of the problem eternal return must address. This problem was 'the will's ill-will toward time and its "It was"'. The connection between this point and the contradiction between

---

20 Recall that very much the same terms occur in the introductory paragraph of *Ecce Homo*, where Nietzsche tells how he affirms his life—in both stances.

the paths 'back' and 'out' is puzzling. I suggest that we think of the will's ill-will at not being able to will the past as expressing the point that the retrospective view has stilled the will, contradicts the will. (Our fondness for the retrospective stance has made our will ill.) So both passages concern the same problem. And I'll try to show that eternal return is so important to Nietzsche because it symbolizes his solution to this problem—to our deep division.

Now in the terms of these 'two paths', what would it be to solve the contradiction between them? It would be for the retrospective stance to somehow 'meet' the stance of willing, and for the latter to meet it as well. This would 'complete the circle' in both directions. But how might each of these stances 'meet' the other? Each, I suggest, must be satisfied with the other in its (the meeter's) own terms, by its own standards. Each must find that it grows (is empowered) through the other. And for us to be able to 'will the past' is for that retrospective stance to serve will.

The retrospective stance steps back from willing to regard things without and despite that willing. In its original form it is the promise to comply with the rules. But in its fullest, most developed form it is the commitment to knowledge and science. What would satisfy it about (the stance of) willing would be to *know it*, and to *make it knowing*. The first of these is accomplished by genealogy, and the latter by a conversion within willing itself. I start with how genealogy knows willing.

For a long time humans have supposed that they know themselves. Even in the early and simple forms of retrospectivity, in which persons bind themselves only to the simplest of social rules, the stance still lets each person find a self, an identity. It's this backward turn from the drives that makes it possible for a person to commit to a 'self' in some independence from them. One commits to certain rules or virtues, for the sake of which one undertakes to overrule the drives. I now 'remember' who I've promised (to myself and others) to be, and believe that this *is* me, much more than those drives my commitments override. Since I seem to *commit* to these rules and virtues in conscious acts, and then to *remember* and follow them consciously, this self or identity seems transparently evident to me.

This confidence in self-understanding is all the greater in persons with developed forms of the will to truth—philosophers, psychologists, other scientists. These take themselves to have an especially overt and considered awareness of their own decisions, thoughts, and feelings. It is the framing point of the *Genealogy* that even here, and here especially, persons fail in the most important points to understand themselves.[21]

We experience our own choice as the determining and responsible factor in our thinking and acting, but in fact this choice—where it is operative at all—merely executes the aim or will embedded in the norms and values by which we choose. Those values, by their design for social purposes, make our choices instrumental for those purposes. The motives for which we think we choose are trumped by ulterior purposes. Indeed even our pride in our self-responsibility serves those purposes: we are the better herd animals when we believe ourselves to be freely setting our own ends in our own interests.

So I have never known my own willing. But genealogy, by exposing the social formation of my values, now makes this possible. It lets my retrospective stance truly understand—bring into view—the forces that really aimed the rules and values to which I commit myself, and with which I identify myself. It reveals the direction of that 'throw' whose momentum my choices merely follow. It even lets my retrospection understand itself, in particular—how its own backwards look originally (and partly still) serves to commit me to rules that oppose my drives. It lets me grasp the ascetic function at work in the 'will to truth' in which I commit myself to know.

Genealogy is the highest achievement of the will to truth, inasmuch as it penetrates to the most hidden, most difficult, and most important truths—the facts of what and why we are (as we are). By pursuing this genealogy we close in on what we really are. We uncover to ourselves more and more of the drives, habits, and values working in us, and more and more of the selective forces that gave them their thrust and tendency. Of course this insight is far from complete. But we do arrive, for the first time, in the proper domain of a genuine self-understanding—we're now looking in the right place, in the right terms.

But even as the highest form of the will to truth, genealogy is in its own right most dangerous. It is most effectively ascetic: it 'cuts into life', by examining and exposing the drives, habits, values, and desires that together constitute our willing. As we diagnose any one of these, and understand the forces that shaped it and the purposes they have made it serve in us, we chill or enervate whichever will we study. We step out of this (projective) will, and expose it to a (retrospective) look that flattens and disenchants it. This is why genealogy can be a route to nihilism, alienating us from our values, one after another.

21 See especially GM Preface 1; also, e.g., GM I 1–2, and the end of GM III 23.

Nietzsche's answer is to show a way to turn this dangerous instrument to a new positive purpose, a purpose that is rooted in the abiding core of our will and drives. There's a certain way we can put genealogy to work, within a life-plan that promises us a superior kind of power and accomplishment. We can use genealogy to make ourselves stronger, by taking an increasing control over the habits and values built into us by social-historical processes. So we put it to the service of our 'will to life', which deeply aims at power and control.

Genealogy exposes ways we have *lacked* such power—ways we have been controlled and used by forces outside us, social forces in whose interest our habits and values have been designed. As we've seen, there is an overall tendency in social selection to evolve values and practices for their *social* utility, i.e., for the way they make society more cohesive and effective. Our deep urges to be like others, to share their values and viewpoints, serve this social function. Moreover, other of our values have been selected to serve the interests of particular social groups or of specific kinds of people. For example, most Christian values are a 'slave morality' in the sense that they have developed and spread because of how they serve the interests of the weak, sick, and suffering. Nietzsche calls me to discover whether my values have been designed in the interest of the kind of person I really am.

By exposing how other forces have made us, genealogy gives us the chance to work, prospectively, towards a kind of control that was never possible before. We can act on these parts and constituents of ourselves for the first time *knowingly*—with an understanding why we have them, and what they are doing (have been aimed to do) in us. By genealogy, we can judge those designed-in purposes of our ways of thinking and acting—and decide whether we favour those purposes. And if we don't favour them, we can try, at least, to redesign those thoughts and acts for different ends. We can try, as Nietzsche often urges us, to reconfigure them so that they suit our individual physiologies—the peculiar mixture of animal drives at the bottom of each of us.

It's this new power, I think, that Nietzsche refers to as 'freedom', and offers as his principal ideal for how a person should be. He thinks he has discovered or invented an importantly new kind of freedom, stricter, fuller, and more genuine than the old. He offers this ideal both as an improvement on the existing value of freedom, and as a best (so far) achievement of the 'power' at which he thinks our most basic and indispensable drives aim.

Thus besides completing the knowing stance, genealogy also enables the projective stance of will to perfect itself. It confers a self-understanding that makes possible a new freedom in that willing, by which will achieves the power and control at which it aspires. This new freedom is more genuine than the kind we have just by our agency—by our being able to promise and commit ourselves. Genealogy corrects the mistake we 'agents' have long made in thinking freedom something possessed and automatic. It constitutes a 'will to responsibility, freedom of the will' (GM III 10).

If freedom is being responsible, i.e., being the principal determiner or explainer of the things I do, then that sovereignty I already have in being able to honour commitments does indeed give me some of this. It lies especially in my sense of power *over my drives*—I identify myself not with them but with those retrospected rules and promises, and pride myself in being able to override those drives to follow those rules.

But as we've seen, there's a major hole in my responsibility: I fail to recognize that the values to which I commit myself have been designed to 'do things with me'. They have been designed, for example, to make me a better herd animal. Since I don't understand what these values are 'for', and what they're doing in me—how I'm being 'used' in my commitment to them—responsibility really slips through and out of me, and belongs to the social forces that made my values by which I am steered. So, for example, I may act out of a habit of pity or benevolence, and cite this as a motive in my choice, yet fail to understand why I have this habit or motive of benevolence—what it is doing in me, what work it was designed to do. And in this case I am in fact being used by those selective forces for those purposes the drive was shaped to play in me.

So by the new (genealogical) insight into this 'what' we are, we now notice a way we *have not been free*. We discover the incompleteness and inadequacy of the freedom we've had by discovering a new constrainedness—how we are subject to the social-historical forces that designed the habits and values we live by. And from this, we can next project and plan out a way to overcome this constraint, and become for once (or more fully) free.

To accomplish this freedom I need more than genealogy. The latter of course belongs to the retrospective, theoretical stance, so the task is now to reflect it in my projective stance of willing. I need to win my freedom not just in theory but in practice. This is what Nietzsche calls 'incorporating' this understanding (see GS 11 and 110). It is to make it effective in how I really do aim myself, moment by moment. My existing values are built into my drives and socialized habits, and I don't annul them just by saying that I do. I need to push genealogical insights down to the very points at which these drives and habits operate. I must build into my everyday responses those countering diagnoses supplied by genealogy, so that I see *why* I will, *while* I will. Willing only really takes up theory into its own projective stance when it takes practical regard of it in its concrete and everyday moments of willing.

Nietzsche thinks that our projective will must *test* and *assess* the genealogical truths by *how far* they can be incorporated. Some truths it will be impossible to import into our practice. For there are limits to how pliable and alterable our drives and habits are. With some drives, we *won't be able* to make their diagnoses effective—they work on us so fundamentally and pervasively that we can't hope to make them self-aware. This means there are limits to the freedom we are capable of. Here see, for example, *Human, All Too Human* I 41 on how 'the motives influencing [a human being] cannot ordinarily scratch deeply enough to destroy the imprinted script of many millennia' (it says this is due to the shortness of human life).

However it is not just truths that get tested by this effort at incorporation—for these truths also test our drives, and those norms and values we identify ourselves with. I test a drive or value by seeing whether I can incorporate the insight why I have it—what it's doing in me. Can I build a diagnosis of my willing into my very willing? Can I, in the very act of willing-valuing X, understand this act? Can I will X while I know why I will it? Many or most of our aims and values, Nietzsche thinks, won't survive this test—or rather they will need to be heavily revised so as to be sustainable, still value-able, in the light of that incorporated diagnosis.

So our insight by genealogy into our willing gives us a new opportunity: to begin knowingly to *redesign* this attitude. Carrying out this redesign is the way to freedom, the truer freedom we're capable of.

At this point it's worth noticing what this shows about Nietzsche's difference from Kant. Nietzsche often seems to be finding 'conditions' of our experience—limitations and biases built deeply into the ways we think and value. But his naturalistic orientation makes these *not* Kantian transcendental and logical 'conditions of the possibility', but quasi-Darwinian conditions-of-selection: what these ways of thinking and valuing have been selected to do. Since the latter are not logical conditions, they are susceptible to that redesign. The upshot indeed is that Nietzsche uses his 'conditions' for an opposite purpose than Kant: Kant identifies conditions to validate them for science, but Nietzsche identifies them in the hope of freeing himself from them.

I've tried to show how the knowing stance completes itself in genealogy, and how the willing stance completes itself in the freedom this makes feasible, which reconciles these two stances with one another. It shows how the retrospective stance, when it *finally finds truths*, can be not a drag and burden to our forward-pushing will, but can in fact help it to find a new kind and degree of power, in the power over our values.

I suggest that Nietzsche means—or partly means—the image of eternal return to express this reconciliation. In this reconciliation the will overcomes its ill-will against the past, i.e., against the retrospective stance that so deflates it, by discovering how the truth about its past empowers it to overcome its past. I think one role of eternal return is to model and remind us of this reconciliation. That these two paths, back (*zurück*) and out (*hinaus*), meet each other and join in a ring symbolizes how these two basic stances of humans, whose conflict has been our grand problem, find their own completion by joining one another.

So eternal return is the emblematic attitude in which one holds together the contrary stances of retrospection (or theory) and will. It symbolizes the fusion of retrospect and prospect, hence of will to truth and will to power (life). It shows life how to face the past, and keep willing. It solves, finally, our problem of the past.

References

Richardson, John, 1996, Nietzsche's System, Oxford: Oxford University Press.
Richardson, John, 2004, Nietzsche's New Darwinism, Oxford: Oxford University Press.

Translations

Untimely Meditations, ed. D. Breazeale, trans. R. J. Hollingdale, Cambridge: Cambridge University Press, 1997.
Human, All Too Human, trans. R. J. Hollingdale, Cambridge: Cambridge University Press, 1986.
Thus Spoke Zarathustra, trans. Graham Parkes, Oxford: Oxford University Press, 2005.
On the Genealogy of Morality, trans. Maudemarie Clark and A. J. Swensen, Indianapolis: Hackett, 1998.
Philosophy and Truth: Selections from Nietzsche's Notebooks of the Early 1870s, trans. D. Breazeale, Atlantic Highlands, NJ: Humanities Press International.

Towards Adualism: Becoming and Nihilism in Nietzsche's Philosophy

Manuel Dries

> More strictly: one must not allow of anything at all that has being [*nichts Seiendes überhaupt*]—because then becoming loses its value and appears downright meaningless and superfluous. (Nachlaß November 1887–March 1888, KSA 13, 11[72])

In this essay, I shall argue that Nietzsche held two doctrines of becoming: one more radical, which he requires to fend off nihilism, and one much more moderate—the ontology of relations he develops under the label 'will to power'. Based on the latter he develops what I wish to call his 'adualistic'—neither monistic nor dualistic—practice of thought, a 'simultaneity-thinking' (*Zugleich-Denken*) that is no longer subject to nihilism. I shall argue further that we can only make sense of Nietzsche's oft-criticized *radical* affirmation of becoming (*Werden*) or impermanence—best defined *ex negativo* that there is no rational, true, benign, systematic, permanent reality for us—if we assume that he saw nihilism not merely as a possibility but as a *real* threat.

For his belief in the *reality* of the threat of nihilism to be intelligible, we have to attribute to Nietzsche at least three assumptions that underpin his entire project. The three assumptions are these:

(1) 'what there is, is becoming (and not being)', and
(2) 'most (if not all) strongly believe in being'.
(3) nihilism is a function of the belief in being.

Everything else can be seen as following directly from these assumptions.

This essay consists of six parts. In part 1, I wish to examine the above assumptions in more detail and show that Nietzsche's move towards becoming is motivated by the anomaly of nihilism. Secondly, I wish to show that Nietzsche not only believes that nihilism is already happening, but also that it is the task of the philosopher to be a *Gegenkraft* ('counter-force'), and this is the reason for the particularly *radical* nature of his affirmation of becoming. In the third part, I will examine whether Nietzsche is himself guilty of reintroducing a problematic dualism between becoming and language that renders all conceptual determinations mere falsifications. This issue will make it necessary, fourthly, to examine Nietzsche's ontology of relations. I will argue that Nietzsche's presentation of becoming as will-to-power relations can be seen as a *less* radical presentation of becoming. In the fifth part, I will show that this only partially removes a problematic dualism of becoming and language. Nietzsche struggles to find a language that captures his views on temporality and identity. Finally, I wish to show that Nietzsche, in order to avoid the above difficulties, at times suggests a non-reductive practice of thought that accounts for both the relative permanence or duration of 'persons' and 'things' and their constantly changing, temporal complexity—in short, the simultaneity of being and becoming.

1. Nihilism as the Function of the Belief in Being

Nietzsche wrestles with the problem of nihilism inherent in the valuation of being and absolute truth, namely 'that the highest values devalue themselves' (Nachlaß Autumn 1887, KSA 12, 9[35]). He perceives a problematic asymmetry between the secularized world-views people have adopted or are in the process of adopting, and their values—moral and other—which are still based on a strong belief in some kind of permanence or essence(s) or being (and are supported by a grammar and language that sustain their metaphysical picture). He is convinced that only when we have successfully erased from our thinking the categories of being that we formerly used to interpret existence—and which in Nietzsche's eyes are still used by everyone (apart from himself)—can nihilism be overcome:[1]

> Assuming we have recognized how the world may no longer be interpreted with these three categories ['unity', 'purpose', 'being', M.D.] and that upon this recognition the world begins to be without value for us: then we must ask where our belief in these three categories came from—let us see if it isn't possible to cancel our belief in them. Once we have devaluated these three categories, demonstrating that they can't be applied to the universe ceases to be a reason to devaluate the universe. (Nachlaß November 1887–March 1888, KSA 13, 11[99])

In addition, the morality of the Western tradition that continues to sustain the paradigm of being (despite the evidence for becoming delivered by

1 As I argue elsewhere, the early Romantics anticipated this view of Nietzsche's (Dries 2007, pp. 127–162).

natural science)[2] must itself be shown to be *an effect* of becoming as will to power, rather than grounded in, or corresponding to, divine being. Only then will our attitude towards the world or 'the whole' (*das Ganze*) change. Instead of incomplete, teleological notions of becoming (such as he attributes alike to Hegel, the young Hegelians, Socialists, and scientific realists) and Schopenhauer's privileging of permanence, Nietzsche wishes—in his eyes for the first time—to do proper justice to becoming. His new 'world conception' calls for the substitution of being with becoming:

> One is necessary, one is a piece of fate, one belongs to the whole, one *is* the whole—there is nothing which could judge, measure, compare, condemn our Being, for that would mean judging, measuring, comparing, condemning the whole ... But there is nothing apart from the whole!—That no one is made responsible any more, that a kind of Being cannot be traced back to a *causa prima*, that the world is no unity, either as sensorium or as 'mind', this alone is the great liberation—this alone re-establishes the innocence of becoming. (TI 'The Four Great Errors' 8)

We see here a threefold shift: towards 'the whole' (*das Ganze*), towards immanence (away from any exogenous stability or teleology), and towards becoming. This is Nietzsche's overcoming of nihilism in a nutshell: by fully embracing becoming, by deracinating the categories which devalued becoming in the name of transcendent being, nihilism can be overcome. The state that Nietzsche wishes to reach is what he calls the 'innocence of becoming' (*Unschuld des Werdens*). But this is only possible by *undoing the belief in being* by teaching that there is no such thing as unity: 'that the world is no unity, either as sensorium or as "mind"' (ibid.) is no objection to it.

In a more formalized way, Nietzsche's argument looks like this: if all there is, is becoming, and most people (if not all and excluding himself) believe in being, then nihilism is the result of the conflict that arises between the two premises—that is, the belief in (the value) being clashes with the realization that all there is, is becoming. I would argue that we need to attribute both the initial assumptions to Nietzsche in order to make sense of pronouncements such as 'nihilism is just around the corner'. If his contemporaries did not strongly believe in being, the confrontation with becoming simply would not have the nihilistic effect Nietzsche predicts.

It emerges, therefore, that for Nietzsche, the danger of nihilism is *a function of the belief in being*: the higher the valuation of being (as Truth with a capital, τέλος, permanence, unity as oneness, and systematicity), the

2 On the impact of the natural sciences on Nietzsche's philosophy, see Moore 2002; Stack 2005; and Emden 2005.

greater the risk and the effect of nihilism. But there is also, I believe, a third assumption which we must attribute to Nietzsche if we wish to make sense of his project, an assumption which derives from his general belief in the efficacy of forces. It can be formulated as follows: the more deeply ingrained a belief, the more radical a force is necessary to overthrow and undo that belief. Consequently, Nietzsche's radicalization of becoming has to be proportional to the *intensity* he attributes to the belief in being of his age. Given his first assumption (that all is becoming), it is not surprising that most of his philosophical project is concerned with undoing the belief in being which he attributes (second assumption) to most of his contemporaries.

2. Becoming as *Gegenkraft*

As many of his notes reveal, Nietzsche believes that the inevitable 'antagonism' between the new paradigm of becoming and the old still dominant paradigm of being is already at work and evident, resulting in a gradual *Auflösungsprozess*, 'a process of dissolution': 'This antagonism—not esteeming what we know [becoming, M.D.] and no longer being permitted to esteem what we would like to pretend to ourselves [being, M.D.]—results in a process of dissolution' (Nachlaß Summer 1886–Autumn 1887, KSA 12, 5[71]). This insight leads him to the conclusion—problematic at best—that the nihilistic process of dissolution should also be accelerated (*beschleunigt*). Nihilism—this time induced by the philosopher who is also a 'physician'—is supposed to play a vital part in its own cure.[3] As he writes in one of his most problematic notes in Spring 1885: 'an ecstatic nihilism could under certain circumstances be unavoidable for the philosopher: as powerful pressure' (Nachlaß May–July 1885, KSA 11, 35[82]). I will return to this passage towards the end of this section.

Let us again look at the argument implied in manifesto-like statements such as the above. Convinced of the inevitability of the dissolution of the paradigm of being through his own belief in the truth of becoming, he wishes everyone else to accept his own change of paradigm. Zarathustra's conviction—that it is the task of the philosopher to 'push that which is already falling' (Z III 'Of Old and New Tables' 20)—issues directly from Nietzsche's belief that nihilism is a function of the belief in being and that it is *actually* a real threat. The task of the philosopher is therefore to accel-

3 'To be the doctor here, to be merciless here, to guide the blade here—this is for us to do, this is our love of humanity' (A 7).

erate this process of dissolution, i.e., actively to undo the belief in being. How does he go about this?

For someone like Nietzsche who thinks in terms of forces and believes in force and counter-force, it is not surprising that he frequently announces his desire to be a counter-force (*Gegenkraft*) himself. As he says in the *Genealogy*, any constitution of *Sinn* ('meaning'), even at an organic level, he sees as the result of successful encounters or 'counteractions': 'Results of successful counteractions. The Form is fluid, but the "meaning" [*Sinn*] even more so' (GM II 12). A counter-force is therefore required to balance, control, or (and I take this to be Nietzsche's intention) overthrow another force if its 'meaning'—its current interpretation—is perceived to be a threat. Nietzsche's choice of 'therapy' is designed to match his belief in the intensity or embeddedness of belief in being. A counter-force of similar magnitude and intensity is called for, because he believes that the belief in being is still metaphysically grounded. This, I think, is the logic behind his questionable and inconsistent *radicalization* of becoming and also the argument that justifies (for Nietzsche) his radical *presentation* of becoming. His late note on the *Birth of Tragedy*, written in Spring 1888, could well be applied to his entire project: what is needed is 'a counter-forcc to all Nay-saying and Nay-doing, a remedy for the great fatigue' (Nachlaß Spring 1888, KSA 13, 14[15]).

Nietzsche's ambiguous views on science also hinge on this argument. According to him, science itself favours the required paradigm shift. While people may regard science merely as useful and unproblematic, they will soon discover, he thinks, that it is really 'die grosse Schmerzbringerin':

> So far it [science] may still be better known for its power to deprive man of his joys and make him colder, more statue-like, more stoic. But it might yet be found the great giver of pain! —And then its counter-force might at the same time be found: its immense capacity for letting new galaxies of joy flare up. (GS I 12)

But science proceeds slowly, by way of hypothesis, experiment, and falsifications, and only over long stretches of time will it have an impact and change a people's self-image. Also, like the senses, science both shows becoming and hides it—under the veil of objectivity—from view. And while the natural sciences might reveal enough to slowly weaken people's belief in being (which is precisely what Nietzsche believes has been the case since the Renaissance), the result is not that they have abandoned the belief in being;[4] rather, people no longer know who they are and what they

4 While future information technology might depend on results in quantum physics (e.g., quantum cryptography), this does not require anyone to change his ontology.

should think. While this is a stage of nihilism that Nietzsche endorses (because its direction is right), he worries that people might become (or are already) stuck in this nihilistic phase in which the belief in being stands against the reality of becoming. As he realizes, a nihilist is

> the man who judges that the world, as it is, should not exist and of the world, as it should be, that it does not exist. Consequently, existing (acting, suffering, willing, feeling) has no meaning: the pathos of the 'in vain' is the nihilist pathos—and at the same time, as pathos, an inconsistency of the nihilist. (Nachlaß Autumn 1887, KSA 12, 9[60])

Nietzsche is impatient and—against his own convictions—he even attempts to control time. While science has the potential to be a *Schmerzbringerin* and bring about the painful paradigm shift, it simply does not do it as quickly as the situation demands. Nietzsche's project becomes that of assisting science and presenting what he thinks is based on the latest results in the natural sciences in such a radical form that it will deracinate the belief in being either immediately, or at least more quickly.[5] He accepts that this may temporarily make matters worse, for a counter-force will take some time to take effect. There will be a period in which the belief in being still applies and functions, although its control over people's world-view and self-image will weaken. If the task of the philosopher is to speed up the process, and if the belief in being is as metaphysically embedded as Nietzsche believes is the case, it follows that the counter-belief he wishes to offer as a remedy must be presented with the same metaphysical intensity. He therefore does much more than simply suggest that the basic belief in being must be denied. He insists that 'one must not allow for there to be anything permanent [*nichts Seiendes überhaupt*] at all' (Nachlaß November 1887–March 1888, KSA 13, 11[72]) and presents a quasi-metaphysical counter-doctrine, namely, a radical, eternally-recurring, infinite becoming without meaning and *τέλος*:

> And do you know what 'the world' is to me? Shall I show you it in my mirror? This world: an immensity of force, without beginning, without end, a fixed brazen quantity of force which grows neither larger nor smaller, which doesn't exhaust but only transforms itself … as a play of forces and force-waves simultaneously one and 'many', accumulating here while diminishing there, an ocean of forces storming and flooding within themselves, eternally changing, eternally rushing back, with tremendous years of recurrence … as a becoming

5 Rex Welshon recently argued that Nietzsche's view 'is nothing more than a philosophically free expression of the contemporary scientific worldview' (2004, p. 159).

> that knows no satiety, no surfeit, no fatigue. (Nachlaß June–July 1885, KSA 11, 38[12])[6]

The point is not to examine the plausibility of radical flux or an eternally recurring becoming; my aim is merely to show why Nietzsche's assumptions require him to attack being in the radical way he does, 'since becoming would otherwise lose its value' (Nachlaß November 1887–March 1888, KSA 13, [11[72]).

We can now return to the second part of the passage on the 'ecstatic nihilism' Nietzsche wishes to induce. It runs as follows:

> An ecstatic nihilism might, under certain circumstances, be unavoidable for the philosopher: as powerful a pressure and hammer which he uses to destroy and do away with degenerating and dying species, for a new order of life; or in order to instil that which is degenerating and that which wants to die with a longing for the end. (Nachlaß May–July 1885, KSA 11, 35[82])

There are many such passages in which he shows his willingness to change the belief in being of his contemporaries at the expense of those who are not able to bear what he sees as the new paradigm. This is a result of his belief that nihilism (a function of the belief in being) will sweep people off their feet, whereas a *gradual* weakening of the belief in being would simply take too long. He accordingly proposes a drastic remedy:

> Bringing joy to humanity by sustaining their illusion, their belief. **Instead my countermovement:** —Domination of humanity for the purpose of its **overcoming**. Overcoming with doctrines through which it perishes, **except those who can bear it.** (Nachlaß Spring–Summer 1883, KSA 10, 7[238])

He seems to suggest that his drastic revelation, though destructive of the weak, will only last for a relatively short period and is therefore more desirable than any protracted and lingering decadence that will ultimately lead in any case to the same, inevitable paradigm shift.

As a physician of culture, Nietzsche believes that the 'casualties' will still be fewer than are caused by leaving things as they are. His goal is a new and cheerful 'innocence of becoming'.[7] If his three assumptions are accepted, his argument is sound. Unfortunately, the same assumptions

---

6 It should also be said that Nietzsche goes against the standards which he derived from his genealogy of being. He, too, employs unsinnliche ideas. Although he believes that there are instances when the senses reveal the truth of becoming—'If the senses show becoming, passing away, change, they do not lie' (TI '"Reason" in Philosophy' 2)—he insists elsewhere on the 'deception by the senses' (Nachlaß Spring 1880–Spring 1881, KSA 9, 10[E93]). We will return to this inconsistency later.

7 This, I think, is Nietzsche's eudaimonism, which I intend to discuss elsewhere.

commit him—despite his pronouncements against *Hinterwelten* (KSA 13, p. 46)—to a new complex of problems. For he also insists that radical becoming cannot be described in language. This opens the way for another 'anomalous' dualism with, as we shall see, a similar nihilistic potential.

3. Becoming versus Language

In an attempt to defend Hegel's conception of becoming, some critics argue that his critique of metaphysics is ultimately more successful than that of Nietzsche. Nietzsche's becoming denotes the absence of any real determination, and his critique of the metaphysical tradition amounts therefore to a mere *inversion* of the traditional opposition of being and becoming. For Nietzsche, one critic argues, 'becoming' denotes an ultimate reality: 'What is real for Nietzsche is "becoming"—flux, multiplicity, change. Nietzsche uses many different terms to denote this flux … But always the meaning is the same: *becoming is restless primordial indetermination*' (Houlgate 1986, p. 49; my emphasis). Nietzsche accordingly abolishes the metaphysical dualism between a true world of being and an apparent world of becoming; 'however, within his own *one* world he has preserved an opposition between what he sees as the fundamental reality and what he sees as mere appearance [within language or sense experience]' (ibid., p. 91). Thus, while he is right to criticize the abstract conception of being which disregards becoming, he is wrong in believing that 'life is becoming without logical form or identity, without "being"' (ibid.). He therefore remains tied to the problematic opposition of being and becoming:

> In criticizing being and *seiend* distinctions, Nietzsche should have gone on to criticize the dichotomy between being and life which turned both into abstractions. This he failed to do; instead of criticizing both the abstractions of being and becoming, he simply played off one against the other. (ibid., p. 95)

Hegel, on the other hand, overcomes this *exclusive disjunction* of being versus becoming ('either being or becoming'). Instead, he 'sees being (at least when it is fully determined as the Idea) as the inherent dynamic form and continuity of becoming itself, and he sees the "apparent" world of linguistic terms and concepts as revealing rather than concealing the character of the reality they describe' (ibid., p. 93).

I am not concerned here with the soundness of this defence of Hegel.[8] What I wish to examine is whether it accurately reflects Nietzsche's posi-

8 Houlgate is unsuccessful in defending Hegel against Nietzsche for at least two reasons. He fails to give an account of Hegel's 'being as becoming' that is free

tion. More than once, Nietzsche does indeed refer to the inability of language to express *Werden*. Despite his Spinoza-inspired attacks on the Platonic-Christian two-world metaphysics,[9] he does seem to introduce a dualism between becoming and language, thereby equating language with 'error' and 'falsification'. Again we encounter one of his basic contradictions: he simultaneously maintains that 'the means of expression of language are not suitable for expressing becoming' (Nachlaß November 1887–March 1888, KSA 13, 11[73]), and also that all philosophy should do (presumably within language) is to express becoming: 'Philosophy, in the only way acceptable to me, as the most general form of history, as an attempt somehow to describe Heraclitean becoming and to abbreviate into signs (so to speak, to translate and mummify it into a kind of illusory being)' (Nachlaß June–July 1885, KSA 11, 36[27]).

This seems clear enough evidence that becoming is Nietzsche's new *Hinterwelt*—not beyond the matter and force of his new 'one' world, but certainly beyond the schematizations of our senses and language. Is he simply demanding the impossible? We seem to have here what I wish to call Nietzsche's version of the 'impossible presentation thesis': his *exclusive disjunction* entails the impossibility of presenting becoming within language, i.e., within a system of signs that 'fixes' meaning by 'expressing' it (*Feststellung*).[10] As in the case of Schopenhauer's being–becoming dichotomy, Nietzsche's own dichotomy between becoming and any kind of determinateness annuls the value of what is given within language.[11]

Attributing this position to Nietzsche, who dedicated his entire migraine-free time to becoming an 'artist of language' (Nachlaß April–June 1885, KSA 11, 34[124]), is, to say the least, problematic. His views on language are indeed more subtle than this. Aware of the double nature of language as both revealing and concealing, Nietzsche also knows that lan-

---

from the kind of systematicity, rationality, and teleology that, as we saw above, Nietzsche finds so problematic. Further, he is mistaken in attributing to Nietzsche a simple inversion of the being–becoming dichotomy. As I shall shortly argue, Nietzsche also considers being and becoming as a unity, albeit not in the 'systematic' and 'continuous' way of Hegel.

9 See, e.g., A 17.

10 The early Romantic philosophers such as Novalis, Schlegel, and Schelling assumed the logical and ontological priority of an Absolute (das Unbedingte) that is never 'present' and can only be represented within reflection and language. This idea leads them to their philosophies of 'infinite approximation' (see, e.g., Frank 1997 and Bowie 2003).

11 Volker Gerhardt points to this inconsistency—thereby tacitly accepting the 'impossible presentation thesis'—when he remarks: 'Nietzsche attempts the impossible, namely, to express the fact of becoming within language' (1996, p. 296).

guage is always both *limitatio* and *conditio*. Anticipating the later Wittgenstein's view, he states, in a note on 'mature artworks' of Spring 1888: 'Any mature art is based on an abundance of conventions: insofar as it is language. Convention is the condition of great art not its prevention' (Nachlaß Spring 1888, KSA 13, 14[119]). In this passage, he at least seems aware of the pitfalls of metaphysical realism. And in another late note he makes it unmistakably clear that it would be wrong to dismiss language for its alleged failure to present or correspond to any extant particulars or entities (*Wesen*):

> The demand for an adequate mode of expression is nonsensical: it's of the essence of a language, of a means of expression, to express only a relation ... The concept of 'truth' is absurd ... the whole realm of 'true', 'false' refers only to relations between entities, not to the 'in-itself' ... Nonsense: there is no 'essence-in-itself', it's only relations that constitute entities, and neither can there be a 'knowledge-in-itself'. (Nachlaß Spring 1888, KSA 13, 14[122])

Here he clearly denies the *viability* of the view that language *aims* at corresponding entities or fundamental truths. The metaphysical realist will always be unable to satisfy the sceptic's doubt regarding the correspondence of *Denken* and *Sein*. Nietzsche is aware of the nihilistic potential of such a metaphysical realism and, anticipating the ontology of current hermeneutics,[12] he inverts what I shall call the 'truth-vector' of language: essences and truths are not to be conceived as the *targets* of intentionality or of interpretations, but should be understood as *results* of intentionality or interpretations.[13]

But what are we to make of his contradictory insistence that there is no 'in-itself', that language *constitutes* entities, and yet, in any description, fails to express becoming?

While Nietzsche might well have been—accidentally or deliberately—self-contradictory (and both interpretations can be found in the secondary literature), I think there is a strong argument, following directly from Nietzsche's assumptions, that would eliminate the above inconsistency. In a nutshell, it runs as follows. We know that Nietzsche endorses the radical doctrine of becoming in the (by his own standards) necessary attack on the belief in being. *This does not, however, mean that he himself actually subscribes to any radical ontology of becoming.* In the light of our earlier discussion, we might speculate that—at least for the later Nietzsche—talk

12 On Nietzsche's importance for philosophical hermeneutics, see Vattimo 1986.

13 On truth as the result of interpretations, see Abel 2003, pp. 4–7; also Abel 1998, p. 326.

about all-pervasive becoming, inexpressible by language, is addressed to those who need to be awoken 'with a philosophical hammer' from what he certainly believed to be their 'dogmatic slumber'. In order to examine this issue further, we need to examine Nietzsche's most detailed description of becoming, his ontology (or phenomenology) of relations, also known by the name of will to power.

4. Becoming as Power Relations

I will try to limit my analysis of the will to power to the one question of relevance to our analysis: is Nietzsche guilty of merely inverting the being–becoming dichotomy, thereby introducing a dualism between a more fundamental reality and language? This question has two parts: does will to power—as a description of becoming—rule out being altogether and denote indeterminacy? And secondly, is will to power a metaphysical theory, or should we interpret Nietzsche as a phenomenologist who is not interested in theories about ultimate reality?

I will begin with the second question. Peter Poellner has recently proposed that Nietzsche should be read primarily not as a metaphysician (as some critics still do)[14] but as a phenomenologist.[15] Anticipating the fundamental reorientation of phenomenology (against philosophy in its traditional orientation towards epistemology or metaphysics), Nietzsche regards the 'the first-personal investigation of how a world can manifest itself in experience, and how, in particular, it does so in human experience, as the fundamental philosophical enterprise' (Poellner 2006, p. 302). Metaphysics and epistemology in their traditional sense are 'while not rejected, at best considered derivative' (ibid.). This, of course, requires the reader to understand the radical nature of this phenomenological turn, and only then is it possible to see, according to Poellner, that none of Nietzsche's physiologi-

---

14 Despite his explicit attacks on foundations, Nietzsche is indeed often interpreted as hypostatizing becoming and making it an ultimate reality. Stambaugh, for example, argued that the novelty of Nietzsche's position lies in his absolute denial of duration: 'The flux of time is in its own way a concealed kind of "substance", for it continuously flows on. The flux is constant, continuous. It always flows, or "is"' (1972, p. 7; see also Danto 1965, p. 96; Poellner 1995, p. 91; and Young 1992, p. 97). In his recontextualization of Nietzsche's ideas within the scientific writings of his contemporaries, Moore also concludes that becoming as will to power is a metaphysical Bildungstrieb (2002, p. 55).

15 Meaning here 'phenomena as they are perceived'. Nietzsche would, of course, reject the idea of any Cartesian 'first philosophy' or fundamental theory.

cal explanations or his accounts of the efficacy of consciousness are referring to 'what *really* is the case in an ultimate ontological sense' (ibid., p. 297). Instead, such descriptions should be interpreted as 'non-metaphysical, practical methods of understanding and acting on the world within the context of a dominant concern with the *phenomenology of the human life-world*' (ibid., p. 298).[16]

What evidence do we have (other than the denial of 'essences in themselves' which we saw earlier) that when it comes to becoming, Nietzsche is *not* trying to present a foundational metaphysics of becoming? I would like to begin with section 370 of *The Gay Science*. Here, Nietzsche attacks not only Schopenhauer's pessimism and Hegel's panlogism for their respective privileging of 'rest', 'stillness', and 'calmness'; he also dismisses those Romantics who desire 'intoxication' (*Rausch*) and try to connect with an ineffable absolute. This passage is important because it shows that the later Nietzsche rejects any interest in some kind of noumenal realm.[17] Dionysian intoxication (*Rausch*) is now listed as a decadent form of 'anaesthesia', as a problematic practice designed to shy away from the *real* contradictions of life—'real' in a phenomenological *sense*, not as an ontological *reference*.[18] In this passage at least, Nietzsche rules out any kind of noumenal beyond[19] and, in the famous genealogy of truth in *Twilight of the Idols* 'How the "True World" Finally Became a Fable. The History of an Error', he demands a world-view that values this world *as* this world: 'The true world is gone: which world is left? The illusory one, perhaps? ... But no! We got rid of the illusory world along with the true one!' This passage

16 'Neither his (implicit or explicit) claims concerning the efficacy of consciousness, nor his advocacy … of "physiological" explanation should be understood metaphysically as theses about what really is the case in an ultimate ontological sense. Rather, both of these approaches should be interpreted as mutually compatible, non-metaphysical, practical methods of understanding and acting on the world within the context of a dominant concern with the phenomenology of the human life-world' (2006, pp. 297–298).

17 See also GM III 5.

18 Poellner invokes Frege's distinction between sense and reference: 'the phenomenologist is only interested in the level of sense (in Husserl's broad understanding of Sinn, whereby all intentional contents, not merely linguistic ones, involve senses). She is not interested, qua phenomenologist, in the level of reference, e.g., in whether some apparent represented object used as a sample really exists. But this temporary suspension of the "natural attitude" is of course not an end in itself, but is engaged for a better understanding of the Sinnstruktur of our actual experiential world' (2006, p. 299).

19 See also Nietzsche's remark on 'secret routes to worlds beyond and false divinities' (Nachlaß November 1887–March 1888, KSA 13, 11[99]).

invites multiple readings, of course, and it does not follow (as Poellner and, for example, Clark argue)[20] that Nietzsche now steers clear of foundational metaphysics and refers to the *human lifeworld* only. We could equally read this passage as referring to his new and metaphysical world of will-to-power becoming, which would be entirely different from (and thereby 'abolish') the Platonic-Christian world formerly considered to be 'true', and also entirely different from (and thereby 'abolish') the world that was 'formerly' seen as mere appearance, namely, sense impressions, things, etc. We come back to the alternative of either phenomenology or metaphysics.

It is easier to see what Nietzsche *rules out* by emphasizing becoming. By his shift towards becoming as will to power, he deracinates the four metaphysical hypostases he regards as most problematic: substantiality, rest, causality, and agency.[21] But what is he *affirming* when he describes becoming as 'will to power'? At first sight, will to power seems like a traditional metaphysical doctrine insofar as it makes a statement about the world as a whole. In the light of our previous discussion of Nietzsche's explicit *denial* of extant particulars as referents for language, we should be cautious about assuming from the start that any description of the whole as will to power corresponds to any 'essence' of what is ontologically real. For now, I shall treat the will to power as an attempt to formulate an explanatory hypothesis, and not, as many passages would certainly allow us to do, as a transcendent principle that controls the movement of totality from outside and to which every phenomenal configuration might be reduced.[22] In one of the most famous passages, Nietzsche describes the will to power as follows:

20 See Clark 1990.

21 As Richardson (2006, pp. 211–212) argues, becoming as will to power seems therefore to imply that change is pervasive, i.e., that there are no substrata exempt from change; that change is constant, i.e., there are no pauses in change; that change is along a continuum rather than by way of isolated causes and effects; and, finally, that change is what there is, i.e., there are no underlying beings that change.

22 In support of this interpretation of will to power, see, e.g., Müller-Lauter 1999a, 1999b. Recent scholarship on the will to power, e.g., Deleuze 1983, 1994; Richardson 1996; Figal 1998; Müller-Lauter 1999b; Smith 2000; Porter 2006, understands 'power' not as an independent state to be reached (Richardson 1996, p. 16). It also rejects the notion of power as self-preservation, because the goal of life as will to power is not the maintenance of power relations but an increase in change, even at the expense of particular forms of successful power (Smith 2000, p. 111).

> My idea is that every specific body [atoms, chemical substances, M.D.] strives to become master over all space and to extend its force (—its will to power:) and to thrust back all that resists its extension. But it continually encounters similar efforts on the part of other bodies and ends up arranging ('uniting') with those that are sufficiently related to it: —thus they then conspire together for power. And the process goes on. (Nachlaß Spring 1888, KSA 13, 14[186])[23]

A 'body' (*Körper*) cannot, however, have any numerical identity because it is not based on parts but on relations, and the number of relations is constantly changing.[24] *Körper*, as any other *Dinge* or 'objects', are themselves best conceived, Nietzsche thinks, not as substances but as 'sums' or 'bundles' (*Summen*) of will to power quanta. Yet even the term 'quanta'[25] shows that he still retains some kind of entities which together form relations. As Nietzsche writes:

> Every thing is a sum of judgements (fears, hopes, some inspire confidence, others do not). Now, the better we know physics the less phantasmal this sum of judgements becomes ... Finally we understand: a thing is a sum of excitations within us: however, since we are nothing fixed [Festes] a thing is also not a fixed sum. And the more stability we attribute to things, – – – (Nachlaß Spring 1880–Spring 1881, KSA 9, 10[F100])

This passage seems to give support to the view that Nietzsche starts out from the kind of phenomenological attitude Poellner suggests, by discussing intentional states such as fear, hope, and trust. But there can be no doubt that he immediately adds weight to his phenomenological 'sum

23 In Henry Staten's reading, this passage denotes the 'overwhelming of others' (1990, pp. 141–142) and Nietzsche's 'fantasy of infinite extension, as though in the case of some monstrous cosmic protozoan' (1990, pp. 141–142). It should be said that Staten omits the second half of the passage in which Nietzsche explains that power is not an independent state to be reached, nor is it the goal of one 'body' to annihilate its relational other. Instead, 'power' denotes the relation (conspirieren zusammen).

24 See the following note: 'And for us, even those smallest living beings which constitute our body (more correctly: for whose interaction the thing we call body is the best simile–) are not soul-atoms, but rather something growing, struggling, reproducing and dying off again: so that their number alters unsteadily, and our living, like all living, is at once an incessant dying. There are thus in man as many "consciousnesses" as—at every moment of his existence—there are beings which constitute his body' (Nachlaß June–July 1885,·KSA 11, 37[4]).

25 'If we eliminate the ingredients, what remains are not things but dynamic quanta in a relationship of tension, whose essence consists in their relation to all other quanta, in their "effects" on these—the will to power not a being, not a becoming, but a pathos, is the most elementary fact, and becoming, effecting, is only the result of this' (Nachlaß Spring 1888, KSA 13, 14[79]).

selves' by referring to physical theories (obviously about processes between quanta underlying—or at least coexisting with—the above phenomenological perspectives) that would support his view.[26]

In another example, Nietzsche again tries to argue for the *reality* of relations, and proposes that even if one of his books existed only in the heads of all the people who had previously read it (of course, at the time, hardly anybody had actually read any of his books), this book would should still be considered as *real*:

> Let us assume that my book existed only in the minds of people, then everything consisted, in a sense, of **their** thoughts and essences—it would be a 'sum of relations'. Yet is it therefore no longer anything? Parable for all things. Just as our 'Neighbour' [*Nächster*]. That a thing dissolves into a sum of relations proves nothing against its being real. (Nachlaß Autumn 1881, KSA 9, 13[11])

Of course, the point Nietzsche is trying to argue is that abolishing any metaphysics of substance (the doer behind a deed, etc.) should not devalue our phenomenological description of it. Again, this shows to what extent many of the arguments hinge on his *belief* in belief in being: even here he is attacking this belief in order to avert nihilism. There are other passages, of course, that seem to imply ontological 'referents' rather than a phenomenological 'sense': 'The law of conservation of energy demands eternal recurrence' (Nachlaß Summer 1886–Autumn 1887, KSA 12, 5[54]). But then again, every *Weltconception*, as Nietzsche had earlier argued, should not be seen as realist but justified instead as a creative act (Nachlaß Summer 1872–early 1873, KSA 7, 19[52]). Against the scientific realism of his times he argues:

> The physicists believe in a 'true world' in their own way: a static atom-systematization that is the same for all entities and follows necessary motions, —so that for them the 'apparent world' reduces to each entity's perspective of universal and universally necessary being … But they are wrong here: this atom they arrive at according to the logic of that consciousness-perspectivism,

26 According to Hales and Welshon (2000), Nietzsche promotes the idea of a 'bundle self' that implies the 'No-Self view' consistent with Buddhism. The self is seen as 'a loosely organized confederation of functional states and dispositions' (p. 159) without a strong notion of diachronic identity. Manfred Frank (2007, pp. 152–170) among others has shown that such a theory of subjectivity has difficulties in accounting for self-consciousness—a serious deficiency in Nietzsche's philosophy of mind (as well as in most post-modern accounts of subjectivity) that has yet to receive proper attention. Paul Katsafanas (2005, pp. 24–25) shows an awareness of the problem.

> —is then itself a subjective fiction … They forgot to make this perspective-positing force part of 'true being'. (Nachlaß Spring 1888, KSA 13, 14[186])

We could play this pseudo-Heraclitean game indefinitely. The textual evidence suggests that Nietzsche tries to write sometimes as a phenomenologist and at other times loses himself in (meta-)physical speculation—or like Lucretius, in (meta-)physical poetry—and we can safely suggest that all three modes are meant to avert the impending and *actual* threat of nihilism. Also, Nietzsche's move is, I think, characteristic of the paradigm of becoming: he shifts from an ontology of substances to an ontology of processes or relations. Becoming as will to power denotes processes involving directional forces and counter-forces, and Nietzsche conceives of such forces as engaged in a process of 'interdetermination' (reminiscent of *Wechselbestimmung*, the early Romantic term for the constitution of consciousness). 'Is will possible without these two oscillations of Yes and No?' Nietzsche asks:

> there must be oppositions, resistances, and thus, relatively, overarching unities... Localized – – –
>
> if A exerts an effect on B, then only as localized is A separated from B. (Nachlaß Spring 1888, KSA 13, 14[80])

Because Nietzsche's process metaphysics requires that there be not just flux, i.e., constantly changing relations between forces, but also, as he says, 'übergreifende Einheiten' ('overarching unities'), Johann Figl proposes (and I agree with him) that we should understand will to power not as radical becoming, but as the irreducible relation of *both* being *and* becoming: 'will to power is then that concept which ties together being and becoming' (1982, p. 85). This seems to provide an answer to our first question: becoming is *not* to be conceived as absolute 'indetermination', or 'structureless thereness' (Danto 1965, pp. 96–97), entirely separate or free from determination. On the contrary, the description of becoming, once untangled, seems much more moderate.

But we have yet to explain Nietzsche's contradictory statements regarding the ability of language to express this (more moderate) becoming.

5. Becoming, Language, and Time

Nietzsche believes (and says) that, on the one hand, 'there is no will: there are points of will constantly augmenting or losing their power' (Nachlaß November 1887–March 1888, KSA 13, 11[73]); and that on the other, 'the means of expression that language offers are of no use to express becoming' (ibid.). I will deal with the former proposition first.

While will to power supposedly designates the mode of being of every configuration in the phenomenal world, Nietzsche warns that 'will' or 'power' (understood as a *single* substance or principle) ultimately does not exist. Every world constitution, conscious or unconscious, is the result of multiple 'volitional' or 'intentional' processes.[27] A 'sum' of dynamic will-points always culminates in a 'power situation' between volitional activities, but as processes, they never arrive at final positions, and continually reconfigure in different relationships of power:

> Struggle of atoms, as of individuals, yet, at a certain difference of force two atoms become one, and two individuals one. And vice versa one becomes two when the internal state effects a disgregation of a centre of force. —Hence against any absolute conception of 'atom' and 'individual'. (Nachlaß Autumn 1885, KSA 11, 43[1])

And as for the supposed inadequacy of language to express becoming, Nietzsche insists that language is unable (*unbrauchbar*) to express will to power, but he then proceeds immediately to deliver a reformulated description of his (non-substantial) idea of will as 'will-points' (*Willens-Punktuationen*). Once again, his assumptions help us to clarify why he holds such a paradoxical position. The subject-object structure of language supports what he believes to be the belief of his contemporaries, namely a metaphysics of substance that carries within itself nihilism as a product of the belief in being. Because of its inherent structure, language cannot capture what he himself regards as true: namely, processes and relations without any essential agents to sustain them. As I indicated at the end of the section 2, we have to assume that he switches at times between, or tries to speak from, and for, more than one paradigm.

When addressing adherents of the paradigm of being, Nietzsche argues (in keeping with the 'impossible presentation thesis') that language *cannot* express becoming. Why is this so? Regardless of what is expressed within the subject-object structure of a language, to someone who *believes in being*, i.e., in isolated and substantive subjects and objects, language's semantic units and grammar will always confirm that structure and with it the paradigm of being itself. Therefore, when he addresses his contempo-

27 What Nietzsche describes as 'dynamic quanta' or 'will-points' (Willens-Punktuationen) seems to have something like the following structure: centre → vector → goal; or alternatively, subject → affect → intentional object. As Welshon (2004) suggests, 'The structure of intentional psychological events: <subject→ affect → intentional object> is … an instance of a more general structure that is plausibly instantiable by non-conscious, non-animated and perhaps even non-living entities. Nietzsche is proposing that psychological events are structured in a manner isomorphic to that exemplified by all efficient causal relations' (p. 173).

raries (who, in his view, believe in being), Nietzsche must necessarily uphold the view that language cannot express becoming, thereby introducing the problematic dualism we have just noted. In short: (i) language cannot express what you [my contemporaries] think the world is essentially, namely being; (ii) fortunately, being does not exist. It is important to understand that Nietzsche's discourse is always *located* or *positioned*, addressing particular people or groups, and, to *some* extent (to make himself comprehensible to them) by using *their* language, and so his whole activity is how to get *them* from their false conception—expressed in a specific linguistic form (which he adopts when speaking to them)—to his *own* views.

But when addressing, as he often does, the future paradigm of becoming, Nietzsche thinks he can indeed express and describe becoming within language. Again, we might ask how this can be so. Will-points *also* follow a teleological structure somewhat similar to that of language ('I need beginnings and centres of motion, starting from which the will reaches out').[28] For someone who has already changed and who accepts Nietzsche's paradigm of becoming, who already believes in processes and relations rather than substances and 'doers behind deeds', etc., language *can indeed* correspond to and express becoming (as plural events between directed quanta of forces, but without any teleology that governs the whole): 'a quantum of power, a becoming, insofar as none of it has the character of "being"'(Nachlaß November 1887–March 1888, KSA 13, 11[73]).

But can the implied dualism really be avoided? The task language would have to master within a paradigm for which something like the will to power serves as its explanatory hypothesis would be to express the simultaneity of two different, yet related, levels of becoming or temporality. As Richardson (2006, p. 225) has recently argued, becoming as will to power firstly denotes a real, pre-conscious background becoming, 'by which perspectivity and meaning arise and evolve' (let this be 'background time'); and secondly, will to power also denotes an ideal, perspectival temporality *for* a perspective, i.e., 'the way time appears to the perspectives' (let this be 'conscious time'). Nietzsche frequently observes that, behind all conscious intentionality and language ('conscious time') lies also an un-

28 Nietzsche admits of a plurality of teleological forces but he wishes to refute any outside, first cause behind such plural events. As he tries to explain in his refutation of any strong notion of causality in 1888: 'Will to power in principle. Critique of the concept of "cause". I need the starting point "will to power" as the origin of motion. Consequently, motion must not be conditioned from outside—not caused … I need beginnings and centres of motion, starting from which the will reaches out' (Nachlaß Spring 1888, KSA 13, 14[98]).

conscious intentionality ('background time'), an *unbewusste Absichtlichkeit* (Nachlaß Autumn 1885–Spring 1886, KSA 12, 1[76]). He can therefore hold that any conscious time (e.g., the way I perceive time) must always already *be* a selection, 'an interpretation that can be false; moreover a simplification and falsification etc.' (ibid.). How are we to understand 'falsification' (*Fälschung*) in this context?

I think that Nietzsche's idea may best be understood by analogy with a spotlight picking out a certain scene on a theatre stage on which many scenes are being performed simultaneously. The spotlight picks out one scene and brings it into focus, while the rest of the actions on and off stage continue, but in the dark. In order to illuminate the complexity, we should have to introduce multiple temporal and intentional 'spotlights'—originating from the point of view of each actor within the diegetic[29] time of his particular scene, thereby overlooking the temporality of his non-diegetic perspective (the person he is in 'real' life), and so forth. Of course, the 'spotlights' not only reveal pre-existing events or objects, to some extent they also create and constitute them. Conceived as such, 'the whole' in Nietzsche's world-conception denotes a continuum of perspectival, interpretative processes, both unconscious 'background time' and 'conscious time': 'The will to power interprets: the development of an organ is an interpretation; the will to power sets limits, determines degrees and differences of power' (Nachlaß Autumn 1885–Autumn 1886, KSA 12, 2[148]). Viewed as such, life is seen as perspectival at all levels: a minimal intentionality or directedness is assumed to be already at work in non-conscious organic life-forms such as 'protoplasm': 'In truth, interpretation is itself a means of becoming master of something (The organic process presupposes constant **interpreting**)' (ibid.). On more complex organic levels, Nietzsche's 'sum selves' acquire their own complicated perspective, composed of inherited and selected drives, sense experiences, incorporated memories, and unconscious and conscious future projects.[30] In his description of perspectivism, Nietzsche attacks the presupposition underlying any subject distinct from the body and sense perception. Such a view of the self is implausible, since it requires a non-

29 Pertaining to diegesis: 'the fictional time, place, characters, and events which constitute the universe of the narrative' (OED).

30 See Nachlaß May–June 1885, KSA 11, 35[58, 59]. Earlier (ibid., 26[272]), Nietzsche insists that even the inorganic must be thought of as having a minimal directedness. Recently, philosophy of mind has started to seriously consider such a 'panpsychist' theory, see, e.g., the responses to Galen Strawson's paper 'Realistic Monism. Why physicalism entails panpsychism' in the collection of essays entitled Consciousness and its Place in Nature (Freeman/Strawson, et al. 2006).

directional 'vision', 'an eye turned in no direction at all, an eye where the active and interpretive powers are to be suppressed, absent, but through which seeing becomes a seeing-something, so it is an absurdity and non-concept of eye that is demanded' (GM III 12).

The fundamental asymmetry, then, is between the 'directional'[31] way in which 'übergreifende Einheiten' view and thereby experience becoming (conscious temporality), and the multi-directional temporality of the whole (background temporality) in which they *become* by participation in what Nietzsche calls the continuum. This leads, according to Richardson, to the following asymmetrical situation: 'Life itself (the organism) views time differently than it lives it. Since becoming lies in the temporal structure of perspectives, and not in how they view time, life tends to miss its own becoming' (2006, p. 215).[32] Both temporalities are, so to speak, at work within us simultaneously. Within a 'reductive' physicalist theory of mind,[33] one might argue that conscious time supervenes on background becoming, implying that the temporality of the whole determines conscious time, even from within a conscious perspective.[34] This, I think, is not the view Nietzsche holds.[35] Rather, his 'sum selves,' from within their perspectival temporality which limits their causal efficacy, determine the continuum, just as background temporality determines the 'sum selves'. I suggest that this type of adualistic 'interdeterminism' is perhaps best conceived along the lines of mutual 'interruptions': at a certain conscious moment, you intend to carry out a certain action, and then, after some 'time' (which you have failed to notice) has passed, you might wonder why you ended up doing something completely different; or at other times, you 'find' yourself engaged in an action you had not been consciously aware

31 The German word gerichtet implies both 'having a direction' as well as a valuation or judgement.

32 Richardson sees proof for Nietzsche's temporal realism in his 'naturalist allegiance to a physical reality, within which these wills have evolved'; Nietzsche therefore 'cannot avoid supposing a time that is independent of those wills—a time in which not just organisms' bodies but all matter interacts, including inorganic matter that does not support perspectival will' (2006, p. 226). Günter Abel, on the other hand, situates Nietzsche's temporal continuum within his general interpretationism, thereby defending Nietzsche's anti-realism against the charge of a new essentialism (2000, p. 438).

33 As opposed to non-reductive physicalist theories that also exist in the 'analytical' tradition (see Strawson et al. 2006).

34 Which leads to an over-determination.

35 I think it is necessary to go further than Leiter's illuminating but reductive readings in 2001, 2002, p. 104, and Leiter/Knobe 2007.

of, and from that 'moment' on you are 'interrupting' and 'determining' this action, thereby taking it in a different direction.[36]

We have finally arrived at a much less radical version of Nietzsche's *Werden*: he allows for instances of being with relative duration and also relative stability; his sums are indeed 'complex forms of relative life-duration [with their conscious temporality, M.D.] within the flux of becoming [within the temporality of the whole, M.D.]' (Nachlaß November 1887–March 1888, KSA 13, 11[73]). So when he states that language falsifies and 'fails' to express becoming, he could be understood as indicating that language cannot afford a God's-eye perspective, and that it falsifies when it presumes[37] to use what modal logicians today call 'rigid designators' that pretend to capture an event once and for all in all possible worlds. This would indeed efface the simultaneity of unconscious background becoming and conscious becoming as it is experienced from within a perspective. It is necessary to use language in such a way that it shows an awareness of the interrelation of both temporalities. But the argument we used earlier still applies: whether or not you understand such a language 'correctly' would depend on your paradigm.

For his descriptions to be true to his belief in becoming as will to power, Nietzsche sometimes tries to express his vision through adualistic descriptions: self-consciousness is, he thinks, better described as *Selbst-bewusst-Werden* rather than *Selbstbewusstsein*. Each 'sum self' has the status of relative being and its own perspective; yet at the same time, it is also the *result(ing)* of a long process of selection. It instantiates and is living its entire evolutionary history that it has incorporated (*einverleibt*):

> Man is not just an individual but the living-on organic totality [*das Fortlebende Gesammt-Organische*] in one particular line. That he exists proves that one species of interpretation (albeit always under further construction) has also kept existing, that the system of interpretation has not switched. 'Adaptation'. (Nachlaß End of 1886–Spring 1887, KSA 12, 7[2])

36 This more complex interdeterminism (see also Richardson 2008) should perhaps be conceived along the lines of interruptions in both directions—the kind of interruption recently suggested by studies into the effect of testosterone levels. After exposure to images of sexual content, those with higher levels of testosterone (measurable through the length of their index fingers) show a higher level of arousal which—for a considerable amount of time—interferes with their ability to make informed decisions.

37 Something any hypothetical adherent of the paradigm of becoming would no longer think possible.

In the light of this passage, Nietzsche's intertwining of being and becoming demands not a monistic but rather an adualistic reading.[38] In his late philosophy, he delivers a theory that corroborates the intuition he had as early as 1872: 'The order in the world, the toilsome and slowest result of horrific evolutions understood as the nature of the world—Heraclitus' (Nachlaß Summer 1872–Beginning of 1873, KSA 7, 19[124]).

I now wish to leave Nietzsche's adualistic intertwining of being and becoming—*both* being *and* becoming, *neither* being *nor* becoming—behind and move from the micro-level of the will to power to the macro-level of Nietzsche's philosophy as a whole. As already indicated, Nietzsche follows two agendas. His reception shows how difficult it is to account for both.

6. Nietzsche's Simultaneity-Thinking

> The logic of our conscious thinking is only a crude and facilitated form of the thinking needed … by the particular organs of our organism. A simultaneity-thinking [*ein Zugleich-Denken*] is needed of which we have hardly an inkling. (Nachlaß April–June 1885, KSA 11, 34[124])
>
> Life no longer dwells in the whole … The whole no longer lives at all: it is composite, calculated, artificial, an artefact. (CW 7)

So far I have attributed to Nietzsche a certain double standard: when he addresses the adherents of the paradigm of being he presents a radical doctrine of becoming in hyperbolic terms; on the other hand, his process ontology (hypothetical or not) of will to power turns out to be much less radical, allowing for stability and duration. But this schematic separation of standards obfuscates the real problem, namely that Nietzsche tries to do both at the same time: shock the believers in being out of their nihilistic assumption *and* prepare for a non-nihilistic, new paradigm. To date, his reception shows that his strategy was successful in so far as it has certainly generated wide-ranging interest, but it also shows that it failed miserably by generating a plethora of (mis-)appropriations. As we know, he has been

38 Abel interprets this passage as follows: 'An adualistic viewpoint is required. Nietzsche advocates such a view. He assumes a continuous spectrum of what exists or happens in one form or another, from the furthest reaches of the inorganic to mental states, consciousness, becoming-self-conscious, cognitive and other mental activities, and planned actions and their executions. The organic appears therefore as the evolutionary-historical and continuous precursor of consciousness. The world of Nietzsche is a world of such continuum-relations (Abel 2001, pp. 6–7).

both celebrated and rejected as the thinker of new values for the select few, for an aristocracy of the powerful against cultural disintegration (Nietzsche uses the term *Disgregation*) and weakness.[39] More recently, he has become the forebear of deconstructive trends in the continental tradition, the thinker of becoming, multiplicity, interpretation, masks, etc.—hailed for his non-totalizing aspects and despised for his laissez-faire relativism (*mere* interpretation). I believe reconstructing Nietzsche's assumptions helps considerably to make sense of this reception, which is puzzling at best.

In this final section, however, I also wish to move from the double standard and the consequences I have just described to a second 'double standard' of a different kind. I wish to show that, at least at times, Nietzsche thinks about unifications, also on an interpersonal and socio-political level, within an adualistic framework. Nietzsche's project of forestalling nihilism requires him to conceive a proper unity (*das Ganze*) as well as difference. As I will show, in some of his remarks on the phenomenology of love, he finds evidence for a notion of *community* for the new paradigm he envisages in his moderate moments (i.e., when he is not speaking as a strong counter-force to the belief in being). For Nietzsche's deconstructive demands exist side by side with his calls for unity,[40] and both issue from his attack on nihilism.

Like several of his predecessors, Nietzsche is very aware of a set of problems that tend to undermine the success of unifications. The three dilemmas that concerned, for example, Schiller in his reaction to Kant—I call them elsewhere the either-or dilemma, the synthesis dilemma, and the relativism dilemma (Dries 2006, pp. 53–58)—also feature prominently in some of Nietzsche's phenomenological observations on unities. He, too, realized at an early stage that most unities suffer from a confusion of unity with oneness. Thus, if the concept of a new unity is necessary in order to attain an affirmative attitude towards life after any two-world metaphysics

39 For passages in which Nietzsche associates 'disgregation' with weakness, see TI 'The Problem of Socrates' 9 and Nachlaß Spring 1888, KSA 13, 14[83, 219]; also May–June 1888, KSA 13, 17[6]. Disgregation is, however, also associated with 'genius', the 'sublimest machine', and Nietzsche equates complexity with Zerbrechlichkeit, 'fragility' (Nachlaß Spring 1888, KSA 13, 14[133]).

40 'But there is no such substratum; there is no 'being' behind doing, effecting, becoming', he writes in GM I 13, making his liquidation (Verflüssigung) of any anthropocentric viewpoint all too apparent. And yet, he clearly has a vision of a new free and durational subject that appears to be—in Quine's terms—an 'entity with identity': 'The freer and the more stable the individual, the more demanding his love: finally it longs for the Übermensch because nothing else satisfies his love' (Nachlaß November 1882–February 1883, KSA 10, 150).

has been abandoned, then this new unity—in order to avoid relapsing into the old belief in being—must be conceived differently.

Let me begin with a note on Goethe in which Nietzsche criticizes two methods of enquiry which he finds equally problematic. Any scientific method that attempts to fuse and combine what should remain separate is seen as problematic and just as unsuccessful as any method that attempts to separate what belongs together (*das Zusammengehörige*).[41] In his evaluation of altruism in 1880, to give another example, he points out that the idea of a unified society *problematically* demands that the oppositions among individuals be reduced to a minimum. The kind of society created by such a homogenization turns out to be uninteresting and unproductive, 'to its palest hue … reduced' (Nachlaß Autumn 1880, KSA 9, 6[58]). Nietzsche also thinks that such lowest-common-denominator reductions fail, because in their attempt to bring about the desired 'sameness' (*Gleichheit*), all productivity stops and the unity *as* unity dies: 'This is euthanasia, utterly unproductive! Just like those men without deep feelings—the kind, calm and so-called happy—are, after all, also unproductive' (ibid.).

But Nietzsche does not only distrust levelling syntheses. Conversely, he also thinks that our traditional practice of oppositional thinking creates the impression that we can *always* select and choose between two sides. He disapproves of this practice of thinking in mutually exclusive, either-or alternatives: 'Just as we have separated dead and alive, logical and illogical etc. To unlearn our mutually exclusive oppositions—this is our task' (Nachlaß July–August 1882, KSA 10, 1[3]).

Aware of the dilemma of *relativism*, the early Nietzsche reminds us that only those things which are not absolutely other and separate can have any effect on each other: 'what is absolutely foreign to each other, cannot have any kind of effect on each other' (PTAG 14). Provocatively, he remarks in *Human, All Too Human* that the ability to 'kill' depends on 'distance':

> We all, indeed, lose all feeling of injustice when the difference between ourselves and other creatures is very great, and will kill a mosquito, for example, without the slightest distress of conscience. (HA I 81)

If distance increases to such an extent that a connection is no longer felt, then annihilation of the other side becomes a possibility. For a community, this means that, at the very moment when one group perceives itself as absolutely self-sufficient, it will be in danger of becoming indifferent to

41 See Nachlaß Winter 1872–73, KSA 7, 24[2].

other groups. Unities can only be successful, according to Nietzsche, if all of the above problems are kept in view.[42]

In this context, Heraclitus emerges as an interesting model. As early as *Philosophy in the Tragic Age of the Greeks*, Nietzsche presents and praises him for his move towards immanence, i.e., his denial of the existence of incompatible metaphysical realms: 'he denied the duality of totally diverse worlds' (PTAG 5); he also approves of his affirmative attitude towards change, so that rather than associating it with suffering, becoming is perceived with 'blissful wonder' (*beglücktes Erstaunen*) (ibid.); and, finally, he admires his adualistic epistemology—a method Nietzsche finds appealing because it lacks the reductive either-or structure and displays a different way of dealing with oppositions (such as whole and part, nature and man, object and subject). In contrast to Nietzsche's reading of Parmenides and Plato, Heraclitus understands oppositions not as mutually exclusive alternatives, and he is therefore able to adopt a different perspective on the world. What Nietzsche finds promising is Heraclitus' attitude towards the world as a whole. He no longer separates absolutely, nor does he unite absolutely; he allows neither dualistic, absolute distinctions nor monistic oneness.[43] Instead, his practice of seeing suspends such a logic of alternatives. For Heraclitus, Nietzsche emphasizes,

> the many perceivable qualities are neither eternal substances nor phantasms of our senses (Anaxagoras is later to imagine the former, Parmenides the latter); they are neither rigid autocratic being [*Sein*] nor fleeting semblance [*Schein*] flitting through human minds. (PTAG 6)

Perceptions are *neither* eternal essences *nor* mere appearances, neither static, independent being (*Sein*), nor fleeting transient illusions of the human mind (*Schein*). Whereas the metaphysical realist holds that what exists, exists in itself, independent of my naming or thinking, and the idealist holds the exact opposite, namely, that what is, is only because of my thinking it, Heraclitus undercuts such a false rigidity. There are two concomitant types of nihilism here: the idealist loses the world (Jacobi's charge against Fichte's subjectivism) and the realist loses the self (Fichte's charge against Spinoza's fatalism). Once the reductive *either-or* is discarded, Heraclitean

42 As we saw earlier, precisely because Nietzsche feels so distant from his contemporaries (who adhere to the paradigm of being which he has left behind), he appears willing to sacrifice some of them along the way to his goal of overcoming nihilism.

43 In Twilight of the Idols, Nietzsche will give a more critical assessment of Heraclitus (TI '"Reason" in Philosophy' 2).

phenomena *erhalten sich*—con-tinue, preserved in (and not devalued by) their hovering between the poles defined by the *neither-nor*.[44]

For Nietzsche, then, Heraclitus' understanding of becoming breaks with our common practice of thinking in our inherited dualistic manner. The latter's active suspension enabled Heraclitus to think and approach the world differently: his neither-nor does not lead to any kind of absolute disintegration or to any levelling synthesis; it leads precisely to 'bliss' and 'astonishment' and a better way of seeing, negating neither the observer nor the observed.[45] To borrow a term from the phenomenologist Bernhard Waldenfels (2002, p. 193), reality becomes, as a *Widerfahrnis*, an implicating 'en-counter' ('ein ein-dringliches Ereignis')—not between two already existing subjects or between an already determined subject and an object; instead, it is from the 'encounter' that both self and world emerge.

Nietzsche understands the self as such a meeting point of will-to-power relations. As we saw, as complex 'sum', it does not disappear altogether; as constant encounter, it is *not* 'redundant' (*überflüssig*) but emerges as real. Again, I think, it is crucial to distinguish between the Nietzsche who announces the 'death of the subject' in his advocacy of becoming as against the belief in being; and the Nietzsche who attempts to think from within, assisting those who have made the leap towards the paradigm of becoming. Did he perhaps envisage those who agree with Quine that physical objects are myths, 'a device for working a manageable structure into the flux of experience' and that 'forces are another example ... nowadays that the boundary between energy and matter is obsolete ... these are myths on the same footing with physical objects and gods, neither better nor worse except for differences in the degree to which they expedite our dealings with sense experiences' (1980, p. 45)?

For those who no longer believe in being, radical gestures will no longer be necessary. Yet Nietzsche's assumptions imply that even then, new ideas will be needed in order to adjust our necessary web of beliefs in

---

44 Zupančič has recently described 'the figure of the two' as Nietzsche's most radical gesture. As in Novalis' understanding of 'illness', she explains the logic of the ascetic ideal as the irreducible doubleness of life and death as follows: 'That which, in a decadent way, turns against life (the "ascetic ideal") is itself something that springs from life ... the opposition of life and death, the tension between them, becomes the very definition of life. Life is two things: it is life and it is death; it is the living edge between them. Therefore, death, in the emphatic sense of the word, is the death of this edge, the end of this tension, the fall into one or the other ... which is always the fall into One' (Zupančič 2003, pp. 18–19).

45 See also Nietzsche's passage on 'the differential [der Unterschied] as the true object of feelings' (Nachlaß Summer 1875, KSA 8, 9[1]).

such a way that life flourishes within the new paradigm. Thinking the whole, but differently, then becomes the vital task. Caught up in his violent rhetorical assaults on the belief in being, and without compassion for those in need of it, Nietzsche only rarely delivers ideas for such a new *community*.

I wish to close with a brief examination of Nietzsche's phenomenology of love. In the aphorism 'Love and Duality', Nietzsche describes love as a special type of unity that is only successful as a unity when it retains its constitutive duality:

> What is love but understanding and rejoicing at the fact that another lives, feels and acts in a way different from and opposite to ours? If love is to bridge these antitheses through joy it may not deny or seek to abolish them. —Even self-love presupposes an unblendable duality (or multiplicity) in one person. (HA II 75)

The unity between two lovers cannot last, Nietzsche observes, when they allow either the one side or the other to become dominant. Differences must be given a positive value and give rise to joy (*Freude*). A unity will only last, Nietzsche holds, if it remains in a state of 'unblendable duality' (*unvermischbare Zweiheit*)—in a state of adualistic togetherness, both together and separate. Similarly, in another passage on 'Love makes the same' (D 532), he ridicules the idea that love demands that we erase the dividing differences. In the attempt to achieve a union without otherness, both give up their idiosyncrasies for the other. Such a false synthesis is again just as problematic as the above either-or.

In his discussion of love, then, Nietzsche—like many thinkers before him, for example, the young Hegel—comes closest to a possible model for his new 'whole' as a community: any false either-or would diminish the other and with it the relation; any false synthesis would ultimately truncate the characteristics of both; and, as we saw earlier, allowing for radically independent domains leads to separation by indifference. In order for a community to be successful, the two (or multiple) parties must avoid the three dilemmas. I interpret Nietzsche's scattered phenomenological observations, informed and supported as they are by his process metaphysics, as pointing in his less aristocratic and selective moments towards a unity that would foster cohesion (*Zusammen-halt*) that would no longer be subject to the confusion of unity with oneness and would thus provide the right kind of model for the paradigm of becoming which he envisages. Once the belief in being has dissolved, Nietzsche clearly wants more than joyful affirmation that 'determines the noncenter otherwise than as loss of centre' (Derrida 1978, p. 292). He also wants new identities and centres, but for that to be a possibility (as a new paradigm of becoming for and from which he thinks he is already speaking), he thinks our logic must first be adapted to the new paradigm (on 'unity', see Gemes 2001, pp. 350–354).

In the fragment 'The New World-Conception' of 1888, Nietzsche seems to offer a new myth; and, while incommensurable to those who believe in being, his conception would ensure what I interpret as a contentious contentment (providing some kind of *haltloser Halt*) for the new paradigm of becoming. This seems to require us to think the world as both becoming and being *and* as neither becoming nor being. Nietzsche imposes a double vision:

> The world persists; it is not something that becomes, not something that passes away. Or rather: it becomes, it passes away, but it has never begun to become and never ceased from passing away—it contains itself in both. (Nachlaß Spring 1888, KSA 13, 14[187])

In *Ecce Homo* 9, he seems to confirm that the task of revaluation requires many abilities, but most of all it requires an adualistic art that would combine—in a kind of alchemy of thinking[46]—*separatio* and *conjunctio*, namely, 'the art of separating without making inimical, to mix nothing, to "reconcile" nothing; a tremendous variety that is nevertheless the opposite of chaos—this was the precondition, the long secret work and artistry of my instinct' (EH 'Why I Am so Clever' 9). Again, it can be shown that this model of simultaneity he proposes in his more compassionate moments issues from his fight against what he assumed to be the real threat of nihilism. In *The Gay Science* 346, he cautions against what he calls a 'terrifying either-or' (*furchtbares Entweder-Oder*) that might come upon future generations:

> Have we not come to mistrust an opposition—an opposition between the world in which until now we were at home with our venerations—and which may have made it possible for us to endure life—and another world that we ourselves are ... and that could easily confront coming generations with the terrible either-or: 'Either abolish your venerations or—yourselves!' The latter would be nihilism; but would not the former also be—nihilism? That is our question mark. (GS 5 346)

For an adualistic practice of thinking to become a reality, he felt—perhaps by reflecting upon himself—that we need to work on the logic we have inherited, incorporated (in our evolutionary temporality), and accepted. As Nietzsche insists: 'a simultaneity-thinking is needed of which we have hardly an inkling' (Nachlaß April–June 1885, KSA 11, 34[124]). More precisely: the simultaneity of that which appears to be mutually exclu-

46 The forgotten German name for alchemy is Scheidekunst.

sive.[47] If we read Nietzsche's 'simultaneity-thinking' as the double standard he considered necessary for the new paradigm of becoming to be proof against nihilism, then his two agendas have to be understood as consistent. The real inconsistency lies, I think, in the violent separation and selection he was happy to accept as part of his cure.

Conclusion

I suggested in this essay that Nietzsche's entire project is motivated by what he sees as the real threat of nihilism. I further suggested that this threat seems to depend on two assumptions we attributed to Nietzsche: firstly, that all there is, is becoming, and secondly, that the belief in being among his contemporaries is all-pervasive. From these two assumptions, we inferred that nihilism is a function of the belief in being, i.e., the stronger the latter, the higher the existential disillusionment when it is confronted with becoming. Nietzsche accordingly attacks the belief in being, since, convinced as he is that all is becoming as will to power, it is only by undoing the belief in being that we can overcome nihilism.

We saw further that, in Nietzsche's eyes, nihilism was already emerging—albeit slowly—and that he regarded it as his task as a philosopher to accelerate this process.[48] In order to undo the belief in being, he makes another assumption: a counter-force is required, and the intensity of becoming as *Gegenkraft* must be equivalent to the intensity of the belief in being. I took this to explain Nietzsche's radical doctrine of becoming. Another example would be his formulation of eternal recurrence as 'the most extreme form of nihilism' (Nachlaß Summer 1886–Autumn 1887, KSA 12, 5[71]).

We have also seen that, faced with the nihilistic implications of the 'impossible presentation thesis', Nietzsche does not himself hold any radical doctrine of flux. His conception of will to power is, when examined closely, much more moderate. But again, he is not consistent here and his tone changes frequently, depending on whether he is trying to attack the belief in being, or whether he is genuinely trying to present plausible ex-

47 For current developments in 'transconsistent' logic, see Priest 1987 and 1995.

48 That he saw himself precisely as both annihilator and donor is clear in Ecce Homo: 'I am by far the most terrible human being there has ever been; this does not mean I shall not be the most beneficient. I know joy in destruction to a degree corresponding to my strength for destruction—in both I obey my Dionysian nature, which does not know how to separate No-doing from Yes-saying' (EH 'Why I Am a Destiny' 2).

planatory accounts of the world's constitution, both phenomenological and metaphysical, within a paradigm that has already abolished the strong belief in being.

But even in his conception of becoming as will to power, he only partially deflates the 'impossible presentation thesis'. His attempts to express within language the simultaneity of a real background temporality and a conscious perspectival temporality ultimately fail to capture and express the processes he assumes in his metaphysics of relations. This failure is not ultimately problematic, since in his theory of truth, truths are not timeless and 'out there' to be discovered, but are always the result of temporal, interpretative processes that can no longer be thought of as presuppositionless. In this respect, Nietzsche's ontology of processes supports his theory of truth: for an organism is always already an interpreting 'unity' before it somehow acquires a conscious, first-person perspective. The first-person perspective is the blind spot not only in Nietzsche's theory of self-consciousness.

Finally, we saw that Nietzsche attempts to conceive of 'the whole' beyond the logic of mutually exclusive alternatives. He tries to account for both being and becoming without privileging either the one or the other. We found additional support for this in his scattered remarks on relationships and love. Again—and this applies to his overall project—Nietzsche's adualism is ultimately motivated by the weight he attributes to the threat of nihilism.

What remains after all this is obvious. An evaluation of Nietzsche's project must depend on an evaluation of his two (or three) basic assumptions. What do we make of his belief in our belief in being? And what do we make of his belief in becoming? And if we really conclude that he had a point, then perhaps we should consider the model of 'simultaneity-thinking' (*Zugleich-Denken*) as an alternative to Nietzsche's own impatient and radical agenda.[49]

---

49 I would like to thank the participants of the 2005 conference 'Nietzsche on Time and History' at Peterhouse, Cambridge, where an early version of this paper was first presented. I am very grateful for the comments and criticism I received from the members of the Cambridge German Philosophy Seminar, Raymond Geuss, Richard Raatzsch, Fabian Freyenhagen, Christian Skirke and Jörg Schaub. Without the extensive discussions with Walter Schmoll, Mark Henderson, Hugh Barr Nisbet, Manolis Simos, Margaret Clare Ryan, and Anna Wehofsits this paper would not have been possible. The usual disclaimers apply.

References

Abel, Günter, 1998, Die Dynamik der Willen zur Macht, Berlin: Walter de Gruyter.
Abel, Günter, 2000, Interpretationswelten, Stuttgart: Suhrkamp.
Abel, Günter, 2001, Bewusstsein–Sprache–Natur: Nietzsches Philosophie des Geistes, in: Nietzsche-Studien, 30, pp. 1–43.
Abel, Günter, 2003, Wahrheit und Interpretation, www.nietzschesource.org/gabel-1.
Bowie, Andrew, 2003, Aesthetics and Subjectivity: From Kant to Nietzsche, Manchester: Manchester University Press.
Clark, Maudemarie, 1990, Nietzsche on Truth and Philosophy, Cambridge: Cambridge University Press.
Danto, Arthur C., 1965, Nietzsche as Philosopher, New York: Columbia University Press.
Deleuze, Gilles, 1983, Nietzsche and Philosophy, trans. H. Tomlinson, New York: Columbia University Press.
Deleuze, Gilles, 1994, Difference and Repetition, trans. P. Patton, London: Athlone Press.
Derrida, Jacques, 1978, 'Structure, Sign and Play in the Discourse of the Human Sciences', in: Writing and Difference, trans. A. Bass, Chicago: Chicago University Press, pp. 278–294.
Dries, Manuel, 2006, 'Friedrich Schiller's Adualistic Conception of Unity', in: Publications of the English Goethe Society, 1/75, pp. 53–58.
Dries, Manuel, 2006, The Paradigm of Becoming, PhD Thesis, University of Cambridge.
Emden, Christian J., 2005, Nietzsche on Language, Consciousness and the Body, Chicago: University of Illinois Press.
Figal, Günter, 1998, Nietzsche: Eine philosophische Einführung, Stuttgart: Reclam.
Figl, Johan, 1982, Interpretation als philosophisches Prinzip: Friedrich Nietzsches universale Theorie der Auslegung im späten Nachlaß, Berlin: Walter de Gruyter.
Frank, Manfred, 1997, 'Unendliche Annäherung': Die Anfänge der philosophischen Frühromantik, Frankfurt a. M.: Suhrkamp.
Frank, Manfred, 2007, 'Non-objectal Subjectivity', in: Journal of Consciousness Studies, 14/5–6, pp. 152–173.
Freeman, Anthony / Galen Strawson, et al., 2006, Consciousness and its Place in Nature. Does Physicalism Entail Panpsychism? Charlotteville, VA: Imprint Academic.
Gemes, Ken, 2001, 'Postmodernism's Use and Abuse of Nietzsche', in: Philosophy and Phenomenological Research, LXII/2, pp. 337–360.
Gerhardt, Volker, 1996, Vom Willen zur Macht, Berlin: Walter de Gruyter.
Hales, Steven / Rex Welshon, 2000, Nietzsche's Perspectivism, Urbana: University of Illinois Press.
Houlgate, Stephen, 1986, Hegel, Nietzsche and the Criticism of Metaphysics, Cambridge: Cambridge University Press.
Katsafanas, Paul, 2005, 'Nietzsche's Theory of Mind', in: European Journal of Philosophy, 13/1, pp. 1–31.
Leiter, Brian, 2001, 'The Paradox of Fatalism and Self-Creation in Nietzsche', in: John Richardson / Brian Leiter (eds.), Nietzsche, Oxford: Oxford University Press, pp. 281–321.
Leiter, Brian, 2002, Nietzsche on Morality, London: Routledge.
Leiter, Brian / Joshua Knobe, 2007, 'The Case for a Nietzschean Moral Psychology', in: Brian Leiter / Neil Sinhababu (eds.), Nietzsche and Morality, Oxford: Oxford University Press, pp. 83–109.

Moore, Gregory, 2002, Nietzsche, Biology, Metaphor, Cambridge: Cambridge University Press.
Müller-Lauter, Wolfgang, 1999a, 'Über Werden und Wille zur Macht', in: Nietzsche-Interpretationen 1, Berlin: Walter de Gruyter.
Müller-Lauter, Wolfgang, 1999b, 'Über Freiheit und Chaos', in: Nietzsche-Interpretationen 2, Berlin: Walter de Gruyter.
Poellner, Peter, 1995, Nietzsche and Metaphysics, Oxford: Clarendon Press.
Poellner, Peter, 2006, 'Phenomenology and Science in Nietzsche', in: Keith Ansell-Pearson (ed.), A Companion to Nietzsche, Oxford: Blackwell, pp. 297–313.
Porter, James I., 2006, 'Nietzsche's Theory of the Will to Power', in: Keith Ansell-Pearson (ed.), A Companion to Nietzsche, Oxford: Blackwell, pp. 548–564.
Priest, Graham, 1987, In Contradiction: A Study of the Transconsistent, Dordrecht: Martinus Nijhoff.
Priest, Graham, 1995, Beyond the Limits of Thought, Cambridge: Cambridge University Press.
Quine, Willard V., 1980, From a Logical Point of View, Cambridge, MA: Harvard University Press.
Richardson, John, 1996, Nietzsche's System, Oxford: Oxford University Press.
Richardson, John, 2006, 'Nietzsche on Time and Becoming', in: Keith Ansell-Pearson (ed.), A Companion to Nietzsche, Oxford: Blackwell, pp. 208–229.
Richardson, John, 2008, 'Nietzsche's Problem with the Past', in: Manuel Dries (ed.), Nietzsche on Time and History, Berlin: Walter de Gruyter, pp. 87–111.
Smith, J. H., 2000, Dialectics of the Will: Freedom, Power, and Understanding in Modern French and German Thought, Detroit: Wayne State University Press.
Stack, G. J., 2005, Nietzsche's Anthropic Circle: Man, Science, and Myth, Rochester, NY: Boydell and Brewer.
Stambaugh, Joan, 1972, Nietzsche's Thought of the Eternal Return, Baltimore: Johns Hopkins University Press.
Staten, Henry, 1990, Nietzsche's Voice, Ithaca, NY: Cornell University Press.
Vattimo, Gianni, 1986, 'Nietzsche and Contemporary Hermeneutics', in: Yirmiyahu Yovel (ed.), Nietzsche as Affirmative Thinker, Dordrecht: Martinus Nijhoff, pp. 58–68.
Waldenfels, Bernhard, 2002, Bruchlinien der Erfahrung, Frankfurt a. M.: Suhrkamp.
Welshon, Rex, 2004, The Philosophy of Nietzsche, Stocksfield: Acumen.
Young, Julian, 1992, Nietzsche's Philosophy of Art, Cambridge: Cambridge University Press.
Zupančič, Alenka, 2003, The Shortest Shadow: Nietzsche's Philosophy of the Two, Cambridge, MA: The MIT Press.

Translations

The Case of Wagner, trans. Walter Kaufmann, New York: Vintage Books, 1967.
Daybreak, ed. Mademarie Clark and Brian Leiter, trans. R. J. Hollingdale, Cambridge: Cambridge University Press, 1997.
Ecce Homo, trans. R. J. Hollingdale, introduction Michael Tanner, London: Penguin, 1992.
The Gay Science, ed. Bernard Williams, trans. Josefine Nauckhoff and Adrian del Caro, Cambridge: Cambridge University Press, 2001.
Human, All Too Human, trans. R. J. Hollingdale, Cambridge: Cambridge University Press, 1986.

On the Genealogy of Morality, ed. Keith Ansell-Pearson, trans. Carol Diethe, Cambridge: Cambridge University Press, 1994.
'Philosophy in the Tragic Age of the Greeks', trans. Marianne Cowan, Washington, DC: Regnery Publishing, 1962.
Thus Spoke Zarathustra, ed. Adrian del Car and Robert B. Pippin, trans. Adrian del Caro, Cambridge: Cambridge University Press, 2006.
Twilight of the Idols, trans. Duncan Large, Oxford: Oxford University Press, 1998.

Part III

Eternal Recurrence, Meaning, Agency

Shocking Time: Reading Eternal Recurrence Literally

Lawrence J. Hatab

In this essay I argue that Nietzsche's idea of eternal recurrence, the identical repetition of life in every detail, can and should be read literally.[1] Readers of Nietzsche usually have been perplexed by his avowal of this conception, and the literal sense of the repetition of life generally has been seen as problematic, if not false, even by Nietzsche's admirers. Yet I assume that Nietzsche was perfectly serious about eternal recurrence and saw it as the heart and climax of his philosophy.

I concede a point that has become something of a standard view, that Nietzsche did not in the end offer eternal recurrence as an objective, scientific, cosmological fact. For many, this means that eternal recurrence should not be understood as a claim about world events, but as an expression of an existential task, a test or a means of coming to affirm the conditions of life (which is the spirit of the published versions in *The Gay Science* and *Zarathustra*). I too stress an existential version, but I add something that seems missing in other versions: unless eternal recurrence is taken 'literally', its existential effect would be lost; one would always be susceptible to the psychological loophole that repetition 'isn't really true'. To avoid the possibility of 'armchair affirmation', I focus on the *literal* meaning of eternal recurrence, without necessarily endorsing its *factual* meaning. This distinction between the literal and the factual has the following advantages. While not presuming a cosmological interpretation of eternal recurrence, we can better understand why Nietzsche did experiment with an objective, descriptive approach to this notion. In my reading, Nietzsche always regarded eternal recurrence as more than simply a hypothetical thought experiment pertaining only to human psychology: he always took it to express something about life and the world as such. A certain extra-psychological literality would better fit the world-disclosive and 'revelatory' spirit of Nietzsche's accounts of eternal recurrence, and it would also not be utterly inconsistent with cosmological experiments in the

---

1 A good deal of this essay is drawn from the text of my book *Nietzsche's Life Sentence: Coming to Terms with Eternal Recurrence* (Hatab 2005). I thank Routledge for permission to publish the essay.

notebooks. In sum, my argument is that eternal recurrence should be seen as the only authentic expression of a Nietzschean life-affirmation *by force of* its literal meaning. In advancing this argument, I challenge those who concede eternal recurrence as some kind of literal claim about the world, but who see such a claim as either false or injurious to other elements of Nietzsche's thought; and those who redescribe eternal recurrence as a metaphorical or symbolic expression of some insight or philosophical position that has nothing to do with literal repetition. To prepare my account I need to set out a few basic elements of Nietzsche's philosophy that are essential for coming to terms with eternal recurrence.

Nietzsche's philosophy, in all its elements, is focused on the question of the meaning of life—not in the sense of finding a decisive answer to 'Why are we here?' but rather the *problem* of finding meaning in a world that ultimately blocks our natural interest in happiness, preservation, knowledge, and purpose. To be precise, the question is not 'What is the meaning of life?' but 'Can there be meaning in life?' So the question that preoccupies Nietzsche's investigations runs: is life as we have it meaningful, worthwhile, affirmable *on its own terms*? No culture, no form of thought has ever denied (how could it?) that our 'first world', immediate existence, is constituted by negative limits—confusion, change, suffering, loss, and death—as checks on all positive possibilities in life. In the end one must confess that life as we have it is tragic, measured against our highest aspirations.

Nietzsche's diagnosis of the Western tradition is that, in one form or another, the answer to this question of meaning in life as we have it has been: No. 'Concerning life, the wisest men of all ages have judged alike: it is no good' (TI 'The Problem of Socrates' 1). Whether in scientific, rationalistic, religious, or moralistic terms, initial conditions of existence have been judged to be deficient, confused, fallen, alien, or base, and thus in need of correction or transcendence altogether. Nietzsche judges all such judgements as implicitly nihilistic, and sees as his task an affirmative revaluation of a necessarily tragic existence, captured in the phrase *amor fati* (GS 276).

It is important to establish that life-affirmation, in response to the *question* of meaning in life, is the core issue in Nietzsche's thought, and that it lies behind and animates all of his supposed 'doctrines', such as will to power, perspectivism, and especially eternal recurrence. Accordingly, Nietzsche's texts cannot be reduced to 'doctrines' or 'propositional content' that call for assessment according to conceptual, empirical, or logical criteria. Nietzsche's philosophical work always bears on the existential *task* of coming to terms with meaning and value in life.

Nietzsche's proclamation of the death of God offers an effective entry point to the force of his thinking. God in the West was the ultimate symbol of foundational transcendence. God's demise, however, reaches far beyond religion because a divine reference had been the warrant for all sorts of cultural constructs in moral, political, philosophical, even scientific spheres. With God out of the picture, all corollary constructs dissolve as well. The death of God, therefore, announces the loss of all traditional truth principles, and we are left with a stark choice: either a nihilistic collapse of meaning or a revaluation of meaning in different terms according to immediate life conditions. Nietzsche takes the latter option and advances what can be called an existential naturalism: not a reductive naturalism confined to scientific categories, but an embrace of the full range of life forces that exhibit a radical becoming. The finite, unstable dynamic of earthly existence—and its meaningfulness—become the new measure of thought, to counter various attempts in philosophy and religion to 'reform' lived experience by way of a rational, spiritual, or moral 'transcendence' that purports to rectify an originally flawed condition (GS 109; TI '"Reason" in Philosophy' 16).

Nietzsche's conception of will to power embodies his naturalistic measure of thought. Traditional philosophy was animated by constructions of binary opposites, with the aim of privileging a positive side over a negative side: being over becoming, eternity over time, constancy over change, good over evil, truth over appearance, etc.—all providing fixed measures that can resolve negative forces confronting human existence. Nietzsche rejects such oppositional structures in favour of mixed conditions, where each side cannot escape, in fact is structurally related to, the other side. Will to power is Nietzsche's counter-concept to binary thinking because it indicates an 'agonistic' force field, where any state is partly constituted by its 'contest' with some counter-force, its drive to overcome resistances. So world conditions emerge in a network of *tensions* that cannot be reduced to stable identities. Will to power depicts in dynamic terms the idea that any affirmation is also a negation, that any condition or assertion of meaning must overcome some 'Other', some obstacle or counter-force. Since power intrinsically requires resistance from an obstacle (Nachlaß Autumn 1887, KSA 12, 9[151]), one's power to overcome is essentially related to a counter-power; if resistance were eliminated, if one's counter-power were destroyed or even neutralized by sheer domination, one's power would evaporate, it would no longer *be* power. Power is *overcoming* something, not annihilating it: 'there is no annihilation in the sphere of spirit' (Nachlaß End of 1886–Spring 1887, KSA 12, 7[53]). Will to power enables Nietzsche to both explain and diagnose various incapacities to affirm the natural course of life. The most dramatic element in Nietzsche's account of life-denying dispositions is his portrayal of the ascetic ideal, where an inability to withstand or engage external contests of power directs power internally and conducts a self-consuming battle against natural drives in the name of a presumed spiritual transcendence.

My aim is to make sense out of Nietzsche's serious interest in eternal recurrence as essential to his philosophical project. In the light of traditional approaches to the problem of time, which exhibit metaphysical, epistemological, and normative objections to sheer temporal conditions, eternal recurrence can be seen as Nietzsche's formula for the 'redemption' of time and becoming. As I read Nietzsche, eternal recurrence is the only authentic alternative to all other conceivable models of time with respect to affirming natural life and its temporal flux. Without qualification Nietzsche calls himself 'the teacher of eternal recurrence' (TI 'What I Owe to the Ancients' 5). With life-affirmation as the central focus, recurrence is called the 'highest formula of affirmation that is at all attainable' (EH III Z 1), and a notebook passage calls it 'the thought of thoughts' (Nachlaß Spring–Autumn 1881, KSA 9, 11[143]).

For Nietzsche, finding meaning is the primal task of life. With the death of God, tragic finitude is now shown to be the 'brute given', and all cultural issues will have to be traced back to this abyssal base. The question now is: *can* meaning be found in tragic finitude? If it can, what could be its measure? I argue that, for Nietzsche, these questions can only be answered adequately in the light of eternal recurrence, which amounts to the only *positive* expression of meaning that is not susceptible to *flinching* from finite becoming.

For Nietzsche, temporal becoming cannot be explained or justified in terms of something 'other' than its immediate conditions. We cannot leap outside becoming and justify in linear terms the present according to the future or the past according to the present. The 'self-justification' of becoming is what Nietzsche calls *necessity* (Nachlaß November 1887–March 1888, KSA 13, 11[72]). Necessity, for Nietzsche, is not equivalent to causal or logical necessity, or the necessary teleological force of an intended beginning or an inevitable end. Such notions account for temporal movement in terms of something *other* than immediate conditions of temporal flux. Nietzschean necessity simply captures the idea of 'no alternative', but *without* recourse to some fixed explanatory scheme that constructs necessity by bracketing temporal events as such. This is why Nietzsche can say in stark terms that 'event and necessary event is a tautology' (Nachlaß Autumn 1887, KSA 12, 10[138]). Necessity also figures in Nietzsche's approach to the meaning question and his call for affirmation, which entails the capacity to say Yes to the necessity of all events in themselves, that is, *only* in terms of how they emerge immediately in time, with no mandate for grounding them in causes, purposes, or fixed references. In *Zarathustra*, Nietzsche specifically plays on the German word for necessity (*Notwendigkeit*), calling it the turning around (*Wende*) of distress (*Not*), distress over the existential trauma of eternal recurrence (Z III 'On Old and New Tablets' 2–3). Necessity for Nietzsche is bound with affirming all elements of existence, including those that exceed rational models of necessity.

My argument is that eternal recurrence is the only way in which Nietzsche's ideal of life-affirmation can be even ascertained, much less expressed. To make headway here, I need to make a clear distinction between life-affirmation and life-enhancement, terms that make various appearances in the texts (*Bejahung* and *Erhöhung*), and that may easily confuse readers. Nietzsche obviously promotes affirmation, while at the same time he discusses beliefs that are called life-enhancing, life-preserving, and life-promoting. The problem is that often these beliefs are attacked as life-denying. What are we to make of this? Life-affirmation is consistent with the agonistic structure of will to power as an affirmation of otherness. Since meaning can only be understood in terms of tensional conflict, affirming one's meaning dictates affirming what runs counter to it as well. Affirmation is anything but comfortable and pleasant, anything but the approval of everything; it entails the difficult capacity to contend with otherness without wanting to annul it. Life-denial stems from an incapacity to affirm the agonistic structure of meaning, which incites various projects of transcending, reforming, resolving, or stilling tensional becoming. Yet Nietzsche can call life-denying projects life-enhancing because they further the interests of certain types of life and cultivate their own forms of power. Although these projects are deemed nihilistic by Nietzsche, they are nevertheless valuable as antidotes to *suicidal* nihilism (GM III 28). So, short of suicide, all forms of life are saying Yes in some way. Nietzschean affirmation is different from selective, enhancing Yesses in calling for a global Yes to all conditions, including those that run counter to one's interests. This is why eternal recurrence can have a 'crushing' effect (GS 341); it allows no escape from the perpetual occurrence of what threatens or negates one's deepest values.

We are now in a better position to fathom the central drama in *Thus Spoke Zarathustra*. The figure of Zarathustra comes to redeem the meaning of the earth from all convictions that look away from natural life out of revenge against time and becoming. Yet Zarathustra is not some sage with an achieved wisdom who delivers the task of affirmation to others; he himself must go through the task and experience its difficulties. Eternal recurrence is the mark and measure of saying Yes to life, and Zarathustra must confront his own revulsion at the recurrence of the small man, and of the very forces that turn away from life. If Zarathustra is to love life, he must will the return of life-denying forces. Far from being a sign of weakness, Zarathustra's struggle is in keeping with the agonistic structure of meaning. Life-affirmation requires the willing of that which opposes one's will. And the repetition motif in eternal recurrence ensures that such a task could not rest with a 'formal' affirmation of the abstractions 'life', 'time', or 'the earth'. Rather, repetition forces us to face the *material* task of engaging the precise events and course of our lives, including the most repulsive occasions. Not even Zarathustra can escape the traumatic shock of recurrence.

The drama of Zarathustra's task of affirmation would seem to lose all its force if recurrence is read as something other than literal repetition. I think this is one reason why Nietzsche was interested in exploring a 'cosmological' account. But I also think that such an account would only amount to a perspectival supplement to the central drama of life-affirmation, and that Nietzsche never would have taken a scientific account, even if cogent, as decisive or sufficient for engaging eternal recurrence. As long as the existential core of recurrence is retained, Nietzsche would not be averse to a 'descriptive' approach to the course of time, particularly because cyclic repetition is inseparable from Nietzsche's sense of affirmation and its challenges.

Although Nietzsche does not explicitly say so, I think there is in his texts an implicit default argument for eternal recurrence with respect to how time, becoming, and meaning are to be construed. In other words, all other conceivable models fail the affirmation test in one way or another, leaving eternal recurrence as the only alternative. Keeping in mind that, for Nietzsche, the concept of time cannot be separated from the existential meaning of temporal events, there seem to be six conceivable alternative models of time and meaning, each of which would be diagnosed by Nietzsche as fugitive evasions of the life-world. I name these alternative models *positivistic, salvational, teleological, cyclical, pessimistic,* and *novelistic.*

The positivistic model of time can be dismissed because it conceives temporal movement in objective terms as the measurable relations between quantified 'points' of past, present, and future 'nows'. Although Nietzsche appreciates the non-teleological element in scientific thinking, he dismisses its detachment from matters of existential meaning (GS 346). Indeed, objective models of time require their own constructions of 'being' (the 'now', and the cognitive permanence of the measuring principle itself) that look away from becoming. The purported *value* of such an outlook stems from the sense of detachment and mastery over temporal events. What is dishonest here is the presumption of a value-free, objective analysis. At least asceticism is honest in responding to temporal life as an existential problem. Scientific approaches to time ignore (suppress) the deep issues of meaning that are intrinsic to temporal finitude.

The salvational model of time is best illustrated in the Christian view, where the temporal world is a once and for all creation with an absolute beginning and end, consummated by a transformation into eternal perfection. The Christian account insists on a linear, irreversible, non-repeatable course of time that finally overcomes temporal life altogether. Such a view obviously fails Nietzsche's affirmation test and would honestly admit as much. The existential significance of linear time is directly addressed by Augustine in book 12 of *The City of God*, where he argues against the Greek model of eternal, cyclic repetition because it entails the return of the fall after the achievement of blessedness (ch. 13). In this same chapter he cites a scriptural passage describing the birth and death of Christ as a unique, unrepeatable event. Cyclic repetition would render the *effort* for salvation meaningless and absurd. Augustine defines religious truth as the 'straight path' to salvation, as opposed to a 'godless' circularity (ch. 20).

Teleological models of time modify the trajectory of the salvational view by staying within temporal movement. But time is still conceived as a direction towards completion that will overcome or resolve the temporal finitude and limitations of earlier or present conditions. Nietzsche diagnoses worldly forms of progress as no less moralistic and fugitive than salvational models. For teleological thinking, conditions of temporal becoming can only be meaningful or bearable in terms of something ahead of, other than, immediate experiences of life.

Certain cyclical views of time seem to avoid the faults of salvational and teleological models by not picturing an end to temporal movement. Yet Nietzsche's proposal of identical cycles would draw out the existential issues that distinguish eternal recurrence from other cyclical views. Stoic philosophy proposed a model of identical cycles, but even this would not measure up to the existential test: for the Stoics, repetition follows from a strict causal necessity and an immanent divine providence; and the posture of Stoic equanimity suggests an avoidance (suppression) of the existential *trauma* intrinsic to the repetition scheme (again, at least the salvational model does *not* evade the trauma).

The pessimistic model of time is well expressed by Schopenhauer. Like the salvational view, pessimism engages the trauma of temporal finitude but rejects the idea of any positive transcendence or transformation. Time simply manifests itself (as appearance) and then ends in nothingness. For Schopenhauer, the tragic finitude of existence, surrounded by nothingness, should prompt one to turn against life as an absurd mistake, and to welcome extinction as the only conceivable release. So of course Schopenhauer would (willingly) fail Nietzsche's affirmation test.

The novelistic model is the most interesting case. With the pessimistic model, we can see why Nietzsche would reject a finale in nothingness, even though it might at first seem consistent with a tragic acceptance of destruction. But given Nietzsche's promotion of creativity, one would think that a repetition scheme would not be his preference. Why not a model of eternal novelty, where time neither begins nor ends and issues forth ever new conditions, never to be transcended, transformed, reformed, completed, or annihilated? Would not eternal novelty be the more Nietzschean choice over the seeming constriction of eternal repetition? It certainly would seem so, and yet the case of eternal novelty is specifically rejected by Nietzsche, which helps show why he was convinced of the unique significance of eternal recurrence with respect to the question of life-affirmation. In a note of 1885, Nietzsche diagnoses eternal novelty as a residue of theological habits that took solace in God's infinite freedom from earthly constraints: 'It is still the old religious way of thinking and desiring, a kind of longing to believe that in some way the world is after all like the old beloved, infinite, boundlessly creative God' (Nachlaß June–July 1885, KSA 11, 36[15]). Nietzsche seems to think that eternal novelty would be a naturalized modification of theological freedom from worldly conditions as they are.

In the light of life-affirmation, it seems to me that eternal novelty is the most plausible alternative to eternal recurrence. Yet Nietzsche was convinced that it could not measure up, because it stems from a 'desiring' and 'longing' that the world be 'boundlessly creative'. Since Nietzsche does not argue on strictly cognitive grounds, since values, interests, and needs are his first-order concerns, his question would not be 'What are your reasons supporting eternal novelty?' but rather, 'Why is eternal novelty *important* to you? Why are you *interested* in such an idea?' The existential response to the prospect of repetition is the baseline issue. Eternal novelty, in Nietzsche's estimation, is still another form of *looking away* from concrete conditions of life. In rejecting repetition, the novelistic model betrays a dissatisfaction with life as it is, masked by its apparent celebration that the world *will* always be different (better?).

Such is the sketch of what I have called the default argument for eternal recurrence. Given the question of life-affirmation, eternal recurrence comes forth as the only conceivable temporal model that does not fall prey to a fugitive gaze *away* from life as lived. Moreover, it is important to recognize that the scheme of identical repetition is essential for the operation of this argument (especially evident in the case of eternal novelty). If eternal recurrence were in any way disengaged from a literal sense of repetition, the force of the default argument would be lost.

As I indicated earlier, I want to argue that Nietzsche's commitment to eternal recurrence can best be explained by reading cyclic repetition literally but not factually. A literal account can stand 'in between' an existential and a factual sense of recurrence. Resistance to a literal reading of recurrence stems not only from a preference for presumably subtler, more interesting (safer?) metaphorical senses; it also reflects epistemological concerns owing to descriptive, referential connotations in common usage: 'literally true' can be equivalent to 'factually accurate'. But a look at the *Oxford English Dictionary* reveals a complex history of usage that opens up many questions about literal language. The primary meaning of 'literal' pertains simply to alphabetic letters, and thus not to a differentiation from metaphor (since metaphors have letters). In this sense, a 'literal mistake' would refer not to a descriptive error but a misprint. Other meanings grow out of and modify this primary sense: word for word transcription (e.g., quotation vs. paraphrase); taking words in their natural or customary meaning; the primary sense of a word or the direct wording of a passage, as distinguished from metaphorical or suggestive meaning; a matter-of-fact, unimaginative disposition; accurate meaning or reference, free of metaphor, exaggeration, or error. We notice here a shift from 'literal' in a *wordly* sense (cf. the German *wörtlich*), pertaining to written words as such, to a *descriptive* sense, pertaining to how a text (or the world) is understood.

The tendency to conflate literal descriptions and factual accounts opens up another complex history of usage. The Latin *factum* originally referred to actions and deeds as opposed to words (the Greek *ἔργον* had the same sense; and consider the German *Tatbestand*). Then fact came to mean an actual event known through direct observation or reliable testimony, as opposed to conjecture, fiction, or misrecollection. My point is that a historical look at language usage shows that current familiar meanings of the literal and the factual have arisen out of linguistic shifts and relations that exhibit anything but clear, direct references, and thus anything but 'literal' or 'factual' sources. The same holds true for the literal–metaphorical distinction, which itself has a philosophical history within Greek thought, when new frameworks emerged for understanding natural speech and mythopoetic language, frameworks which generally presumed deficiencies in these prior modes of discourse. The problem is that even 'metaphor' is metaphorical; *μεταφέρω* in Greek means 'to carry over', and so the familiar sense of metaphor—as 'combining' one sense with another owing to a similar, though non-equivalent, likeness—is itself a metaphorical trope. As we have seen, the connotations of 'literal' as univocal (non-combined) meaning and descriptive accuracy are themselves 'carried over' from a reference to written letters. One can say then that the former extended senses are no longer 'literally' literal (in the primary wordly sense). A similar metaphorical process can be located in the meaning of 'factual' as carried-over from doing-versus-speaking. So the familiar connotation of the literal–factual dyad, which presumes a secured, unambiguous actuality (as is), is itself an ambiguous extension of even more direct meanings (as written, as done). Consequently, it is hard to avoid the conclusion that indirect, non-literal forces such as metaphor go all the way down in human language, thus undermining the presumed privilege bestowed on the literal–factual dyad, and also subverting the notion that a metaphor cannot have a literal sense.

It is well known that Nietzsche insists upon and celebrates a non-foundational perspective on language. For Nietzsche, language by its very nature is a formative, creative shaping of the unstable flux of experience. Language cannot be understood as a representational description of non-linguistic 'facts' presumed to be independent of metaphorical, rhetorical, and imaginative forces in language. Nietzsche's linguistic theory and his own textual practices presume and portray a radical undecidability between literal and figurative meanings in philosophical language. But there is evidence in Nietzsche's writings that the groundlessness of language is on occasion associated with an irreducible immediacy in linguistic presentations, which I take as a source for my venture into the literal sense of recurrence.

If the literal–factual dyad cannot be sustained as a baseline reference, there opens up the possibility of a different sense of the literal that is not equivalent to, or suggestive of, factuality. I want to understand the literal in a functional and performative sense rather than a descriptive sense. I begin by calling the literal *as written* in place of the descriptive *as is*. But this is not enough. In addition to *what* a text presents, I need to include *how* language and texts are engaged and received. This brings us to certain historical questions and particular remarks in Nietzsche's writings that will help shape what I want to call *mimetic literality*.

The story of the literal–factual dyad cannot be told apart from the complex history of orality and literacy in the Greek world. The emergence of philosophical reflection in ancient Greece was intrinsically connected to shifts from an oral mode of culture to one influenced by reading and writing. Oral culture was shaped according to structures of poetic production and audience reception that in retrospect exhibit a non-reflective immediacy: poets were 'inspired' vehicles for sacred transmissions, and audiences were 'enchanted' recipients of enthralling poetic performances. The sheer graphics of writing permitted an isolation of texts from such performance milieus, and the fixity of written words permitted a host of reflective operations that greatly altered *how* the linguistic resources of Greek culture would be understood.

I bring this up only to highlight the 'literal' effect of graphic letters in crafting a reflective departure from an *oral* sense of 'literalness' that has nothing to do with familiar connotations of rational truth, but rather the *immediate disclosive force* of poetic language in performance. As is well known, Plato criticizes poets and rhetors because they were 'out of their minds' when performing their creative and oral functions. Their inspired condition overtook self-control and was incapable of reflective analysis of what they were saying and why they were saying it. What is not always recognized is how this critique of poetic psychology figured in Plato's discussion of imitation (*μίμησις*) in the *Republic*. In addition to Plato's concerns about the content of traditional poetry (particularly its tragic world-view), he also targeted the form of oral performance and its effects on both performers and audiences. The 'representational' sense of *μίμησις* (copying a natural object) was not Plato's primary concern (see 603B ff.); rather, it was the psychological effects of mimetic identification, wherein performers and audiences would be captured by, and immersed in, oral presentations, thus losing reflective self-control and being enraptured by the 'reality' of poetic speech and disclosure. Particularly dangerous for Plato was the mimetic force of empathic identification with the suffering of tragic heroes (605).

Nietzsche occasionally discusses what I am calling mimetic psychology, especially in his reflections on Greek art. *The Birth of Tragedy* contains several relevant treatments. Apollonian and Dionysian forces are exhibited in nature herself, *before* the mediation of artistic works (BT 2). Forming and deforming powers are intrinsic to nature's very course, and dreams and intoxicated states (both of which exceed conscious control) are preconditions for the more cultivated manifestation of Apollonian and Dionysian powers, particularly those of language and music. Artists are said to 'imitate' such primal natural energies, which could not mean representational simulation, but rather the more performative sense of 'impersonating' these energies in artistic practices (impersonation being one of the meanings of *μίμησις* in Greek). Singing and dancing, for example, exhibit an enchanted, ecstatic elevation, a quasi-divine transformation where one is not really an artist because one 'has become a work of art' (BT 1).

In many respects Nietzsche associates the Dionysian with music (BT 6, 17), especially its immediate emotional force that 'overwhelms' conscious individuation. The Apollonian is associated with poetic language and theatrical technologies that shape a more individuated world. But since music and language are coordinated in tragic drama (BT 21), immediate disclosive force still operates in its performances. Poetic metaphors are thus not 'symbolic', they possess a living power to disclose (BT 8). For Greek audiences, dramatic fiction was not a departure from reality, it produced on stage powerful scenes of 'a world with the same reality and irreducibility that Olympus with its inhabitants possessed for the believing Hellene' (BT 7). Tragic drama produced a Dionysian effect of mimetic identification, originally embodied in choral impersonation, where one acts 'as if one had actually entered into another body, another character' (BT 8). Tragic appearances have a *reality* because they tell us: 'Look there! Look closely! This is your life' (BT 24).

The problem with Euripidean drama, as Nietzsche saw it, was that it brought the critical 'spectator' on stage (BT 11). Particularly problematic was the effect of the Prologues in Euripides' plays, where the context and course of the drama was laid out in advance for the audience. The effect was to preclude or diminish mimetic identification, so that the audience would no longer 'become completely absorbed in the activities and sufferings of the chief characters or feel breathless pity and fear' (BT 12). The modern 'aesthetic' audience has been thoroughly schooled in the mode of critical reflection, where art is meant to be understood by way of interpretive tools beyond the immediate presentation of the work, beyond the 'powerful artistic magic' that should 'enrapture the genuine listener' (BT 22).

Nietzsche admits that 'the meaning of tragic myth set forth above never became transparent in conceptual clarity to the Greek poets', which is one reason why tragedy did not have the strength to survive (BT 17). This is why Nietzsche found favour with Kant and Schopenhauer in *The Birth of Tragedy*: they made it possible for *philosophy* to confront tragic limits and thus expand the sense of tragedy beyond its original artistic forms. Indeed, Nietzsche designates the tragic turn in philosophy as 'Dionysian wisdom apprehended in concepts' (BT 19). In gathering these issues, I want to argue that eternal recurrence can be understood as a tragic-mythic concept, a formation meant to engender a literal, immediate disclosure that yet is not construed as a cosmological fact.

Nietzsche naturally assumed that eternal recurrence would prompt reflection, and he did mean it to have philosophical significance. But it can be said that reflection is a second-order disposition derived from the immediacy of mimetic identification. I do not think we can say, for instance, that Zarathustra was engaged in a 'philosophical analysis' of eternal recurrence; he was responding to its world-disclosive impact directly in terms of his own life and experience of meaning. There is still nothing wrong with a reflective stance towards eternal recurrence. Yet, recalling Nietzsche's complaint about an exclusively 'critical audience' in drama, I think Nietzsche would question a philosophical audience that is exclusively critical, that engages recurrence solely in terms of philosophical adjudication rather than existential impact. He would not object to his audience *becoming* critical, but rather *arriving* critically.

In *Ecce Homo* Nietzsche relates the 'history' of *Zarathustra* and its 'fundamental conception' of eternal recurrence, which 'came to' him during a walk through the woods in August 1881 (EH III Z 1). *Zarathustra* is described as having 'invaded' and 'overtaken' him. Then Nietzsche offers an account of inspiration that clearly articulates the advent of the text as he experienced it (EH III Z 3). He speaks of 'revelation', where 'one is merely incarnation, merely mouthpiece, merely a medium of overpowering forces'. The process is involuntary: 'like lightning, a thought flashes up, with necessity, without hesitation regarding its form—I never had any choice.' The presentation of image and parable is also involuntary: 'one no longer has any notion what is an image or a parable: everything offers itself as the nearest, most obvious, simplest expression.' Alluding to Zarathustra, he suggests that parables show a kind of irreducibility, 'as if the things themselves approached and offered themselves as parables'. A text from *Zarathustra* (III 'The Homecoming'), which Nietzsche then quotes, also contains a description of parabolic speech as being 'upright' (*aufrecht*), 'straightforward' (*aufrichtig*), and issuing 'straight talk' (*gerade redet*). What is important here is the recognition of a non-reflective original base out of which eternal recurrence takes shape. And I am suggesting that Nietzsche aimed to offer his readers the possibility of a comparable non-reflective engagement with recurrence and its immediate disclosive force.

Let me close with a few questions. The kind of mimetic immediacy I am suggesting raises an interesting problem concerning the relationship between eternal recurrence and life-affirmation. Surely affirmation is central to the meaning of eternal recurrence, but the inspirational milieu Nietzsche hints at makes me cautious about assuming recurrence to be 'based' in life-affirmation in some way, in the sense that affirmation was Nietzsche's primary message and recurrence served a kind of instrumental function, to draw out or test affirmative dispositions. This is a widely held view and there is certainly something right about it. But I am not sure Nietzsche was simply searching for devices that could express or serve the kind of life-affirmation that marks his thought (recall the distinction between affirmation and life-enhancement). I ask: could we, or even Nietzsche, have imagined his strict sense of life-affirmation apart from or before the presentation of eternal recurrence? I'm not sure. Could it be that absent or prior to the stark sense of eternal recurrence, 'affirmative' postures towards life would simply be confined to, or delimited by, affirming various kinds of life-*enhancement*, which are not equivalent to Nietzsche's agonistic/holistic conception of affirmation? Life-denial was Nietzsche's target from the beginning of his thinking. But everyone claims to be pro-life! Was it the vision of eternal recurrence that first clarified for Nietzsche what genuine affirmation of natural life would have to *mean*? And did recurrence first give him a way to weed out and smoke out all other selective forms of life-enhancement that parade as pro-life yet always aim away from it? Could even interpretations of eternal recurrence that aim away from its literal sense be part of this parade?

References

Hatab, Lawrence J., 2005, Nietzsche's Life Sentence: Coming to Terms with Eternal Recurrence, New York: Routledge.

Translations

The Birth of Tragedy, in: *Basic Writings of Nietzsche*, trans. and ed. Walter Kaufmann, New York: Viking Press, 1966.

Ecce Homo, in: *Basic Writings of Nietzsche*, ed. and trans. Walter Kaufmann, New York: Viking Press, 1966.

The Gay Science, trans. Walter Kaufmann, New York: Vintage Books, 1974.

On the Genealogy of Morality, in: *Basic Writings of Nietzsche*, ed. and trans. Walter Kaufmann, New York: Viking Press, 1966.

Thus Spoke Zarathustra, in: *The Portable Nietzsche*, ed. and trans. Walter Kaufmann, New York: Viking Press, 1954.

Twilight of the Idols, in: *The Portable Nietzsche*, ed. and trans. Walter Kaufmann, New York: Viking Press, 1954.

The Will to Power, trans. Walter Kaufmann and J. R. Hollingdale, New York: Random House, 1967.

Suicide, Meaning, and Redemption

Paul S. Loeb

1. The Problem of Suicide

In the famous opening sentences of his 1942 collection of essays, *The Myth of Sisyphus*, Albert Camus writes:

> There is but one truly serious philosophical problem, and that is suicide. Judging whether life is or is not worth living amounts to answering the fundamental question of philosophy. All the rest—whether or not the world has three dimensions, whether the mind has nine or twelve categories—comes afterwards. These are games; one must first answer. And if it is true, as Nietzsche claims, that a philosopher, to deserve our respect, must preach by example, you can appreciate the importance of that reply, for it will precede the definitive act. These are facts the heart can feel; yet they call for careful study before they become clear to the intellect. (Camus 1991, p. 3)

Now, against this startling declaration, Nietzsche scholars would probably want to cite his claim in *Twilight of the Idols* that the value of life cannot be estimated by the interested party of the living. Nietzsche also says there: judgements of value concerning life, for it or against it, can in the end never be true and are in themselves stupidities. But they are worthy of consideration as symptoms, and for a philosopher even to see a *problem* in the value of life—much less to adopt a *negative* attitude to life—is thus an objection to him, a question mark concerning his wisdom, an un-wisdom (TI 'The Problem of Socrates' 2).

It might seem, then, that Nietzsche would not consider Camus' problem an authentic problem, much less the single truly serious and fundamental problem of philosophy. He might even think that Camus' declaration is a question mark concerning his wisdom. But there is a sense, of course, in which Camus' declaration poses a *meta*-problem that is at the heart of Nietzsche's entire philosophy: namely, why is it that so-called sages since Plato have seen a problem in the value of life and contemplated the necessity of practising what they preach?[1]

In the third essay of the *Genealogy*, Nietzsche offers a partial solution to this meta-problem by pointing to the peculiar character of philosophers themselves: their instinct for an optimum of favourable conditions under which they can expend all their strength; as well as their need for ascetic cloaks as a result of the emergency conditions under which philosophy arose and survived at all. But Nietzsche soon widens his enquiry to consider the causes of the dissatisfaction with life felt by every human animal. The answer, he proposes, is that human animals are unable to find any meaning for their lives and are therefore inherently inclined to suicide. What has so far prevented them from *acting* on this inclination is their invention of a meaning in the ascetic ideal—that is, in the rebellion against life and its most fundamental presuppositions (the animal, the material, the senses, appearance, change, becoming, desire, death) (GM III 28). Indeed, Nietzsche is amused to tell us, it was secretly the love of life that prompted the human animal to find its meaning in this ascetic ideal. For it was only by *rebelling* against life that the human animal was able to find the meaning it needed to continue living, willing, and suffering. As he expresses it in his introductory and concluding formula, the human animal preferred to will nothingness rather than not will at all.

Thus, when philosophers such as Camus declare that the question of suicide is a fact that the heart can feel, Nietzsche would point, first, to the peculiar character of the philosopher's heart. Second, to the nihilistic meaning of the ascetic ideal that was implanted millennia ago into the heart of the human animal so that it might survive. And, finally, but most fundamentally, to the death-wish that dwells in the heart of every human animal because it secretly suspects that there is in fact no meaning at all to its willing and suffering (not even the nihilistic meaning of the ascetic ideal). Camus' problem is thus absolutely worthy of consideration for Nietzsche: not as a question to be answered, but as a symptom of the profound and ineradicable drive to suicide built into the human animal.

What Nietzsche might find surprising, however, is Camus' *method* for solving his problem of suicide—what he calls 'absurd reasoning' (Camus 1991, p. 3). This is because it is a point of honour for Camus to presuppose, and directly confront, the complete absence of meaning in all human life. Thus, shortly after posing his problem of suicide and the worth of life,

1 Recall here Plato's claim at the start of the *Phaedo* that philosophers should be willing and ready to die. Although they should not help themselves, they should eagerly wait for someone else to benefit them.

Camus notes that dying voluntarily implies that you have recognized, even instinctively, the uselessness of suffering and the absence of any profound reason for living. So the true subject of his essay, he tells us, is 'precisely this relationship between the absurd and suicide, the exact degree to which suicide is a solution to the absurd' (ibid., p. 6):

> One must brush everything aside and go straight to the real problem. One kills oneself because life is not worth living, that is certainly a truth—yet an unfruitful one because it is a truism. But does that insult to existence, that flat denial in which it is plunged come from the fact that it has no meaning? Does its absurdity require one to escape it through hope or suicide—this is what must be clarified, hunted down, and elucidated while brushing aside all the rest. Does the absurd dictate death? This problem must be given priority over others, outside all methods of thought and all exercises of the disinterested mind. (ibid., pp. 8–9)

In keeping with this statement, Camus nowhere seeks refuge in any aspect of the ascetic ideal that according to Nietzsche has injected meaning in all human life up to now. As he works his way towards his solution, and against existentialist philosophers such as Jaspers and Kierkegaard, Camus insists on the question whether it is possible to find life worth living *without appeal* to any of the ascetic concepts that devalue our life in this world: God, afterlife, ideal other world, immortal soul, eternal freedom, morality, sin, and guilt. All such appeals, he writes, involve a sacrifice of the intellect, a suicide of thought:

> My reasoning wants to be faithful to the evidence that aroused it. That evidence is the absurd ... There can be no question of masking the evidence, of suppressing the absurd by denying one of the terms of its equation. It is essential to know whether one can live with it or whether, on the other hand, logic commands one to die of it. I am not interested in philosophical suicide, but rather in plain suicide. I merely wish to purge it of its emotional content and know its logic and integrity. Any other position implies for the absurd mind deceit and the mind's retreat before what the mind itself has brought to light. (Camus 1991, pp. 49–50)

2. Conscious Revolt

What would *not* surprise Nietzsche, I think, is Camus' *solution* to his problem of suicide, his idea of 'conscious revolt' against the absurd (1991, p. 53). Although Camus rejects as escapist the conclusion of other existentialist philosophers that there is after all a meaning to life (through the concept of God, for example), he himself paradoxically concludes that a life *can* have meaning if, and only if, it is lived in *full consciousness of, and revolt against*, the absence of meaning. As Camus writes: 'That revolt gives life its value. Spread out over the whole length of life, it restores its majesty to that life' (ibid., p. 55). For this reason, Camus asserts, the problem of suicide is reversed: 'It was previously a question of finding out whether or not life had to have a meaning to be lived. It now becomes clear, on the contrary, that it will be lived all the better if it has no meaning' (ibid., p. 53). Since suicide is the extreme acceptance of death and its absurdity, life lived in conscious revolt against the absurd actually dictates the *rejection* of suicide:

> It is essential to die unreconciled and not of one's own free will. Suicide is a repudiation. The absurd man can only drain everything to the bitter end, and deplete himself. The absurd is his extreme tension, which he maintains constantly by solitary effort, for he knows that in that consciousness and in that day-to-day revolt he gives proof of his only truth, which is defiance. (ibid., p. 55)

Lest Camus' existential solution seem dated or irrelevant today, let me point out that it is not so far from the solution most contemporary scholars attribute to Nietzsche. For example, Brian Leiter even cites Camus when he concludes his commentary on the third essay of the *Genealogy* with Nietzsche's supposed view that there is *no* meaning or justification for human suffering. But whereas Camus says that we can find meaning in the *conscious revolt* against the absurdity of our lives, Leiter interprets Nietzsche as saying that the highest human beings can avoid both suicidal nihilism *and* asceticism by finding meaning in the *affirmation* of their lives' absurdity: 'This is the attitude of existential commitment, through brute force of will, to carry on in the absence of such a meaning or vindication, to give up, in effect, asking "Suffering for *what*?"' (Leiter 2002, p. 288).

But let me continue for now with Camus' full solution. As I said, I think that Nietzsche would recognize this solution. Because Camus insists that life is essentially absurd, and because he finds meaning in conscious revolt against the absurd, we may infer that he finds meaning in living a life of rebellion against life. Hence, although he no longer finds himself guided by the concepts of the ascetic *ideal*, Camus is still proposing the kind of ascetic or life-denying *life* that Nietzsche calls a contradiction in terms. As Nietzsche writes in the third essay of the *Genealogy*:

> This is all in the highest degree paradoxical: we stand before a discord that *wants* to be discordant, that *enjoys* itself in this suffering and even grows more self-confident and triumphant the more its own presupposition, its physiological capacity for life, *decreases*. 'Triumph precisely in the last agony': the ascetic ideal has always fought under this hyperbolic sign; in this

> riddle of seduction, in this image of torment and delight, it recognized its brightest light, its salvation, its final victory. (GM III 11)[2]

But Nietzsche would not rest content with this analysis. He would argue further that, from the perspective of physiology rather than psychology, the self-contradiction of 'life against life' which Camus seems to represent can be only apparent. In truth, it must be the case that Camus embodied a degenerating life, an unsatisfied instinct and will to power, a sickliness, an exhaustion that longed for the end.[3] But the deepest protective and healing instincts of life which remained intact in Camus found a new means and a new invention, played a new trick, so as to ward off his suicidal instinct: namely, the existentialist philosophy of conscious revolt. Considered more closely, then, Camus' absurd reasoning is highly instructive. Because Camus chooses to live his life in conscious *revolt* against the absurdity of death, he is committed to rejecting suicide. But death is a precondition of life. So if Camus had chosen instead a conscious *embrace* and *affirmation* of life, his logic would have dictated suicide.

3. Life-Affirming Suicide

Here, then, is a striking Nietzschean reversal of Camus' own reversal of the problem of suicide. It is ironic that in this context Camus cites Schopenhauer as a philosopher who should be judged poorly because he did not admit his logic to the point of taking his own life. As careful readers know, Schopenhauer actually rejected suicide and recommended instead the sort of 'self-depletion' endorsed by Camus. Indeed, Schopenhauer's reasoning, which Nietzsche studied closely, anticipated Camus' absurd reasoning by over a century. Here is the relevant passage from the first volume of *The World as Will and Representation*:

> Far from being denial of the will, suicide is a phenomenon of the will's strong affirmation ... The suicide wills life, and is dissatisfied merely with the conditions on which it has come to him ... He wills life, wills the unchecked existence and affirmation of the body; but the combination of circumstances does not allow of these, and the result for him is great suffering. (Schopenhauer 1969, §69)

---

2 Here and throughout this essay I have consulted the translations of Nietzsche's writings by Kaufmann, and the translation of GM by Kaufmann and Hollingdale.

3 Similarly with Socrates. According to Nietzsche, Socrates was incurably sick and knew it, secretly wanted to die, and in the end committed suicide by forcing Athens to give him the poison cup (TI 'The Problem of Socrates' 12).

Nietzsche, of course, was interested in reversing Schopenhauer's valuation and was therefore interested in looking at the problem of suicide from the perspective of life-*affirmation*. Thus, already in Zarathustra's speech on the preachers of death, he had expressed the view that we should welcome the wish of those who would like to be dead. To those who say that life is only suffering, Zarathustra replies: 'See to it, then, that you cease ... And let this be the doctrine of your virtue: "Thou shalt kill thyself!"' (Z I 'On the Preachers of Death'). So it is no surprise to find Nietzsche later, in the *Twilight of the Idols*, offering the same advice to Schopenhauer—advice I think he certainly would have extended to Camus:

> It is not in our hands to prevent our birth; but we can correct this mistake—for in some cases it is a mistake. When one does away with oneself, one does the most estimable thing possible ... with it one almost earns the right to live. ... Society, what am I saying! life itself derives more advantage from this than from any 'life' of renunciation, anaemia, and other virtues—, one has liberated the others from one's sight; one has liberated life from an objection. Pessimism, pur, vert, [pure, green] first proves itself through the self-refutation of the pessimists: one must advance a step further in its logic, not negate life merely with 'will and representation', as Schopenhauer did—one must first of all negate Schopenhauer. (TI 'Reconnaissance Raids' 36).

Notice Nietzsche's explicit claim here—in agreement with the logic of both Schopenhauer and Camus—that the decadent or pessimist who longs for death performs a *life-affirming* and *life-enhancing* act by committing suicide.[4] Life itself, he says, derives an advantage from the pessimist's suicide, and the pessimist's suicide liberates life from an objection. Although the pessimist does not have the right to live, he almost earns this right through his life-affirming and life-enhancing suicide.[5]

Now, if we equate life-*affirmation* with life-*preservation* and *self-preservation*, we will certainly find strange and contradictory Nietzsche's recommendation of the decadent's suicide. But this equation misses his distinction between what benefits life *in general* and what benefits *particular* living beings.[6] Certainly, he would agree, there are life instincts within

4 See Zarathustra's speech on free death for similar advice to those whose advancing age leads them to 'hang withered wreaths in the sanctuary of life' (Z I 'On Free Death').

5 Compare this with Julian Young's assertion that Nietzsche has no answer to Camus' question of suicide and that Nietzsche assumes without telling us why that suicide is a non-option for those who reject life and ought to commit suicide (2003, pp. 103–104).

6 This interpretation differs from Lawrence Hatab's: 2005, pp. 44ff.

the decadent that keep him alive despite his longing to die. But these forces are residual and limited in the economy of the whole. Although they keep the decadent alive, they cannot eliminate, or even ameliorate, the predominant illness, impoverishment, and fatigue that render him constitutionally unable to affirm life. In order to survive his death-wish, the decadent has no choice but to live in revolt against the life for which he is not suited. The residual life-preserving forces that keep alive this *particular* living being thus invariably work against the advantage of life *in general*. For they keep in existence someone who will always object to life in general and whose very existence constitutes an objection to life in general. The only way for the decadent to affirm life *in general* is to deny, that is, end, his own *particular* life.

Nietzsche draws some fairly radical conclusions from this line of reasoning—conclusions that belie the usual, and quite innocuous, understanding of his insistence on life-affirmation. I want to focus here on three of these conclusions. In the first place, Nietzsche's claim that the decadent has only two choices, life-denying survival or life-affirming suicide, means that it is impossible for decadents to *live* in a life-*affirming* manner. We always hear that Nietzsche promotes the life-affirming *life* for everyone, and that he advises pessimists to *change* or *improve* the life-denying aspect of their lives. But this cannot be right. Although Nietzsche does want to promote life-affirmation even with respect to decadents, what this means is that he wants to encourage and intensify their innate urge to self-destruction. This is why he insists, in the same section of *Twilight of the Idols* cited above, that 'the highest interest of life, of ascending life, demands that degenerating life be ruthlessly pushed down and aside'. Or, why he writes a little later in the *Antichrist*: 'The weak and ill-constituted shall perish: first principle of our philanthropy. And one shall help them to do so' (A 2).

Nietzsche's second conclusion has to do with his quest for a counter-ideal [*Gegen-Ideal*] to the ascetic ideal. For consider his claim that the decadent cannot affirm life even though the ascetic ideal gives his life meaning. Thus, meaning is a precondition of survival, but does not guarantee life-affirmation. A counter-ideal must therefore provide the kind of meaning that supports life-affirmation. However, since Nietzsche thinks that suicide is the *sole* life-affirming option *for decadents*, this counter-ideal must embody a new meaning that will overwhelm any residual life-preserving forces and help the decadents *give in* to their dominant suicidal instincts. This means that the usual interpretation of the counter-ideal cannot be right. According to Leiter, for example, the alternative non-ascetic ideal 'must be able to bear the burden of answering the question, "Suffering for what?" and thus blocking "suicidal nihilism", for that is the existential task the ascetic ideal discharges' (2002, p. 287). Given Nietzsche's reasoning above, however, it is actually the case that the counter-ideal must *oppose* the ascetic ideal's ability to *block* the suicidal nihilism of degenerating life. Or, put differently, the counter-ideal must *oppose* the ascetic ideal's ability to *preserve* the impoverished life of decadents who constitute a disadvantage and objection to life. Whereas the ascetic ideal was a trick played by life in its struggle against the decadents' imperative death-wish, the counter-ideal must expose that trick *and at the same time* reinforce and justify the decadents' *appropriate* desire to do away with themselves.

4. The Suicidal Human Animal

Nietzsche's third conclusion is by far the most radical and the one I want to discuss in the rest of this essay. Recall his claim in the third essay of the *Genealogy* that sickliness is the *norm* in humankind:

> That this [ascetic] ideal was able to attain power and dominate humans to the extent which history demonstrates, particularly wherever the civilization and taming of the human was set under way, is the expression of a great fact: the sickliness of the type of human which has existed so far, of the tamed human at least, of this human's physiological struggle against death (more precisely: against disgust with life, against exhaustion, against the desire for the 'end') ... For the human is more sick, more uncertain, more mutable, less defined than any other animal, there is no doubt about that—he is the sick animal ... how should such a courageous and well-endowed animal not also be the most endangered, the most chronically and deeply sick of all the sick animals? (GM III 13)

Nietzsche's references here to the influence of the 'civilization and taming of the human' and to the sickliness of 'the tamed human' allude back, of course, to his thesis in the second essay that society led to the worst sickness ever contracted by the human animal—its suffering from itself. This is a sickness, he writes, from which the human animal has not yet recovered, and that makes the human animal the sickest animal on earth. Society, he speculates, involved a kind of incarceration in which the human animal could not externally discharge its hostile, cruel, and destructive instincts. Instead, the human animal had to turn these instincts inward and against itself—thereby becoming what Nietzsche calls 'the master of self-destruction' (GM III 13):

> from lack of external enemies and resistances and forcibly confined to the oppressive narrowness and punctiliousness of custom, [it] impatiently lacerated, persecuted, gnawed at, assaulted, and maltreated himself; [it] rubbed itself raw against the bars of its cage as one tried to 'tame' it; this deprived creature,

> racked with homesickness for the wild, had to turn itself into an adventure, a torture chamber, an uncertain and dangerous wilderness. (GM II 16)

According to Nietzsche, then, the human animal *as such*—the internalized animal with a vast inner world called the 'soul'—is forced to inflict constant suffering upon itself and therefore embodies a *self-destructive* and degenerating life, an unsatisfied instinct and will to power, and a depression, lethargy, and fatigue that longs for the end. So it is the human animal *as such* that is inherently suicidal and unable to live a life-affirming life. As Nietzsche puts it in the third essay, the human animal was 'suffering from itself in some way, suffering physiologically in any case, like an animal locked in a cage, uncertain as to why and wherefore, desiring reasons [as a relief]'. Finally, the ascetic priest provided *reasons* for its suffering in the form of the ascetic ideal (GM III 20). Of course, these reasons—interpreting the human animal's suffering as punishment for its past misdeeds—were false reasons, and they actually led to even more self-inflicted suffering. But, nevertheless, the human animal found in the ascetic ideal a *pretext* to keep on living despite its self-destructive instincts—so that now it survives even to this day as an objection to life and as a constant source of objections against life.

But this means that Nietzsche's recommendation of a life-affirming suicide, and his quest for a new counter-ideal that will enforce this recommendation, is intended not just for the individual pessimists like Schopenhauer and Camus, nor even just for the weak and sick majority-herd of humankind that he everywhere deplores, but for humankind itself. Whereas the hitherto reigning life-denying ideal gave suicidal humankind an illegitimate reason to live, the new life-affirming counter-ideal must give it a legitimate reason to die. Whereas prior to the ascetic ideal the void of meaning prompted humankind to suicide, the new counter-ideal will offer humankind a meaning and justification for its suicide.

5. Zarathustra's Counter-Ideal

In the third essay of the *Genealogy*, Nietzsche only hints at this last, and most radical, conclusion.[7] In this third essay, he spends most of his time dissecting and criticizing the reigning ascetic ideal, and dismissing seemingly plausible candidates for a new counter-ideal (especially modern sci-

7 But see his remark in the second essay: 'the mass of humankind sacrificed to the flourishing of a single *stronger* species of human—now that would be progress' (GM II 12).

ence). But he never actually proposes or explains what he thinks should be the new counter-ideal. Accordingly, many scholars have found his concluding discussion dissatisfying, evasive, or simply more evidence that Nietzsche is better at tearing down than building up. However, here I think we need to take more seriously his *Ecce Homo* claim (EH III BGE) that his post-*Zarathustra* books were *supposed* to be purely critical and merely preparatory to the constructive solutions he had already proposed in *Thus Spoke Zarathustra.*[8] The reason for this, he tells us at the end of the second essay of the *Genealogy*, is that he himself belongs to his decadent age and is therefore not healthy enough, or strong enough, or courageous enough, or free enough, to speak in his own voice on behalf of the needed counter-ideal to the ascetic ideal. Instead, he says, he needed to imagine a man of the future who would possess the required sublime wickedness and self-assured intellectual malice. Here Nietzsche acknowledges the immoralism of the needed counter-ideal and admits that he is not courageous or free enough to advocate it himself. Only a man belonging to a stronger future age, he says, can liberate us as much from the previous ideal as from what was bound to grow out of it, from the great nausea, from the will to nothingness, from nihilism. Since Nietzsche says elsewhere that 'the great nausea' is induced by the decadence of humankind (GM III 14), and that 'nihilism' is the weariness of humankind over its own decadence (GM I 12), he hints again here at the life-affirming need to facilitate humankind's suicide.

In his *Ecce Homo* review of the *Genealogy,* Nietzsche comments that the reason the ascetic ideal has been so powerful despite being so harmful is that it has been the only ideal so far and has had no rival: 'Above all a *counter-ideal* was lacking—until Zarathustra.' According to Nietzsche himself, then, the counter-ideal to the ascetic ideal does not belong to him and is therefore nowhere proposed or explained by him in any of the works written in his own voice (such as the *Genealogy*). As he writes at the end of the second essay, when he abruptly cuts off his own exuberant account of the counter-ideal's effects, the only fitting thing for him to do is to keep silent. Otherwise he would interfere with something that only the stronger, more godless Zarathustra is at liberty to do. He would *interfere*, presumably, because if he were writing in his own voice, he would be too cowardly to communicate the sublimely wicked and intellectually malicious aspects of the needed counter-ideal. We therefore need to turn at this point to the work in which Nietzsche imagined how a future anti-nihilist and conqueror of nothingness, a redeemer of great love and contempt, would

8 I discuss this claim at length in Loeb 2005.

bring into the world the first rival ever to the ancient and all-powerful ascetic ideal.

Now, as soon as we open *Thus Spoke Zarathustra,* we find Nietzsche's imagined man of the future commanding humankind to overcome itself and to will its own downfall (*Untergang*). In an inaugural refrain that is repeated at many key moments throughout the book, Zarathustra proclaims that humankind is something that should be overcome and that humans themselves must do this. Speaking first as the redeemer of great contempt, Zarathustra calls on humankind to finally achieve its greatest experience, the experience that will enable it to overcome itself. This is the hour of the great contempt towards itself, the hour when humankind recognizes that all it has valued most highly about itself as its reason for being—its wisdom, its soul, its happiness, its reason, its virtue, its justice, its pity—is nothing but poverty and filth and wretched contentment. Speaking next as the redeemer of great love, and alluding to the Gospel's beatitudes, Zarathustra proclaims his love for all those self-destructive humans who want to perish and who find the necessary means to gladly do so.[9] And speaking finally as what he is not, a preacher of repentance, Zarathustra warns humankind of the consequences of preserving itself, and of the despicable smallness, impoverishment, and sterility that will accompany its long-term survival.

To paraphrase Nietzsche himself (EH 'Why I Write Such Good Books' 1), most readers of Zarathustra's first public speech simply cannot believe their own ears (or eyes), and so spend a lot of time and energy trying to explain away Zarathustra's inaugural command. Some scholars debate the meaning of Nietzsche's terms, *Untergang* or *überwinden*; others claim that Zarathustra's command is merely metaphorical, or spiritual, or psychological; others claim that there is a retraction, and even deconstruction, of Zarathustra's inaugural command as the book progresses; and yet others dismiss the entire book as abandoned by Nietzsche when he entered the 'mature' period of his 'true' masterpieces such as the *Genealogy*. But it seems to me that these various scholarly stratagems are an expression of hermeneutic denial, and that for the most part they have been refuted along the way. So I do not propose to discuss them any further here. Instead, I am simply going to assume that Nietzsche does indeed envision his future philosopher as commanding the self-destruction of humankind. I want to ask now what further reasons Nietzsche gives in *Zarathustra* for thinking

9 See also Zarathustra's praise to the perishing tightrope walker (a symbol of humankind) who tossed away his pole and plunged into the depths: 'You have made danger your vocation: there is nothing contemptible in that. Now you perish of your own vocation: for that I will bury you with my own hands' (Z Prologue 5).

this command necessary, how he proposes to ensure that this command is obeyed, and what he imagines will be left if this command is indeed obeyed. In order, my replies to these questions will be: humankind's impotence with respect to the past; the unveiling of the truth of eternal recurrence; and the splitting of the history of humankind in two.

6. The Meaningless Human Past

In trying to understand further Nietzsche's idea that humankind is something that must be overcome, it is important first of all to get past his diagnoses of the decadence that afflicts certain *ages* or certain *segments* of humankind. In the *Genealogy*, Nietzsche pronounces upon the incurable illness of the Jewish slaves, Christian history, modern socialists and anarchists, the priestly caste, the majority herd, and even the sick half of the human race. By contrast, he praises as exceptionally healthy and strong the ancient Greek nobles, the Romans, Renaissance history, Napoleon, the ascetically cloaked philosophers, and humankind's strokes of luck, its rare cases of psychic and physical health. The real question, then, is what kind of incurable affliction Nietzsche finds even in all members of the second, otherwise privileged, list.

We have already seen the start of the answer to this question in Nietzsche's claim that social enclosure led to the human animal's self-inflicted suffering and to its anguished search for the meaning behind this suffering. Nietzsche returns to this point in the last pages of the *Genealogy*, where he emphasizes the *existential* plight of the human animal, its experience of itself as 'a leaf in the wind, a plaything of absurdity, of the absence of meaning':

> If we put aside the ascetic ideal, then the human, the *animal* human, has had no meaning up to now. Its existence on earth has lacked a goal: 'why does the human exist at all?'—was a question without an answer; the *will* for the human and earth was missing; behind every great human destiny rang the even greater refrain: 'In vain!' For the meaning of the ascetic ideal is none other than *this*: that something was missing, that the human was surrounded by a gaping *void*—it did not know how to justify, explain, affirm itself, it *suffered* from the problem of its meaning. (GM III 28)

Notice Nietzsche's point here that there was an absence of meaning even behind every *great* human destiny. Even those he admires most—for example, the ancient Greek nobles, or Napoleon—suffered from the problem of meaning. More particularly, and alluding back to his second essay's claim of the internalized animal's innate suffering from itself, Nietzsche writes that the problem of meaning for the sickly human animal has always been 'the absence of an answer to his questioning cry: "Why do I suffer?"' The meaninglessness of suffering, he observes, has been the curse which has hung over humankind up to now (GM III 28). Although Nietzsche's genealogical hypothesis points to the sudden and violent manner in which the human animal's instinct for freedom was forcibly made latent (GM II 17), this does not provide a reason or a meaning for the resulting self-inflicted suffering. Indeed, speaking of the conquering predatory animals that first created society, Nietzsche says that they 'arrive like fate, without motive, reason, consideration, pretext, they arrive like lightning' (GM II 17).

So Nietzsche only fully confronts the problem of meaning, the most tremendous problem there is for humankind, in the last pages of his *Genealogy*. And this confrontation itself is actually only a lead-in to the proper confrontation, and solution, in his previous book, *Thus Spoke Zarathustra*. This is why at the end of the second essay Nietzsche describes the man of the future, Zarathustra, in precisely the same terms as the passage I have just quoted. So far, he says, human existence on earth has lacked a goal, and so far the will for the human and earth has been missing, but Zarathustra shall liberate or release the will and give the earth its goal and humankind its hope. Nietzsche's idea that the human will needs *liberating*, and that this liberation will succeed in giving meaning to human existence and suffering, refers of course to Zarathustra's well-known speech about the creative will's imprisonment in the unchangeable past:

> The will itself is a still a prisoner. Willing liberates; but what is it that puts even the liberator itself in fetters? 'It was': that is the name of the will's gnashing of teeth and most solitary affliction. Powerless against that which is done—it is an evil spectator of all that is past. The will cannot will backwards; and that it cannot break time and time's desire, that is the will's most solitary affliction ... That time cannot run backwards, that is its wrath: 'That which was'—that is the name of the stone that it cannot roll away. (Z II 'On Redemption')

As predicted in the *Genealogy*, Zarathustra simply presupposes the death of God and with it the end of all belief in some external, transcendent, or metaphysical meaning such as that proposed by the ascetic ideal. Nietzsche's problem of meaning thus becomes a purely *immanent* one.[10] Since

10 Since Nietzsche denies the kind of transcendent meaning provided by the ascetic ideal, and since he suggests that the absence of such meaning is only a problem for those who have overreacted to the death of God (Nachlaß Summer 1886–Autumn 1887, KSA 12, 5[71]), it might seem that he also thinks there is no problem regarding the absence of human self-given immanent meaning. As I argue in this essay,

human creative willing is the only possible source of meaning in the world, how is it possible for the human animal to create for *itself* and for *its own existence* a genuine, this-worldly, and life-affirming meaning? How is it possible for the human animal to set for *itself* a goal other than nothingness? Zarathustra's reply here is that it must first find a way to give such a meaning or goal to its own *past* existence. As he says in the same speech: 'All "it was" is a fragment, a riddle, a dreadful accident—until the creative will says to it, "But thus I willed it." Until the creative will says to it, "But thus I will it; thus shall I will it."' (See also Z III 'On Old and New Tablets' 3.)

But this cannot be done. Although the human animal may creatively will a meaning or goal for its *future*, time does not run backwards and the human animal cannot will this same meaning or goal backwards into its past. This means that the human animal must always leave its past existence without meaning or goal—in Zarathustra's words, as a fragment, a riddle, and a dreadful accident. And since the past determines the present and the future, this meaninglessness undermines even the human animal's goal for its future. So human existence must remain meaningless after all.

7. Providential Reasoning

Most Nietzsche scholars think there is a fairly straightforward resolution to this seemingly intractable problem. Although there are still quibbles here and there, it has been agreed for a long time now, at least since Nehamas' 1985 *Life as Literature*, that there is a perfectly obvious way—one Nietzsche intended—in which the human creative will *can* extend backwards into the past and give it meaning.[11] All that is needed, as Nehamas puts it, is to notice Nietzsche's literary assumption that '[t]he events of the past are necessarily located through and within a narrative, and [that] different narratives can generate quite different events'. From this it follows that my present retrospective willing can indeed have a creative influence on my past that grants it meaning. Here is Nehamas' argument:

> [E]very one of my past actions is a necessary condition for my being what I am today. How I see my present self affects crucially the very nature of my

however, Zarathustra's speech on redemption and the concluding section of GM III show otherwise. Indeed, it is precisely the death of God that for the first time frees humans to consider how they might be able to give themselves meaning (Z II 'On the Blessed Isles').

11 For recent formulations of this interpretation, see Anderson 2005, and Richardson 2006.

> past. If I am even for a moment such as I would want to be again, then I would accept all my past actions, which, essential to and constitutive of the self I want to repeat, are now newly redescribed. By creating, on the basis of the past, an acceptable future, we justify and redeem everything that made this future possible; and that is everything … To accept the present is then to accept all that has led to it. It is in this sense that one can now say of what has already happened, 'Thus I willed it'. (1985, p. 160)

The broader literary framework behind Nehamas' argument is nicely explained and summarized by Julian Young in his recent book, *The Death of God and the Meaning of Life*. Citing *Gay Science* 277, Young argues that a human being's life story and task is according to Nietzsche not given as part of the furniture of the world but is rather always an act of free interpretation and the product of 'our own practical and theoretical skill in interpreting and arranging events' (2003, pp. 94–95). It follows that 'Nietzsche's response to the question of how one is to render one's life meaningful … is to construct one's life as if it were a well-constructed work of literature [a *Bildungsroman*] with oneself as its "hero"' (2003, p. 94). Citing *Gay Science* 299 (and also GS 78), Young observes that we therefore need to learn from artists in order to become the poets of our own lives, and this means we must learn the art 'of viewing our life as if from its end—grasping ourselves as a completed totality, as something "past and whole"' (2003, p. 87). But an essential element in such art is knowing how to construct a redemptive state, 'one that makes sense of, and makes up for, the sufferings and imperfections that have preceded it' (ibid., p. 90). Such a state should enable us to love the whole of our unalterable past and everything that has happened to us—so much so, that we would want to live it all over again, down to every last detail. Citing *Gay Science* 277 again, Young concludes that we must therefore be able to see a 'personal providence' (2003, p.91) in things and reach a 'high point' in which we see, as Nietzsche writes,

> how palpably always everything that happens to us turns out for the best. Every day and every hour, life seems to have no other wish than to prove this proposition again and again. Whatever it is, bad weather or good, the loss of a friend, sickness, slander, the failure of some letter to arrive, the spraining of an ankle, a glance into a shop, a counter-argument, the opening of a book, a dream, a fraud—either immediately or very soon after it proves to be something that 'must not be missing'; it has profound significance and use precisely for us. (GS 277)

As Nehamas himself acknowledges, however, there is a 'grave difficulty' in this literary interpretation of Zarathustra's speech on redemption. For self-deception, he writes, 'may convince us that we are approaching this [ideal affirmative] relationship to life and to the world when in fact we are not'. As he puts it, 'I might be willing, for example, to repeat my life only because I do not let myself see it for what it is, because I do not allow myself to see in the proper light, or to see at all, large and objectionable parts of it' (1985, pp. 162–163). This is indeed a difficulty, but it is grounded on a further, and graver, difficulty. Although Nietzsche does indeed think that my past is a *necessary condition* for my being what I am today, it does not follow from this that my past has my present state as its *meaning* or *goal*. Certainly, I can want to affirm my present state as a peak in my life and I can accordingly redescribe my past so that it appears to have this present state as its inevitable meaning and goal. But this does not mean that my redescription is true or that it is supported by any evidence. In fact, given my wish to affirm what I am today as a high point, and to see what I am today as the inevitable goal of my past, it is much more likely that my redescription of my past is falsification and wishful thinking.

Ironically enough, the *Gay Science* 277 passage cited by Young and others as support for the literary interpretation of Zarathustra's redemption speech makes just this point. Far from recommending that we must learn how to see a personal providence in our past that leads up to our present high point, Nietzsche argues here that this ability is now actually 'the greatest danger of spiritual unfreedom', 'our hardest test', and our most 'dangerous seduction'. The truth which we have worked so hard to understand, and which we are now in danger of falsifying, is that existence is beautiful chaos devoid of all providential reason and goodness. So when we find, Nietzsche writes, that 'the idea of a personal providence confronts us with the most penetrating force, and the best advocate, the evidence of our eyes, speaks for it', we should 'rest content instead with the supposition that our own practical and theoretical skill in interpreting and arranging events has now reached its high point'. Even when at times 'we are excessively surprised by a wonderful harmony created by the playing of our instrument—a harmony that sounds too good for us to dare to give the credit to ourselves'—even then, Nietzsche writes, we should suppose that it is not providence that guides our hand, but rather 'good old chance'—'the wisest providence could not think up a more beautiful music than that which our foolish hand produces then'.

According to Nietzsche himself, then, it is chance and chaos that determines our past as well as the path from our past to our present state. Although we may wish to read our present state back into our past as its meaning or goal, this cannot be done without falsification. So when Zarathustra says that the only way for the human animal to give itself meaning is for it to recreate all of its 'it was' into a 'thus I willed it', this cannot simply mean—as the current consensus has it—some kind of redemptive redescription of the past. Although I might now *say* to my past, 'thus I *willed* it', this does not make it so. In fact, since time does not run backwards, it is impossible for my *present* willing to influence my *past* willing in such a way that the latter shapes the arc of my life as inevitably leading up to my present state. Again, I can *say*, as Zarathustra teaches I also need to say: 'thus I *will* it', and 'thus I *shall* will it'. But neither of these can actually be true. Neither my present nor my future willing can have any power over my past—much less the power to shape my past in such a way that it has my present state as its meaning or goal. Indeed, how could my present or future willing shape the past that made this willing possible in the first place? Thus, it is impossible for the human animal to creatively will a meaning or goal for its own past. Since there is no other source of meaning, and since the past determines the present and the future, it would appear again that human existence must remain meaningless.

Against this line of argument, Nehamas and Young might want to recall their interpretive claim that according to Nietzsche the past is an act of free interpretation and the product of our own practical and theoretical skill in interpreting and arranging events. But we have just seen that this is a misreading of Nietzsche's remarks in *Gay Science* 277, where he is concerned rather to show that the accidental reality of the past is *falsified* by this kind of providential interpretation.

Or, leaving aside his broader literary framework, Nehamas might want to respond that for Nietzsche the nature of the past lies in its relationship to the future, and since the future is yet to come, the nature of the past is not yet settled (1985, pp. 160–161). But this reading is quite clearly contradicted by Zarathustra's emphasis on the unchangeability of the past. And it is contradicted as well by Nietzsche's *Genealogy* analysis of the human animal as the mnemonic animal. In this analysis, Nietzsche argues that, before society bred a faculty of memory, the prehuman will actively forgot anything outside the present moment that could confine its activity. But the new memory faculty suspended or disconnected this active forgetting and represented an active will not to let go, to keep willing that which it once willed in the past. With the advent of society and its mnemotechniques, some things were impressed upon moment-centred animal affects and desire so that they remained there—inextinguishable, omnipresent, fixed—just as they once were. The human will was therefore forced to recognize for the first time an entire arena of possible willing—much more extensive than the sphere outside society—that was completely and forever outside of its reach: namely, that which was willed and can now never be unwilled, deeds that can never be undone, in short, the past, the 'it was'. Nowhere in this analysis does Nietzsche define the nature of this remembered fixed past as determined by its relation to the future, or as somehow *still remaining* to be fixed by a future that is yet to come, or as somehow transformable by its relocation within different narratives.

8. The Hammer of Eternal Recurrence

Although Zarathustra does not explicitly introduce his teaching of eternal recurrence in his redemption speech, he obviously points in this direction when he concludes by asking how the creative will might be taught to will backwards so that it might will something higher than any reconciliation with time. According to the consensus reading I have been concerned to criticize, what Zarathustra means by this is that my affirmation of my present state should teach me how to 'will *backwards*'—that is, to will the exact repetition of my entire *past*. And I would not be able to will in this way unless I had successfully redescribed my past as directed inevitably towards my present state. As Nehamas writes: 'If I am even for a moment *such as I would want to be again*, then I would accept all my past actions, which, essential to and constitutive of the self I want to repeat, are now newly redescribed' (1985, p. 160, my emphasis). Or, as Julian Young writes, the key is to '"write" my life so that I not merely like it, but like it so much that I can will its recurrence for ever and ever, down to every last detail' (2003, p. 91), and to do this we must be able to see the kind of personal providence in things that Nietzsche describes in *Gay Science* 277. Hence eternal recurrence, which Nietzsche describes in *Ecce Homo* as 'the highest formula of affirmation that is at all attainable' (EH III Z 1), turns out to be the new counter-ideal and the means whereby the human animal can for the first time will a life-affirming meaning for its past and thereby for its entire existence.

However, this invocation of eternal recurrence does nothing whatsoever to eliminate the kind of self-deception pointed out by Nietzsche himself in *Gay Science* 277. Saying to my past, 'Thus I will it to be repeated', or, 'Thus I will it to be repeated eternally', does not add any more meaning to my past than saying to it merely, 'Thus I will it.' Just as my present willing can do nothing whatsoever to direct my past inevitably towards my present state, so too my present willing can do nothing whatsoever to ensure the repetition of my vanished past. Indeed, this is precisely Nietzsche's point about the irreversibility of time: ever since its acquisition of a faculty of memory, the human animal has nostalgically willed the return, and even eternal repetition, of its most treasured and joyous past moments. The problem is *not* that the human animal has never before willed this, and that it should start doing so now. The problem, rather, is that the human animal has *always* willed this, and that this willing has been of no avail whatsoever. Indeed, it is precisely this impotence, this confrontation with the immovable stone 'it was', that has caused the human animal to find its existence and suffering devoid of meaning, to take revenge on life, to formulate the ascetic ideal, and to live its life against life.

When, therefore, Nietzsche describes eternal recurrence as the highest formulation of life-affirmation that is at all attainable, his point is *not* that the human animal should aim to will this eternal recurrence and thereby achieve the highest life-affirmation possible.[12] Instead, his point is that the human animal can *never* will this and that therefore the human animal can *never* achieve such life-affirmation. Indeed, no matter how strong and healthy he deemed himself, perhaps the strongest and healthiest of all of his contemporaries, Nietzsche found himself unable to will his life's eternal recurrence. Writing in his notebooks in 1883, Nietzsche exclaims: 'I do not want life again. How have I borne it? What has made me endure the sight? the vision of the superhuman who affirms life. I have tried to affirm it myself —alas!' (Nachlaß November 1882–February 1883, KSA 10, 4[81]).[13]

So, rather than conferring meaning upon the human animal's past, the thought of eternal recurrence actually multiplies and intensifies the meaninglessness of this past to a new and devastating degree. This is why Nietzsche writes in his 1887 notes that the thought of eternal recurrence *builds upon* the nihilism that follows the demise of the ascetic ideal—that is, upon the most paralysing thought of 'continuing with an "in vain", without aim and purpose'. Eternal recurrence, he writes, is the most extreme form of nihilism because it is the 'thought of existence as it is, without meaning or goal, *but inevitably recurring, without any finale into nothingness*' (Nachlaß Summer 1886–Autumn 1887, KSA 12, 5[71], my emphasis). It is not just the thought of meaninglessness, but the thought of meaninglessness *eternally*. Previously the human animal sought solace in the idea of death and nothingness as an escape from the meaninglessness of its life and suffering. But the thought of eternal recurrence closes off all such escape and condemns the human animal to eternal meaninglessness. As Nietzsche writes in his 1883 notes: 'everything becomes and recurs eternally— escape is impossible!" (Nachlaß Winter 1883–1884, KSA 10, 24[7]).[14] Nor can it be replied to this that, because death no longer affords an escape to life's suffering, there is actually less point to suicide. For in the third

12 Compare Reginster 2006, pp. 201ff.

13 This is the reason why Nietzsche would argue (pace Simmel 1986, p. 178; or Berkowitz 1995, pp. 209–210) that his doctrine of eternal recurrence is not itself a comforting self-deception of the kind he criticizes in GS 277.

14 In Loeb 2006 I criticize the argument (Soll 1973) that eternal recurrence precludes any continuity of consciousness and therefore cannot lead to any feeling of accumulated meaninglessness.

essay of the *Genealogy* Nietzsche argues that the thought of death *as escape* keeps humans from giving in to their pre-existing death-wish by imparting meaning to life's suffering and offering the hope of a different or better life. Because eternal recurrence undermines both this meaning and this hope, nothing remains to keep humans from giving in to their inherent suicidal instincts.[15]

Properly understood, then, Zarathustra's doctrine of eternal recurrence is the *counter*-ideal to the ascetic ideal that has so far helped humankind to survive in the face of its longing for death. It is the means whereby Zarathustra *compels* humankind to will its own downfall. In *Gay Science* 341, Nietzsche describes this thought causing the life-impoverished human to gnash his teeth and throw himself down to be crushed under its heaviest weight. And Zarathustra himself, in the midst of his deadly combat with the dwarf that is a symbol for the small human animal, warns his arch-enemy that he will not be able to bear the deadly blow (*Schlag*) of the heavy weight of his death-wielding hammer (*Todtschläger*) of eternal recurrence. Similarly, in his 1883 notes, Nietzsche has Zarathustra say about his teaching of eternal recurrence: 'The human is something that must be overcome. I hold here the hammer that will overcome him!' (Nachlaß Autumn 1883, KSA 10, 21[6]).

In all these places, Nietzsche incorporates his metaphorical conception of the remembered past as a kind of heavy weight. Already at the start of his second *Unfashionable Observations*, he had written of the remembering human being who 'braces himself against the great and ever-greater burden of the past' that 'weighs him down or bends him over, hampers his gait as an invisible and obscure load'. Unlike other animals that are able to forget, the human being is perpetually confronted with the 'it was' that brings him suffering and instils in him a longing for oblivion through death (UM II 1). Later, as we have seen, Nietzsche argues that the human animal's mnemonic essence renders it impotent towards the crushingly heavy stone of the past that it cannot move, and that this impotence causes it to turn its destructive instincts inward against itself. Hence, confronted with the eternal recurrence of this immovable and meaningless past, and also with the loss of its previous hope for some *escape* in death and nothingness, the degenerate human animal has no choice now but to act on its

15 See also Nietzsche's contemporaneous note: 'Inevitably there emerges contempt and hatred for life. Buddhism. The European energy [*Thatkraft*] will push towards a mass-suicide. Thereto: my theory of recurrence as the most terrible burdening' (Nachlaß Summer–Autumn 1882, KSA 10, 2[4]).

exponentially intensified drive to self-destruction.[16] This is why Nietzsche writes in his 1885 notes that eternal recurrence acts as a 'pessimistic teaching and way of thinking, an ecstatic nihilism, [that] can under certain conditions be indispensable precisely to the philosopher—as a mighty pressure and hammer with which … to implant into that which is degenerate and desires to die a longing for the end' (Nachlaß May–July 1885, KSA 11, 35[82]).[17]

9. Superhuman Meaning

The interpretation I have just outlined stands in stark contrast to the usual 'existentialist' reading according to which Nietzsche hopes that eternal recurrence, a doctrine that intensifies this meaninglessness to the highest degree, will enable the human animal to become strong and healthy enough to accept, affirm, and even thrive on this meaninglessness. As I mentioned earlier, Brian Leiter—taking his cue from Camus—proposes this kind of reading when he writes that 'the "meaning" embodied in affirmation of the eternal return is precisely that *there is no meaning or justification for suffering*'. According to Leiter, and I think this a view widely attributed to Nietzsche today, 'to affirm the doctrine of eternal return is to recognize that there is no such meaning': 'This is the attitude of existential commitment, through brute force of will, to carry on in the absence of such a meaning or vindication, to give up, in effect, asking "Suffering for *what*?"'(2002, p. 288). Although Leiter speculates that Nietzsche thinks some higher human beings might be able to do this (something I have argued against here), he also endorses the more usual view that Nietzsche thinks this is something that only the superhuman will be able to do.

Now, it is true that Nietzsche does think the superhuman will be able to will eternal recurrence and affirm life. He says as much in the quote I just

16 There is an interesting link here to Freud's later association, in *Beyond the Pleasure Principle*, between the compulsion to repeat and the death drive.

17 There are many other similar statements in Nietzsche's contemporaneous notes: 'The great noon as turning point—the two paths. The hammer to overpower the human: highest development [*Entfaltung*] of the individual, so that it must perish of itself (and not, until now, of dietary mistakes!) (How death came into the world!)' (Nachlaß Autumn 1883, KSA 10, 21[3]). And: 'The hammer: a doctrine which, by unleashing a death-longing pessimism, brings about a selection of those most suitable for life' (Nachlaß Autumn 1885–Autumn 1886, KSA 12, 2[100]). And: 'the hammer —a danger that can shatter the human' (Nachlaß April–June 1885, KSA 11, 34[78]).

cited, where he bemoans his own human *inability* to do either, but says that he has nevertheless endured life through his vision of the superhuman that *will* be able to do both. However, contrary to Leiter and the usual existentialist reading, Nietzsche nowhere says that the superhuman accepts, affirms, or thrives on *meaninglessness*. Nor does Nietzsche say that the superhuman, like the human, finds meaninglessness in the thought of eternal recurrence. Indeed, my argument so far shows that his view has to be quite the reverse. Although Nietzsche denies that meaning guarantees life-affirmation, he believes that the only way to affirm life is to find meaning in it. So it must be the case that he envisions the superhuman as being able to give meaning to its own existence. And since Nietzsche describes the highest formula of affirmation as the willing of eternal recurrence, it must be the case that he envisions the superhuman as finding meaning in the thought of eternal recurrence.

Which brings me finally to my third and last question: what does Nietzsche imagine will be left once humankind has been led by the thought of eternal recurrence to will its own downfall? In several places, Nietzsche describes eternal recurrence as the thought about time that will split humankind, or the history of humankind, in two.[18] This is in keeping with Nietzsche's thinking about time and history, and in fact represents what he regarded as his most profound and consequential thinking about both. The split Nietzsche has in mind has horizontal and vertical dimensions. Horizontally, the history of humankind will be split into what came *before* the thought of eternal recurrence and what comes *after* it. Vertically, humankind itself will be split in two by the selective action of the thought of eternal recurrence. The lower segment, and great majority, of humankind will be driven by its lower will to power to crushing despair and suicide. But the higher segment, and very small minority, comprising the ancestors of the superhuman, will be driven by its higher will to power to an initial self-emancipation from its mnemonically imposed fetters. This minority's new freedom will in turn enable it to begin giving meaning to its own existence and thereby to begin affirming life.

Elsewhere I have articulated what I believe is Nietzsche's vision of the revolutionary development that will take place *after* the advent of the thought of eternal recurrence and *within* the highest and smallest and most powerful segment of humankind (see Loeb 2001, 2002, 2005, 2006). This is a complex vision and I do not have the space here to rehearse all of its details, nor the many compelling objections to this vision, nor the resources I find in Nietzsche's texts for meeting these objections. However,

18 See for example his 10 March 1884 letter to Overbeck (KSB 6, p. 485).

if I have been successful here, I hope to have at least shown that the standard metaphorical interpretation of Zarathustra's teaching of backwards-willing is incapable of explaining how the superhuman will is supposed to gain control over, and give meaning to, its own past. The only way to explain this, I have argued in these other essays, is to recognize that Nietzsche really does intend Zarathustra's doctrine of eternal recurrence—which he defines in *Ecce Homo* as a teaching of the unconditioned and endlessly repeated circular course of all things—to be understood literally and cosmologically.[19] Although the direction of time is still irreversible, and although it is still impossible to *change* the past, the cosmological truth of circular time allows—indeed, *requires*—present and future willing to have a causal *influence* on past willing. But this means that it is in fact possible for the superhuman will to shape its own past existence in such a way that it has its own completion and perfection as its meaning or goal.

In *Thus Spoke Zarathustra*, Nietzsche illustrates this possibility through a narrative in which the cosmic truth of eternal recurrence reveals that Zarathustra's life is a self-enclosed circular course. Because Zarathustra ceases to exist in the time observed by others between his death and his recreation, he experiences no break or ending at the moment of his death and finds his last conscious moment immediately succeeded by his reawakening consciousness (Nachlaß Spring–Autumn 1881, KSA 9, 11[318]). Zarathustra's life is thus a ceaselessly forward-flowing ring in which the endpoint eternally turns back to become the starting point. This means that the completed and perfected Zarathustra is able to 'backwards-will'—that is, to store mnemonic messages, commands, and reminders (such as, 'It is time!' 'It is high time!') which are addressed to himself at an 'earlier' stage in his life. This mnemonic willing is then buried in his 'younger' self's subconscious and manifests itself in the form of precognitive dreams, visions, omens, and voices. This is why Zarathustra hears disembodied cries, whispers, and laughter calling to him, admonishing him, and commanding him at critical times in his life when he is tempted away from himself or does not feel adequate to his destiny. Because this voice is easily identified as that of his own self at a 'later' point in the narrative, we are led to understand that Zarathustra possesses a recurrence-conscience that enables him to keep promises to his perfected self and to become who he is. The superhuman Zarathustra is thus able to transmit his completion and perfection backwards throughout his entire past existence so as to guarantee it meaning, necessity, and wholeness. What appeared to

19 And even 'factually' in Lawrence Hatab's sense of the term in his recent study (Hatab 2005, pp. 91ff.).

be riddles in his past are solved by the underlying hidden meaning that is his backwards-willed destiny. What appeared to be accidents in his past are necessitated and preordained by his backwards-willing perfected future self. And what seemed to be fragments in his past are unified by the backwards-willed destiny towards which they are directed by his future completed self. As a self-propelled wheel that wills its own will, Zarathustra is quite literally the artist-creator of his own life, and his aesthetic creation may be said to have a meaning that is affirmable for all eternity.

Thus, when Nietzsche says that eternal recurrence will cause a split in human history between what came *before* and what came *after* this thought, he does not mean that there is a *unidirectional* flow of influence from this before to this after. Indeed, given his conception of the superhuman as the type of being that has gained control over time, Nietzsche is actually more concerned with the backwards-willing influence that extends from the future superhuman to its past human ancestry. As we have seen, the human animal, whose essence is defined mnemonically, is impotent with respect to its own past and therefore cannot give its own existence meaning. But it does not follow that human existence is meaningless. As Zarathustra says in his inaugural speech: 'Behold, I teach you the superhuman. The superhuman is the meaning of the earth. Let your will say: the superhuman s h a l l b e the meaning of the earth.' And a bit later, when he is reflecting on his inability to communicate his teaching, Zarathustra says: 'Human existence is uncanny and still without meaning: a jester can become the human's fatality. I will teach humans the meaning of their existence—the superhuman, the lightning out of the dark cloud of the human' (Z Prologue 7). What these pronouncements mean is that the human animal, not being able to give itself meaning, and not being able to derive meaning from the cosmos or its past, must instead receive this meaning from its future progeny—namely, the superhuman that has gained power over time and is therefore able to give meaning to a past existence that *includes its human ancestry.*

10. The Myth of Sisyphus

Let me now conclude by returning to Camus, and in particular, to his discussion of the myth of Sisyphus. Here is Camus' synopsis of this myth: 'The gods had condemned Sisyphus to ceaselessly rolling a rock to the top of the mountain, whence the stone would fall back of its own weight. They had thought with some reason that there is no more dreadful punishment than futile and hopeless labour' (1991, p. 119). Although neither Nietzsche nor Camus mentions it, scholars have rightly found in this myth a fruitful representation of the doctrine of eternal recurrence—that is, of the endless and eternal repetition of the same with no hope of a final culmination or even of any escape into death or nothingness. Indeed, Nietzsche even seems to allude to this myth in the dwarf's reply to Zarathustra about the philosopher's stone being thrown high in the air only to fall back down on him. In this myth, as with eternal recurrence, we find an extreme intensification of the meaninglessness and absurdity that Camus finds in all of human life. So it is quite surprising to see Camus simply *reapplying* here the philosophy of conscious revolt that he outlined in his introductory essay on suicide. According to Camus, Sisyphus is still the absurd hero, and this is because he is able to achieve a moment of consciousness and lucidity at the top of the mountain when he 'watches the stone rush down in a few moments toward that lower world whence he will have to push it up again toward the summit' (1991, p. 121):

> It is during that return, that pause, that Sisyphus interests me … I see that man going back down with a heavy yet measured step toward the torment of which he will never know the end. That hour is like a breathing-hour of consciousness. At each of those moments when he leaves the heights and gradually sinks toward the lair of the gods, he is superior to his fate … If this myth is tragic, that is because its hero is conscious. Where would his torture be, indeed, if at every step the hope of succeeding upheld him? … powerless and rebellious, [Sisyphus] knows the whole extent of his wretched condition: it is what he thinks of during his descent. The lucidity that was to constitute his torture at the same time crowns his victory. There is no fate that cannot be surmounted by scorn … If the descent is thus sometimes performed in sorrow, it can also take place in joy. This word is not too much … One must imagine Sisyphus happy. (Camus 1991, pp. 121–123)

Let us forget for a moment that Sisyphus is supposed to be immortal, and let us follow Camus in imagining his plight to be representative of the *human* plight. In that case, I think there can be little doubt that Camus' response is profoundly dissatisfying. As Julian Young comments, we certainly do *not* imagine Sisyphus happy: 'Surely our response to his predicament is not admiration but rather *pity*.' To make him relevant to the question confronting humans, Young writes, Sisyphus has to be given the option of suicide, and 'given the option it is far from clear that he should not take it' (2003, pp. 166–167). Why does Camus think he would not take it? Well, because Sisyphus finds meaning in his scorn and rebellion against the gods who have punished him with his futile and hopeless task. Thus, insofar as the gods and their punishment are a symbol for life itself, Camus' response to the myth of Sisyphus confirms once again Nietzsche's idea that the human animal cannot *live* its meaningless life except by scorning it and rebelling against it, except by ascetically willing nothingness. But this time Camus seriously underestimates the new intensity and scale of self-destructiveness induced by the prospect of an *eternally recurring* meaninglessness. To a far greater degree than before, Camus' own logic dictates that a mortal Sisyphus has no choice but to take the life-affirming option of suicide. For the human animal, eternal recurrence is the hammer which crushes any remaining life-preserving instincts that would still rather *will nothingness* than not will at all.

Given our discussion of Zarathustra's doctrine of backwards-willing, it is noteworthy that Camus thinks that Sisyphus' mountain-top moment of conscious revolt somehow injects meaning into his futile and hopeless task. According to Camus, '[a]t that subtle moment when man glances backwards over his life, Sisyphus returning toward his rock, in that slight pivoting he contemplates that series of unrelated actions which becomes his fate, created by him, combined under his memory's eye and soon sealed by his death' (1991, p. 123). For Camus, then, Sisyphus' retrospective glance transforms his task into a fate that is 'his' and 'created by him'. Here we see the idea, endorsed by most Nietzsche commentators, that Sisyphus can find meaning in his endlessly repeated climb up the mountain if he retrospectively becomes aware of this climb as leading inevitably to the peak where he has successfully achieved his task. Thus, Ivan Soll (1973, pp. 338–339) and Maudemarie Clark (1990, pp. 270–272), for example, both liken eternal recurrence to the myth of Sisyphus, and both argue that this doctrine allows the meaningful to be repeated eternally as long as each cycle ends by reaching an ideal state or realizing a goal. As Soll writes: 'The entire development of each cycle would exhibit a meaningful pattern. And one might derive great metaphysical comfort from the thought of the successful pattern of fulfilment displayed and repeated in each cycle'.

But in *Gay Science* 277, as we have seen, Nietzsche himself explains how it is that our highly developed practical and theoretical skill in interpreting and arranging events may lead us to precisely this kind of *self-deceptive* comfort. In fact, the task performed by Sisyphus is imposed on him by the gods and is not in any way his or created by him. Nor, despite the providential reasoning Camus attributes to Sisyphus, is this task in any way meaningful. The gods deliberately select an arbitrary stone, an arbitrary mountain, and the completely meaningless task of pushing the stone up the mountain. That the stone always rolls back down after it has been pushed up shows that there is nothing meaningful or successful in Sisyphus reaching the mountain top, or in pushing the stone up there, or even less in achieving a moment of lucidity there. Because the gods who selected Sisyphus' task *intend* precisely *the reverse*, it is only self-deception that allows him to believe that his climb up the mountain has any meaning at all. Similarly, according to Nietzsche, the human animal may derive self-deceptive comfort from the providential thought that its life displays a successful pattern of fulfilment. But in reality, the human animal has no power over its remembered past. The human past does not belong to it, is not created by it, and does not have any later peak moment as its meaning or goal. For the human animal that is unable to achieve *even one single* meaningful cycle, it does not matter that eternal recurrence allows the meaningful to be repeated eternally. Since it cannot give its own existence meaning, the human animal must always find in Zarathustra's doctrine the eternal repetition of *meaninglessness*.

Let us now imagine instead the *superhuman* Zarathustra in Sisyphus' situation. There is still the stone, there is still the mountain, and there is still the task of pushing the stone up the mountain. But because Zarathustra has liberated his own will and gained control over time, he is able to choose the stone, to choose the mountain, and to choose the task of pushing the stone up the mountain. Whereas the human Sisyphus can only deceive himself into believing that the task of pushing the stone up the mountain is his and is created by him, the superhuman Zarathustra is right to believe this. From the mountain top where he has completely achieved his goal, the perfected and ripened Zarathustra—a kind of *circulus vitiosus deus* who makes himself necessary (BGE 56)—commands and helps his imperfect and still-ripening self to achieve just this goal. Whereas the human Sisyphus is only able to attain a peak moment of awareness in which he looks back on his past life and scorns the very conditions of his existence, the superhuman Zarathustra is able to achieve a peak moment of awareness in which he has not only come to terms and learned to get along with whatever was and is, but wants to have it, just as it was and is, repeated into all eternity, shouting insatiably *da capo* (BGE 56).[20]

References

Anderson, R. Lanier, 2005, 'Nietzsche on Truth, Illusion, and Redemption', in: European Journal of Philosophy, 13/2, pp. 185–226.

Berkowitz, Peter, 1995, Nietzsche: The Ethics of an Immoralist, Cambridge, MA: Harvard University Press.

Camus, Albert, 1991, The Myth of Sisyphus and Other Essays, trans. Justin O' Brien, New York: Vintage International.

Clark, Maudemarie, 1990, Nietzsche on Truth and Philosophy, Cambridge: Cambridge University Press.

20 I would like to thank Keith Ansell-Pearson, Daniel Blue, Manuel Dries, Mark Jenkins, and participants in the 2005 Nietzsche in New York conference for their helpful comments on an earlier version of this paper.

Hatab, Lawrence, 2005, Nietzsche's Life Sentence: Coming to Terms with Eternal Recurrence, New York: Routledge.
Leiter, Brian, 2002, Nietzsche on Morality, New York: Routledge.
Loeb, Paul S., 2001, 'Time, Power, and Superhumanity', in: Journal of Nietzsche Studies, 21, pp. 27–47.
Loeb, Paul S., 2002, 'The Dwarf, the Dragon, and the Ring of Eternal Recurrence: A Wagnerian Key to the Riddle of Nietzsche's Zarathustra', in: Nietzsche-Studien, 31, pp. 91–113.
Loeb, Paul S., 2005, 'Finding the Übermensch in Nietzsche's Genealogy of Morality', in: Journal of Nietzsche Studies, 30, pp. 70–101.
Loeb, Paul S., 2006, 'Identity and Eternal Recurrence', in: Keith Ansell-Pearson (ed.), A Companion to Nietzsche, Oxford: Blackwell, pp. 171–188.
Nehamas, Alexander, 1985, Nietzsche: Life as Literature. Cambridge, MA: Harvard University Press.
Reginster, Bernard, 2006, The Affirmation of Life: Nietzsche on Overcoming Nihilism, Cambridge, MA: Harvard University Press.
Richardson, John, 2006, 'Nietzsche on Time and Becoming', in Keith Ansell-Pearson (ed.), A Companion to Nietzsche, Oxford: Blackwell, pp. 208–229.
Schopenhauer, Arthur, 1969, The World as Will and Representation, vol. 1, trans. E. F. J. Payne, New York: Dover.
Simmel, Georg, 1986, Schopenhauer and Nietzsche, trans. H. Loiskandle, D. Weinstein, and M. Weinstein, Amherst: University of Massachusetts Press.
Soll, Ivan, 1973, 'Reflections on Recurrence: A Re-examination of Nietzsche's Doctrine of Die Ewige Wiederkehr des Gleichen', in: Robert Solomon (ed.), Nietzsche: A Collection of Critical Essays, Garden City, NY: Doubleday, pp. 322–342.
Young, Julian, 2003, The Death of God and the Meaning of Life, New York: Routledge.

Translations

Ecce Homo, in: *Basic Writings of Nietzsche*, ed. and trans. Walter Kaufmann, New York: Viking Press, 1966.
The Gay Science, trans. Walter Kaufmann, New York: Vintage Books, 1974.
On the Genealogy of Morality, in: *Basic Writings of Nietzsche*, ed. and trans. Walter Kaufmann, New York: Viking Press, 1966.
Thus Spoke Zarathustra, in: *The Portable Nietzsche*, ed. and trans. Walter Kaufmann, New York: Viking Press, 1954.
Twilight of the Idols, in: *The Portable Nietzsche*, ed. and trans. Walter Kaufmann, New York: Viking Press, 1954.
Unfashionable Observations, trans. Richard T. Gray, Stanford: Stanford University Press, 1995.

Nietzsche and the Temporality of (Self-)Legislation

Herman W. Siemens

1. Introduction

The concept of (self-)legislation or *(Selbst-)Gesetzgebung* is well known as a central category in both Nietzsche's descriptive-phenomenal philosophy of power, and his prescriptive-affirmative ethics. But in the research literature, the temporal articulations and insertions of this category have not received sufficient attention. This paper aims to make good this deficit by examining the concept of (self-)legislation in its multiple intersections with the themes of time and history.

Within the framework of Nietzsche's pluralistic 'ontology' of becoming, (self-)legislation has a central and profoundly ambivalent place. It is central, because for Nietzsche the character of becoming is to be an incessant *Fest-setzen*, a multiple fixing (*Feststellen*) or positing (*Setzen*) of being.[1] It is ambivalent because, on the one hand, *Gesetzgebung* names the positing of the moral law, immutable and universal, in radical contradiction with the pluralism and dynamism of life-as-becoming. On the other hand, however, it also names the only resource we have against the moral law, as can be seen in Nietzsche's repeated attempts to formulate a naturalistic counter-model of legislation that enhances the dynamic and pluralistic character of life.

Legislation is a central theme in Nietzsche's thought from his early *Auseinandersetzung* with the pre-Socratics, the *Kulturphilosophie* of the *Untimely Meditations*, the natural histories of morality from *Human, All Too Human* on (especially the phase of the 'ethics of custom'), through to *Zarathustra*'s 'Old and New Tables', and the question of the transvaluation of all values dominating his later writings. Equally constant is Nietzsche's

---

1 See, e.g., Nachlaß Autumn 1887, KSA 12, 9[91]: 'All that happens, all movement, all becoming as a fixing of relations of degree and force, as a s t r u g g l e.' See also: Nachlaß April–June 1885, KSA 11, 34[88, 89]; Nachlaß Summer–Autumn 1884, KSA 11, 26[359]; Nachlaß August–September 1885, KSA 11, 39[13]; Nachlaß Autumn 1885–Autumn 1886, KSA 12, 2[139]; UM III 3; GS 370; A 58.

view that legislation has its sources in the individual and is to be understood as a function of individuated power. In this respect, Nietzsche's thought conforms to the psychologization of power in the nineteenth century.[2] Because Nietzsche sees legislation as a function of individuated power, his attention is directed towards the legislator as a *type*, or rather, towards a variety of different legislator-types and their exemplifications, including, the Greeks (in the early 1870s); Schopenhauer and Wagner as legislative types in cultural crisis of the present (1874–1875); Zarathustra (1883); and finally, the legislators of the future, the 'Gesetzgeber der Zukunft' (1884–1886). The problematic of legislation is therefore best studied by tracing the sequence of legislator-types across Nietzsche's writings and reconstructing the systematic relations between them. In this paper Nietzsche's ambivalent relation to legislation will be examined by way of some key moments in this diachronic typology of the legislator: Schopenhauer, Wagner, and Zarathustra. I will examine these legislator-types, and their temporal articulations, as efforts to solve the problem of life-affirmative and/or life-enhancing legislation, attempts (*Versuche*) that are successively problematized and superseded by the next type.

Among the different legislator-types generated by the Nietzschean thought process of self-critique and renewed attempt (*Kritik-Versuch*), there are also a number of constants. In the first place, it is characteristic of all these contexts that the thematics of legislation and self-legislation are hard to separate; this is not just sloppy thinking, but a consequence of Nietzsche's conviction that true legislators must at the same time be self-legislators (e.g., Nachlaß Autumn 1885–Autumn 1886, KSA 12, 2[57]). Secondly, what emerges clearly from all these contexts is that the creative and evaluative moments of legislation are central for Nietzsche. And thirdly, that in the forms of legislation affirmed or sought by him, both the descriptive, theoretical meaning and the prescriptive, evaluative meaning of the term 'law' are taken up, so that Nietzsche's affirmative notion of legislation can be seen as an attempt to overcome the categorical separation of *is* from *ought* in modern philosophy, to synthesize the theoretical and the moral domains. This goes emphatically for Nietzsche's own performative instantiations of philosophical legislation (as in, e.g., the formulation 'Gesetzt … '). Although I will not do so in the confines of this essay, these should be considered within the thematic of legislation.

2 See, e.g., Nachlaß Summer 1875, KSA 8, 6[40]; cf. Nachlaß Spring–Summer 1875, KSA 8, 5[170]; Nachlaß Beginning of 1880, KSA 9, 1[68]; Nachlaß April–June 1885, KSA 11, 34[88]. See also Gerhardt 1996, pp. 76–78.

Throughout the essay, attention will be paid to 'legislation' as a *word* in Nietzsche's vocabulary, not just as a concept. A striking feature of Nietzsche's engagement with the problem of legislation is that his unrelenting criticisms of traditional and existing conceptions do not on the whole issue in calls to excise the word 'legislation' from our (moral) vocabulary, nor in efforts to replace it with alternative words. Instead, he attempts time and again to invest the word with new meanings and values. In this regard, the topic of legislation presents a rich case study in Nietzsche's sense of *linguistic finitude*. Just as the moral law can only be overcome from within, through an effort to rethink the concept of law in the context of its genealogy ('die Selbstaufhebung der Moral' (D Preface 4), so too our only recourse against the word 'law' is the effort to invest it with new meanings and values. In what follows I hope to shed light on some of ways in which Nietzsche engages in this task.

2. Schopenhauer and Wagner as Legislators

2.1. 'Schopenhauer as Educator' (UM III)

'Schopenhauer as Educator' is the first published text of Nietzsche's in which legislation and self-legislation are absolutely central. Despite its complex, transitional character, this text is of fundamental importance for understanding the nature and sources of Nietzsche's demand for legislation throughout his work. Yet in this, its first appearance, the concept of legislation becomes enmeshed in an aporia of time that Nietzsche is unable to resolve. 'Richard Wagner in Bayreuth' (UM IV) is the first of several subsequent attempts to come to grips with the temporal aporia of legislation.

The demand for self-legislation appears in the very first section of the text. This opening section reads as a manifesto of radical individualism against conformism with prevailing tastes and fashions:

> In his heart every man knows quite well that, being unique, he will be in the world only once and that no imaginable chance will for a second time gather together into a unity so strangely variegated an assortment as he is: he knows it but he hides it like a bad conscience—why? From fear of his neighbour, who demands conventionality and cloaks himself with it. But what is it that constrains the individual to fear his neighbour, to think and act like a member of a herd, and to have no joy in himself? Modesty, perhaps, in a few rare cases. With the great majority it is indolence, inertia, in short that tendency to laziness. (UM III 1, KSA 1, p. 337)

Nietzsche goes on to describe artists as the non-conformists *par excellence*:

> Artists alone hate this sluggish promenading in borrowed fashions and appropriated opinions and they reveal everyone's secret bad conscience, the law that every man is a unique miracle; they dare to show us man as he is, uniquely himself to every last movement of his muscles, more, that in being thus strictly consistent in uniqueness he is beautiful, and worth regarding, and in no way tedious. (UM III 1, KSA 1, pp. 337f.)

Value is clearly located in the (potential) uniqueness of each of us, or more precisely: in the unique convergence of multiplicity, chance, and necessity that makes each of us what we are, but is buried or suppressed in oblivion by the ease of conformism. As a way to free ourselves from conformity towards our own, unique existence, Nietzsche goes on to propose radically individual self-legislation. Our measure of happiness can only be attained by exercising freedom, that is, gaining control over our lives, by taking responsibility for our own existence, so that we exhibit what its meaning is. We must, in short, find a way to 'live according to our own law and measure' (UM III 1, KSA 1, p. 339).[3]

In this opening section of the text, then, self-legislation articulates a response to two moral impulses driving Nietzsche's thought. The first is the *moral particularism*[4] that finds expression in Nietzsche's demand for radi-

---

3 'The fact of our existing at all in this here-and-now must be the strongest incentive to us to live according to our own law and measure: the inexplicable fact that we live precisely today, when we had all infinite time in which to come into existence, that we possess only a short-lived today in which to demonstrate why and to what end we came into existence now and at no other time. We are responsible to ourselves for our own existence; consequently we want to be the true helmsman of this existence and refuse to allow our existence to resemble a mindless act of chance' (UM III 1, KSA 1, p. 339).

4 Nietzsche's moral particularism, in turn, is to be understood as part of the project to (re-)*naturalize morality*, formulated with increasing clarity in Nietzsche's later thought, and is grounded in two features of that project: (1) his pluralistic 'ontology' of diverse life-forms, the uniqueness of each, and its particular life-conditions (*Lebens-* or *Existenz-bedingungen*); and (2) his naturalized concept of values as means for a given life-form to meet its life-conditions (Nachlaß November 1887–March 1888, KSA 13, 11[118]; Nachlaß Spring 1888, KSA 13, 14[158]). On the (re-)naturalisation of morality, see Nachlaß Autumn 1887, KSA 12, 9[86]: 'my task is to translate the the seemingly emancipated and nature-less moral values back into their nature—that is, into their natural "immorality"'; or, more bluntly: 'basic principle: to be like nature' (Nachlaß Spring 1884, KSA 11, 25[309]). On the physiology of morality, see: Nachlaß End of 1876–Summer 1877, KSA 8, 23[87]; D 174; Nachlaß November 1882–February 1883, KSA 10, 4[90]; Nachlaß Spring–Summer 1883, KSA 10, 7[76]: 'In truth we follow our drives, and morality is only a sign-language of our drives?'); also Nachlaß Spring 1884, KSA 11, 25[460] (cf. GS 162; also Nachlaß Summer–Autumn 1884, KSA 11, 26[38]); Nachlaß Autumn 1887, KSA 12, 10[157]; Nachlaß Spring 1888, KSA 13, 14[105];

cally individual legislation. This is the first (published) articulation of what Volker Gerhardt has described as Nietzsche's sustained attempt to develop a radically individual morality, a morality of *das individuelle Gesetz* ('the individual law'), often placed by Nietzsche in polemical opposition to moral universalism, especially the Kantian morality of the universal law.[5] The second moral impulse behind self-legislation motivates Nietzsche's rejection of conformism or *heteronomy*, our self-subjection to prevailing norms and conventions.[6] At stake here is a *positive perfectionist* impulse to extend one's attainments and enrich the range of human possibilities. Nietzsche's articulation of this impulse in 'Schopenhauer as Educator' owes much to the powerful influence of Emerson. As Cavell, Conant,[7] and others have shown, Nietzsche's Emersonian perfectionism is fuelled by an aversion to heteronomy; or, in Emerson's idiom, to the complacent, conformist acceptance of who we are. Nietzsche's lifelong opposition to heteronomy is here expressed in Emersonian terms as a practice of aversion that is bound up with a positive perfectionist impulse.

In 'Schopenhauer as Educator', then, self-legislation is a response to the twin moral impulses of *particularism* and *perfectionism*. But these impulses do not simply stand on their own. They constitute an affirmation

Nachlaß Spring 1888, KSA 13, 14[158]; see also Nachlaß Summer 1880, KSA 9, 4[67]; BGE 188. On legislation from the perspective of the body, see Nachlaß Spring–Summer 1883, KSA 10, 7[126, 150].

5 'What? You admire the categorical imperative within you? This "firmness" of your so-called moral judgement? This absoluteness of feeling, "here everyone must judge as I do"? Rather admire your selfishness here! And the blindness, pettiness, and simplicity of your selfishness! For it is selfish to consider one's judgement a universal law, and this selfishness is blind, petty, and simple because it shows that you haven't yet discovered yourself or created for yourself an ideal of your very own—for this could never be someone else's, let alone everyone's, everyone's!' (GS 335; cf. A 11). Gerhardt (1992, p. 41) takes the expression 'das individuelle Gesetz' from Georg Simmel's 'Das individuelle Gesetz. Ein Versuch über das Prinzip der Ethik' (Simmel 1968), but it also occurs in a note of Nietzsche's, where it is opposed to the 'ewiges Sittengesetz' (Nachlaß Spring–Autumn 1881, KSA 9, 11[182]).

6 'The virtues are as dangerous as the vices, to the extent that one allows them to rule as authority and law from outside instead of generating them from within oneself, as is right: as the most personal self-defence and necessity, as a condition of precisely our existence and benefit, which we know and acknowledge, regardless of whether others grow with us under the same or under different conditions' (Nachlaß End of 1886–Spring 1887, KSA 12, 7[6]). For Nietzsche's opposition to heteronomy and self-subjection, see also D 108; and Nachlaß Spring 1880, KSA 9, 3[159].

7 See Cavell 1990; Conant 2001; and Conway 1997, esp. pp. 52–56.

of what Nietzsche will come to describe as the pluralistic character of life as will to power and its intrinsic dynamic of self-overcoming and intensification. In the context of 'Schopenhauer as Educator' itself, Nietzsche's particularist and perfectionist ethic is a response to the critical diagnosis of the present generated by his practise of aversion. It is a diagnosis that bears striking similarities to the account of *contemporary nihilism* that fuels Nietzsche's demand for self-legislation in later years. It also bears striking similarities to the problem of modernity, as perceived by Schiller, the young Hegel, and others of his generation. As Habermas describes it in *The Philosophical Discourse of Modernity* (1987, pp. 23–44), modernity is marked by the pervasive and radical sense of disorientation brought on by the collapse of traditional authorities at the hands of modern, Enlightenment critique. In the absence of credible rules or models from the past, modernity is thrown back on itself and must find ways to orient and guide itself in its own terms. In 'Schopenhauer as Educator' the need for self-legislation comes from the moral bankruptcy of the present, denounced as a 'low tide of all moral powers', incapable of generating values; we live instead on a dwindling capital of inherited morality.[8] The line from here to Nietzsche's later account of nihilism, as a bankruptcy of transcendence, is obvious. The same goes for another feature of Nietzsche's *Zeitdiagnose* in 'Schopenhauer as Educator': the pervasive conflict (*Vernichtungskrieg*)[9] of forces and values, resulting in a condition of diremption, atomistic disgregation, and alienation (see UM III 4, KSA 1, p. 367). For Hegel, this is the second major problem of modernity, stemming from the loss of the unifying powers of religious belief and worship brought on by Enlightenment critique. Modernity must find ways both to unify and orient itself.

8 '—then one finally asks oneself: where are we, scholars and unscholarly, high placed and low, to find the moral exemplars and models among our contemporaries, the visible epitome of all creative morality in our time? What has become of any reflections on questions of morality—questions that have at all times engaged every more highly civilized society? There is no longer any model or any reflection of any kind; what we are in fact doing is consuming the moral capital we have inherited from our forefathers, which we are incapable of increasing but know only how to squander' (UM III 2, KSA 1, p. 344).

9 For the similarities between early and later texts on nihilism as a pervasive conflict of forces, compare the following two texts from 1873/74 and 1887: 'Now is lacking that which binds all fractional forces: and thus we find everything antagonistically opposed and all noble forces engaged in a mutually destructive war of extermination [*Vernichtungskrieg*]' (Nachlaß Autumn 1873–Winter 1873/74, KSA 7, 30[8]). And: 'that the synthesis of values and goals (on which every strong culture rests) dissolves, so that the individual values wage war on each other: disintegration' (Nachlaß Autumn 1887, KSA 12, 9[35]).

For Nietzsche, then, self-legislation is to be both the source of *orientation* and a *unifying* power in modernity. And yet: at crucial junctures in the text a profound ambiguity enters into Nietzsche's concept of legislation. It is unclear whether it serves to overcome *the times* by provoking the actualization of better selves in a perfectionist dynamic of intensification; or whether it serves the overcoming or transcendence of *time itself*, in a metaphysics of being (*Sein*). I shall try to indicate three such moments.

a. The first intimation comes in Nietzsche's account of self-knowledge in 'Schopenhauer as Educator' 1. In a move that recurs again and again in the later writings, Nietzsche's thought moves from the demand for self-legislation to the need for self-knowledge. In section1 he describes an indirect, Emersonian route to self-knowledge by way of others—the objects of our love and reverence, who serve to reveal (not who are, but) a higher, unattained but attainable self.[10] The telos of self-knowledge is described by Nietzsche as 'the fundamental law of your own true self', but also as something 'completely incapable of education or formation'. Here the term *Gesetz* serves to connect and synthesise moral legislation and self-knowledge, a self-knowledge that points to something fixed and binding, a sense of necessity or *Müssen* as the ground of the *Sollen* of legislation. This dual usage is a good example of how Nietzsche's affirmative use of *Gesetz* cuts across the prescriptive and the descriptive meanings of the

10 See in particular UM III 1: 'But how can we find ourselves again? How can man know himself? He is a thing dark and veiled; and if the hare has seven skins, man can slough off seventy times seven and still not be able to say: "this is really you, this is no longer outer shell". Moreover, it is a painful and dangerous undertaking thus to tunnel into oneself and to force one's way down into the shaft of one's being by the nearest path. A man who does it can easily so hurt himself that no physician can cure him. And, moreover again, what need should there be for it, since everything bears witness to what we are, our friendships and enmities, our glance and the clasp of our hand, our memory and that which we do not remember, our books and our handwriting. This, however, is the means by which an inquiry into the most important aspect can be initiated. Let the youthful soul look back on life with the question: what have you truly loved up to now, what has drawn your soul aloft, what has mastered it and at the same time blessed it? Set up these revered objects before you and perhaps their nature and their sequence will give you a law, the fundamental law of your own true self … your true being lies not buried deep within you, but immeasurably high above you or at least, above that which you normally take as your I. Your true educators and formative teachers reveal to you that the true, original meaning and basic stuff of your nature is something completely incapable of being educated or formed and is in any case something difficult to access, bound and paralysed; your educators can be only your liberators' (KSA 1, pp. 340–341).

term. But is this 'something fixed' and the constraint it exercises any more than a name for being?

b. This ambiguity is expressed again in a passage from 'Schopenhauer as Educator' 4 where, under the sign of the 'Schopenhauerian human', Nietzsche describes his own practice of critical aversion and its sources in a perfectionist longing to extend human life. The specific question addressed in this passage is: what is to be the standard of critique for the aversive practice of the Schopenhauerian(-Emersonian) human?

> But there is a kind of denying and destroying that is the discharge of that mighty longing for sanctification and salvation and as the first philosophical teacher of which Schopenhauer came among us desanctified and truly secularized men. All that exists that can be denied deserves to be denied; and being truthful means: to believe in an existence that can in no way be denied and which is itself true and without falsehood. That is why the truthful man feels that the meaning of his activity is metaphysical, explicable through the laws of another and higher life, and in the profoundest sense affirmative: however much all that he does may appear to be destructive of the laws of this life and a crime against them. (UM III 4, KSA 1, p. 372)

But what exactly is the status of this 'other and higher life' and 'the laws' or standard of critique that it offers? Is it transcendent and unattainable, a vision of pure being like Plato's sun, a metaphysical ground or standard of judgement that distends the untimely critique of the present into a total negation of becoming? Or is it immanent, the vision of a possible form of life, whose law or standard makes possible a *transformative* critique of the present in favour of a better life,[11] a 'transfigured *physis*' (UM III 3, KSA 1, p. 363)?

c. The ambiguity of legislation takes its most insidious form in section 3, where Nietzsche first presents Schopenhauer as the philosophical legislator-type:

> Let us think of the philosopher's eye resting upon existence: he wants to determine its value anew. For it has been the proper task of all great thinkers to be law-givers as to the measure, stamp and weight of things. (UM III 3, KSA 1, p. 360)

But Nietzsche does not go on to offer a faithful account of Schopenhauer. Instead, the passage involves a reflection on, and *correction* of, Schopenhauer's problem: what is the value of life? What is life worth? Nietzsche takes up the question of the value of life, but then corrects it by asking the

11 Barbera (1994, pp. 229–230) has argued persuasively for the influence of the young Schopenhauer (before *World as Will and Representation*) on 'Schopenhauer as Educator', in particular his notion of the 'bessere Bewusstsein'.

prior question: What are conditions for a 'fair' or 'just judgement' (*ein gerechtes Urteil*) of the value of life? The argument developed in the passage can be put as follows.

With Schopenhauer, Nietzsche agrees that life *as it is* cannot be affirmed. He does not, however, go on to draw Schopenhauer's practical conclusion: that life therefore *ought not to be.* Instead he questions the standpoint of this judgement of life, and argues that a just judgement requires first a *transformation* of the evaluating perspective, and then a transformation of life itself into something better. The problem is, these transformations bring practical and aesthetic resources into play, which compromise the philosopher-legislator's truthfulness (*Wahrhaftigkeit*). The philosopher gets torn between 'the reformer of life and the philosopher, that is: the judge of life' (*Reformator des Lebens, Richter des Lebens*). He gets caught in 'the discord between the wish for freedom, beauty and greatness of life, and the drive for truth which asks only: what is existence worth?'[12]

The worry expressed in these lines is whether the legislator can be truthful like Schopenhauer and still affirm existence, or whether *illusion* is necessary to affirm existence. There is, in other words, a tension for Nietzsche between the truthfulness of the philosopher and the possibility of life-affirmation. The background to this tension in 'Schopenhauer as Educator' is a conflict of loyalties in Nietzsche's mind between Schopenhauer and Wagner, and specifically, between their respective concepts of genius. For Schopenhauer, the genius is primarily a theoretical figure, a thinker devoted to truth, dwelling on the margins of culture; his exceptional insights culminate in the realization that life is worthless. For Wagner, by contrast, the self-modelled genius is an ecstatic, affirmative figure at the very heart of culture. His orientation is primarily practical and he devotes himself to the creation of life-serving illusions (*Wahngebilde, edle Täuschungen*). What troubles Nietzsche is the Wagnerian linkage between illusion and affirmation: what, after all, is an affirmation of life worth, if it is based on an illusory vision of life? It is not until *Human, All Too Human* that Nietzsche will break with this equation decisively and shake off the Wagnerian

12 See UM III 3: 'Let us think of the philosopher's eye resting upon existence: he wants to determine its value anew … A modern thinker will, to repeat, always suffer from an unfulfilled desire: he will want first to be shown life again, true, red-blooded, healthy life, so that he may then pronounce his judgement on it … But this longing also constitutes their d a n g e r: there is a struggle within them between the reformer of life and the philosopher, that is to say the judge of life. Wherever the victory may incline, it is a victory that will involve a loss' (KSA 1, pp. 360–362).

figure of genius in favour of the 'free spirit'. But in 'Schopenhauer as Educator', he is genuinely torn between Wagner and Schopenhauer. His equivocation is laid bare by a telling slippage in his portrait of Schopenhauer, one that culminates in an ambiguity in his use of the word *Zeit*. The slippage is between Schopenhauer's rejection of *the timely* (*das Zeitgemässe*) and his rejection of *time itself* as a 'false, vain, unworthy mother'. The slippage trades on the ambiguity of Nietzsche's use of the word *Zeit* ('time') when he speaks *first* of Schopenhauer's struggle against *die Zeit* (meaning: the times, the present), and *then* of Schopenhauer's struggle against *Zeit* (meaning: time, temporal existence, becoming).

Consider the following three lines from the closing passage:

> and in the end the supposed child of his time [*Kind der Zeit*] proves to be only its stepchild. (UM III 3, KSA 1, p. 362)

Here *die Zeit* means the times, the present. The passage continues as follows:

> Thus Schopenhauer strove from his early youth against that false, idle and unworthy mother, the times [*jener falschen, eiteln und unwürdigen Mutter, der Zeit*]. (ibid.)

Again: *die Zeit* means the times, the present. But then there is the line:

> and when he had conquered his age in himself [*die Zeit in sich*] he beheld with astonished eyes the genius in himself. The secret of his being was now revealed to him, the intention of that stepmother time [*die Absicht jener Stiefmutter Zeit*] to conceal his genius from him was frustrated, the realm of transfigured *physis* was disclosed. (ibid.)

Here *Stiefmutter Zeit* (without the definite article) can only mean: time itself. If this is correct and Schopenhauer's struggle against the times becomes a struggle against time itself, that is, against becoming in the name of being, then we have to ask: does the affirmation of being (against becoming) constitute an affirmation or a negation of life? Can one possibly negate becoming and affirm life? In what sense is being in exclusion of becoming alive at all? In his effort to escape the Wagnerian equation of life-affirmation and illusion, Nietzsche falls back into Schopenhauerian metaphysics. For in this passage, there is an unmistakable alignment of philosophical legislation (qua *Wahrhaftigkeit*) with being *against* becoming.

2.2. 'Richard Wagner in Bayreuth' (UM IV)

'Richard Wagner in Bayreuth' represents an effort to rethink genius and specifically the legislation of genius in its temporal character, its articulations with being and becoming. In the figure of Wagner, self-legislation is once again identified with emancipation from the present. In this text, the word *Gesetz* is used for the 'evil' external power of society and convention (on Burckhardt's *böse Macht*, see Gerhardt 1996, pp. 71–76, 104–112). Nietzsche speaks of 'power, law, ancestry, contract and the whole order of things' (UM IV 4, KSA 1, p. 451) and their 'apparently unconquerable necessity' (ibid.) or *ἀνάγκη* (Nachlaß Summer 1875, KSA 8, 11[20]). But *Gesetz* is also used for the 'good' power of the individual *Kulturkämpfer* ('fighter for culture') and his striving for freedom. As Gerhardt (1996, pp. 99–100) points out, Wagner's life story is dramatized between these two poles of power, from conformity to the former in his early ambitions for 'honour and power', through to a Wotan-like renunciation of external power, in which he gives himself over to his free creative force. In this text, Nietzsche looks for a standpoint or ground of legislation that is sufficiently removed or distant from 'power, law, ancestry, contract and the whole order of things' to allow for radically individual legislation; yet one that also resists the transcending of becoming or temporality towards being that we saw in 'Schopenhauer as Educator'. Thus Wagner is portrayed as caught in a *Kreuzung der Empfindung*—between hatred and rejection of the present and a yearning need for love and community with his contemporaries.[13] Nietzsche's quest for a ground of legislation that is removed from yet immanent to becoming—or at least too indeterminate to signify being—culminates in a kind of 'homelessness', *a legislation from nowhere*, as when he writes that

> the uncanny and exuberant sensation of surprise [*Befremdung*] and amazement at the world is coupled with the ardent longing to approach this same world as lover. (UM IV 7, KSA 1, p. 471)

Or again:

> For it is, to be sure, a life full of torment and shame, to be a homeless wanderer in a world to which one nonetheless has to speak and of which one has to make demands, which one despises and yet is unable to do without—it is the actual predicament of the artist of the future. (UM IV 10, KSA 1, p. 500)[14]

But on its own this is evidently unsatisfactory for Nietzsche, and he looks to characterize Wagner's actual insertion in the present in positive terms. He wants to describe the real necessity for Wagnerian art in modernity, a necessity that is opposed to the 'apparently unconquerable necessity' of

13 For *Kreuzung der Empfindung* see UM IV 7: 'But this sensation becomes a peculiar hybrid [*eigenthümlich gekreuzt*], when to the brightness of this exuberance is joined quite a different impulse, the longing to descend from the heights into the depth, the living desire for the earth, for the joy of communion.'

14 See also UM IV 10 on saving 'this homeless art' for the future (KSA 1, p. 504).

convention opposed by genius. He does so in the language of 'law', describing 'true music' as 'a piece of fate and primal law' that speaks through the mysterious 'primordially determined nature [*ur-bestimmte Natur*]' (UM IV 6, KSA 1, p. 465). This mysterious necessity linking Wagner's art to the present is explained in section 9:

> his appearance in the history of the arts is like a volcanic eruption of the total undivided artistic capacity of nature itself after humanity had accustomed itself to seeing the arts isolated from one another as though this were an eternal rule. (UM IV 9, KSA 1, p. 485; cf. UM IV 7, KSA 1, p. 468)

Wagnerian art is here cast as an organizing force that brings together and synthesizes the chaotic multiplicity of individual arts in its works. What Nietzsche here says of art is part of the broader explanation offered for the real necessity of Wagner and Wagnerian legislation in the present: namely, as a unifying, organizing force able to bind the centripetal atomistic forces of modernity. As Gerhardt argues, Wagner's later self-realization as a free, (self-)legislative genius described in section 9 is to be understood above all as a 'new form of exercising power': as the organizational genius with his 'sovereign disposal over conflicting forces' (Gerhardt 1996, pp. 100–101). Legislation is here thought of as inseparable from freedom or emancipation (UM IV 9, KSA 1, pp. 493–495), as a unifying instance that binds and holds together the most disparate elements. This goes for the personal domain (the forces within Wagner); for the domain of art, for the particular arts and elements of his drama; but also—what Nietzsche especially values—for the political domain, the 'connection between state, society and art' (Nachlaß January–February 1874, KSA 7, 33[7]).[15] This organizational legislation is made possible by Wagner's eye for 'relations' or 'large-scale connections'. It is often presented in Heraclitean terms as 'unity in diversity' ('Einheit im Verschiedenen') (Nachlaß Beginning of 1874–Spring 1874, KSA 7, 32[12]; also January–February 1874, KSA 7, 33[7]), or as the 'unity of justice and enmity' ('Einheit von Gerechtigkeit und Feindschaft') (UM IV 9, KSA 1, p. 494). This goes in particular for Wagner's dramas, which are described as the realization of the 'inner lawfulness' of his life story, namely of his 'One will' and the 'strictly individuated passion' that enforces itself over 'a plurality of passions' and the 'confusing multiplicity of claims and desires' (ibid.; cf. UM IV 2, KSA 1, pp. 435–436). What these expressions reveal, however, is not a genuine Heraclitean balance between the One and the Many (or 'unity in diversity'), but a preponderance of the One *over* the Many. Nor do they ex-

15 Compare also Nachlaß Summer 1875, KSA 8, 11[51]; Nachlaß January–February 1874, KSA 7, 33[4]; Nachlaß Beginning of 1874–Spring 1874, KSA 7, 32[10].

press a Heraclitean affirmation of conflict, but a desire for peace. This can be seen from the following series of expressions taken from section 9:

> the compelling force of a personal will [*die zwingende Gewalt eines persönlichen Willens*]
>
> an overpowering symphonic understanding, which gives birth continuously to concord out of war [*ein übermächtiger symphonischer Verstand, welcher aus dem Kriege fortwährend die Eintracht gebiert*]
>
> that we have before us conflicting particular streams, but also, in force over all of them, a stream with One mighty direction [*dass wir widerstrebende einzelne Strömungen, aber auch über alle mächtig, einen Strom mit Einer gewaltigen Richtung vor uns haben*]
>
> to assert One will across a confusing multiplicity of claims and desires [*durch eine verwirrende Mannichfaltigkeit von Ansprüchen und Begehrungen, Einen Willen durchführen*] (UM IV 9, KSA 1, pp. 493–494)

Such distortions of Heraclitus for Wagner's sake unwittingly exhibit Nietzsche's insight into the tyrannical absolutization of power as Wagner's true tendency. Nietzsche's worries about the egocentric absolutization of power into the tyrannical in Wagner are strongly attested in the Nachlaß, even before the writing of 'Richard Wagner in Bayreuth'.[16] With time this perception gains the upper hand, as can be seen in the later reinterpretation of Wagner in terms of the *décadent* (Nachlaß Spring 1888, KSA 13, 15[88]), and a tyrannical absolutization of bad taste (Nachlaß Summer 1880, KSA 9, 4[221]).

3. Zarathustra as Legislator-Type: Nietzsche's Agonal Model of Self-Legislation

The Wagnerian model of legislation may or may not avoid the metaphysical pitfalls of Schopenhauer's model, but it certainly fails the test of plu-

16 See Nachlaß Beginning of 1874–Spring 1874, KSA 7, 32[32]/7: 'The "false omnipotence" produces something "tyrannical" in Wagner. The feeling of being without heirs —this is the reason why he tries to give his idea for reform as much room as possible, as if to propagate himself by adoption. Striving for legitimacy. The tyrant will not allow for any other individuality but his own and that of his confidants. The danger for Wagner is great when he does not accept Brahms etc.: or the Jews' (cf. HA I 577). Also 32[34]: 'The tyrant-sense for what is colossal. There is no respect coming his way, the true musician regards him as an intruder, as illegitimate.' And again 32[61]: 'Here lies Wagner's significance: he attempt tyranny with the help of the theatre-masses [*die Tyrannis mit Hülfe der Theatermassen*].'

ralism required for a life-affirming form of legislation. It is in response to this problem that Nietzsche develops the third type of legislator I shall examine. At stake here is a *pluralistic*, *egalitarian*, and *dynamic* model of legislation inspired by the signature institution of archaic Greek culture: the agon or contest. Already in 'Homer's Contest', written in 1872 (three years before 'Richard Wagner in Bayreuth'), Nietzsche's abhorrence of Wagner's absolutization of power is hard to miss in his description of 'the Hellenic notion of the contest: it abhors the rule of one [*Alleinherrschaft*] and fears its dangers; it desires, as a protection against genius—a second genius' (CV 5, KSA 1, p. 789).[17] This same sentiment is expressed eleven years later in the *Zarathustra* Nachlaß of 1883 with even greater clarity:

> To rule? dreadful! I do not want to enforce my type. My joy is diversity! Problem! (Nachlaß Summer–Autumn 1883, KSA 10, 15[21]; cf. Nachlaß Autumn 1883, KSA 10, 16[86])

It is to this period, and the pluralistic model of self-legislation developed in a series of notes from 1883 (Nachlaß Summer–Autumn 1883, KSA 10, notebooks 15 and 16), that I now turn. In these notes, the three moments of the problematic of legislation in 'Schopenhauer as Educator' are very prominent; that is, his opposition to the *heteronomy of self-subjection* and to *moral universalism* in the context of the necessity to respond to *contemporary nihilism*. What is distinctive about these notes is how the claims of moral particularity are articulated in a pluralistic context of collective self-legislation. This implies a shift towards a more political concept of legislation and suggests that Gerhardt's thesis of Nietzsche's radically individual morality needs to be modified in the direction of moral laws that bind collectively across particular communities (Conway 1997, pp. 29–30).[18]

In these notes, the task of legislation and new goals is not taken on by Nietzsche himself; nor is it ascribed to Zarathustra as such. Instead the task is pluralized around the figure of Zarathustra, who serves as the legislator for further legislators (not subjects). In response to the question: 'What sense does it have to give laws?' Nietzsche writes:

17 This line needs to be read together with the remark by Kelterborn, Nietzsche's student, that the latter honoured Wagner 'in the first instance not just as the genial musician und *dramatist*, but above all else as a cultural force, a *fellow fighter next to him* (not above him) in the struggle for a higher German culture [einen *Mitstreiter neben*, (nicht über ihm)]' (original emphasis). Dated 1875, from the Nachbericht, BAB 4, p. 351.

18 See also Siemens (2002) where I use the concept of taste to develop this sense of ethical laws.

> the general type of law-giver, who is the herald for **many** law-givers. Main doctrine: to achieve completeness and a **sense of well-being** on every level — no jumping! (Nachlaß Summer–Autumn 1883, KSA 10, 15[10])

According to this lapidary response, law or law-giving must meet two basic demands, if it is to have any sense: *first*, the law must be provisional and not final, not the law to end all legislation; *secondly*, the law must be responsive to diversity, a law *for* many, not a law that subjects the many to One. In these terms, it is clear that Nietzsche's concept of law is radically opposed to the moral law in its traditional claims to eternity and universal validity. But Nietzsche's underlying motives are positive, not negative: the provisional, pluralistic qualities of Nietzsche's law are intended to *take the side of life* against the moral law, as part of the Nietzschean project to naturalize morality. This is clearly expressed in two subsequent notes. In note 15[19], he writes:

> Zarathustra offers the model for **how** one has to behave towards the law, insofar as he supersedes [*aufhebt*] the law of laws, morality, with higher [ones] (Nachlaß Summer–Autumn 1883, KSA 10, 15[19])

It is as an example of the *dynamic character of life* that Zarathustra's attitude to the law becomes exemplary: like him, we are to engage in the dynamic of destruction and creation, in the processes of self-overcoming and intensification intrinsic to life.[19] And in the second note (cited above), it is Zarathustra's joy in the plurality of life-forms, in 'the spectacle of many others', that first brings him face to face with the problem of legislation.

The problem is, then, whether law and law-giving can be rethought in a way that reflects and enhances the *dynamic* and *pluralistic* qualities of life against the rigidity and universalism of the moral law. Nietzsche's solution, adumbrated in the next line, is to rethink law and law-giving on the model of the agon or 'contest for power':

> To call for the agon! Precisely those who would dearly like to hide themselves, the still ones, the pious ones,—competition for mastery! (Nachlaß Summer–Autumn 1883, KSA 10, 15[21]; cf. Nachlaß Autumn 1883, KSA 10, 16[86])

In order to reconstruct Nietzsche's agonal solution, we must return to note 15[10] and Nietzsche's demand that law-giving be somehow counter-final and pluralistic for it to be meaningful. In order to meet this demand, Nietzsche indicates three lines of thought:

---

19 See also Nachlaß Autumn 1883, KSA 10, 16[84], where the 'law-giver' is literally placed in between the destruction of existing laws and the clarion call for new laws: 'The destroying of tablets. The ideal "law-giver". Clarion call [*Heroldsruf*]'.

1. It is to be a 'law for law-givers', not for passive subjects or 'supplicants': a 'law for law-givers / From supplicants [*Betenden*] we must become blessers [*Segnende*]!' (Nachlaß Summer–Autumn 1883, KSA 10, 15[58])
2. It is to be productive and pluralistic, the creative source of many ideals and not a subjection of many to the ideal or law of One: 'Not One ideal of the wise man, instead a hundred ideals of the fool I wish to erect!' (Nachlaß Autumn 1883, KSA 10, 16[86]).
3. It is to be law or law-giving that blesses, completes, and fulfils 'on every level', not a leveller that demands or pleads that we all 'jump' to the same One level.

The first two points are best understood by way of the third. What Nietzsche means here is explicated in the note 15[19] (cited above) on Zarathustra's exemplary, destructive-creative attitude to the law, which continues as follows:

> the fulfillability greater than before (accessible to the individual's interpretation). NB. it **must** be **fulfillable** and from the fulfilment a higher ideal and its law **must** grow! (Nachlaß Summer–Autumn 1883, KSA 10, 15[19])

In these lines Zarathustra figures as the counterpart or counter-exemplar to Paul and his tortured relation to the law that *cannot be fulfilled* (see D 68). But again, Nietzsche's thought must be grasped from a positive perspective in his concept of life: the demand for fulfillable laws, in the sense of laws that are accessible to individual interpretations, addresses the claims of *particularism*, understood as the ethical articulation of the radical plurality and diversity of life forms. This demand, in turn, is presented as the key to the development of further, higher laws; that is, to a dynamic sense of law-giving that replicates the self-overcoming and intensification intrinsic to life. But how exactly is a higher law to 'grow' from the individual interpretations and fulfilment of the law? This thought is developed in note 16[86] that takes up the demand for fulfillable laws:

> Demand: the new law must be fulfillable—and from the fulfilment the overcoming and the higher law must grow. Zarathustra gives the attitude towards the law, insofar as he supersedes the 'law of laws', morality. Laws as backbone. To work on them and create, insofar as one carries them out. Hitherto slavishness before the law! (Nachlaß Autumn 1883, KSA 10, 16[86])

With the image of laws as a 'backbone' (*Rückgrat*) Nietzsche brings us back to the notion of a 'law for law-givers' (1 above). Laws that are subject to individual interpretation and fulfilment break, for the first time, our slavish subjection to eternal, immutable laws, what Nietzsche elsewhere calls our 'fear of commanding', that 'one would rather obey an available law than create a law for oneself, than command oneself and others. The fear of commanding—Rather subject oneself than react' (Nachlaß End of 1886–Spring 1887, KSA 12, 7[6]).

At the centre of Nietzsche's dynamic conception of laws is the notion of creativity: where laws are subject to interpretation and fulfilment—according to the model of a 'backbone' or provisional framework—individuals are placed in a position to work and rework them so as to create new and better laws of their own. With this notion of a 'law for law-givers', we come up against the central paradox for Nietzsche's reinterpretation of law: how to overcome our 'slavishness before the law' through law-giving? How to legislate active self-legislation, rather than passive obedience or prostration? Under what conditions does law-giving cease to be coercive and become instead productive—a stimulant towards individual self-emancipation, towards autonomy in the sense of radically individual self-legislation?

Such questions go to the heart of Nietzsche's problematic relation to the law. For with them, the challenge is posed: can the gestures of tyranny, coercion, subjection, and making-fast, can the functions of unity and universal validity so closely wedded to the concept of law, be overcome? Or must we abandon the concept of law altogether in the name of life? It is these questions that Nietzsche's agonal model of the law is designed to address. What makes the agon interesting as a solution is that it does not attempt to reinvent the concept of law from scratch. Rather, it exploits the traditional meanings and functions of law—coercion/subjection, rigidity/eternity and universality—in order to transform the meanings and functions of law. By situating the traditional concept of law within a unique constellation of forces, Nietzsche turns it against itself, so as to generate a dynamic, pluralistic, and emancipatory sense of law.

Nietzsche's concept of the agon turns on a reinterpretation and re-evaluation of *resistance* in the context of conflicting or competing powers. The resistance offered by an opponent need not be experienced as a negative, inhibiting force, as pain, loss, or a diminution of power to be avoided at all costs. It can also be a stimulant that one seeks out, an obstacle that provokes one to exercise, extend, and measure one's own resources in the effort to achieve mastery (see CV 5, KSA 1, p. 789; cf. EH 'Why I Am so Wise' 7; also Nachlaß Spring 1888, KSA 13, 14[173, 174]). When placed in the context of a 'contest for power', law takes on precisely this meaning, as an obstacle or stimulus that provokes others to resist and surpass it with their own, better laws:

> The rights that I have conquered for myself I will not give to the other: rather, he ought to rob them for himself! like me—and [he] may take them and wrest them from me! To this extent there must be a law which emanates from me, as if it wanted to make all into my likeness [*zu meinem Ebenbild mache*]:

> so that the singular individual finds himself in contradiction with it and strengthens himself ... Whoever appropriates a right will not give this right to the other—but will be an opponent to him insofar as he appropriates it for himself: the love of the father who clashes with his son. The great educator, like nature: he must pile up obstacles, so that they are overcome. (Nachlaß Autumn 1883, KSA 10, 16[88])

On one level, these lines leave the traditional, problematic meanings of the word *Gesetz* untouched. Indeed, they *exploit* these meanings in order to undo them. For *only* if Zarathustra legislates *as if* he wanted to coerce others and make a claim on all will he evoke their resistance, provoking each one (*der Einzelne*) to extend himself and seek its own law, thereby destroying Zarathustra's law and its universal claim in a plethora of self-legislation.

On another level, however, the law does undergo significant reinterpretations. Most importantly, the relation of law to rights is reversed. The traditional (liberal) concept of legislation as a *giving* of (equal) rights to others and/or safeguarding of those rights is rejected in favour of a symmetrical regime of power, in which rights are claimed, conquered, or usurped by dint of one's deeds, not given.[20]

The central problem raised by this model of law is how to stimulate others to conquer their own rights. Nietzsche's agonal solution is to raise one's own law as if (*als ob*) it were universally binding, thereby usurping all rights and forcing others who oppose one to discover and assert their own capacities in reclaiming their rights—like the love of a father who intentionally clashes with his son. Even here, where the relation of laws to rights is reversed and tied to the concept of love, the coercive character of law remains the key to rethinking law in non-coercive terms.

Whether Nietzsche's agonal solution works is questionable. What, after all, is to distinguish 'as if' universal claims designed to provoke conflict and self-legislation from those that simply pacify and subject? What separates the father who clashes with his son out of love, from tyrannical, overbearing fathers like Kafka's? The agonal regime of power presupposes conditions—creative resources, and a resilience on the part of many—which, on Nietzsche's own diagnosis, are scarcely imaginable in the present. As far as I can tell, Nietzsche does not ask what, for contemporary

20 Compare 'My Concept of Freedom' (TI 'Reconnaisance Raids' 38) where Nietzsche argues that the safeguarding of liberal values under liberal institutions has had the effect of producing unfreedom: they turn equality (*Gleichheit*) into sameness (*Gleichmachung*). Against this, Nietzsche defines his concept of freedom as the *struggle for* rights, that is, as the exercise of illiberal—agonal—capacities in the struggle for liberal institutions.

democratic sentiments, are the most urgent questions: under what conditions does law-giving cease to be coercive and become instead productive—a stimulant towards individual self-emancipation? And how can these conditions be promoted and extended across social life? Such questions seem rather to arouse his suspicion,[21] and these notes culminate in a very different line of thought: the *deferral of legislation to the future.*[22] In this, the next phase of Nietzsche's thought on law, the agon migrates to particular communities, to a class or caste of 'legislators of the future', charged with the task of transvaluation. Nonetheless, the notes on agonal law are valuable as an effort to retain the language of law, while reinvesting it with naturalistic meanings that undo the problematic features of the traditional concept of law. What is more, the agonal model of power they deploy brings with it an insight of fundamental importance to the problematic of law and legislation: that legislation is irreducibly *relational* in character. This means that moral particularism, and its realization in radically individual legislation, is unthinkable without relations of resistance—attraction to concrete others, a 'new love'[23] that includes relations of ten-

---

21 See, e.g., BGE 259: 'To refrain from mutual injury, mutual violence, mutual exploitation, to equate one's own will with that of another: this may in a certain rough sense become good manners between individuals if the conditions for it are present (namely if their strength and value standards are in fact similar and they belong to One body). As soon as there is a desire to take this principle further, however, and if possible even as the fundamental principle of society, it at once reveals itself for what it is: as the will to the denial of life, as the principle of dissolution and decay.'

22 See Nachlaß Summer–Autumn 1884, KSA 11, 26[407]; Nachlaß April–June 1885, KSA 11, 34[33, 199, 207, 212]; Nachlaß May–July 1885, KSA 11, 35[9, 39, 45, 47]; Nachlaß June–July 1885, KSA 11, 37[14]; Nachlaß Autumn 1885–Autumn 1886, KSA 12, 2[57].

23 See Nachlaß November 1882–February 1883, KSA 10, 4[83]: 'The dissolution of morality leads in practical consequence to the atomistic individual, and then to the division of the individual into multiplicities—radical flux. This is why now more than ever a new goal is necessary and love, a new love.' See also note 4[89]: 'On the morality of higher human beings. Everything that is otherwise morality, has here become love. But there now begins a new "thou shalt"—the knowledge of the free spirit — the question of the highest goals'. These notes reiterate the problems of disorientation and atomistic disgregation in modernity motivating Nietzsche's call for self-legislation in 'Schopenhauer as Educator'. But they do so in a radicalized form that indicates why a strictly individual morality on its own is insufficient. The problem of 'atomistic Chaos' (UM III 4, KSA 1, p. 367) is now radicalized into a boundless conflict of forces and values that threatens not just relations between 'atomistic individuals', but the very constitution of individuals. In line with this heightened perception of the threat, Nietzsche's demand

sion and antagonism. A radically individual morality cannot, in other words, be achieved in isolation, but is inseparable from the task of founding the kind of ethical community that makes it possible.

References

Barbera, Sandro, 1994, 'Ein Sinn und unzählige Hieroglyphen', in: Tilman Borsche / Federico Gerratana / Aldo Venturelli (eds.), 'Centauren-Geburten': Wissenschaft, Kunst und Philosophie beim jungen Nietzsche, Berlin: Walter de Gruyter, pp. 217–233.

Cavell, Stanley, 1990, 'Aversive Thinking: Emersonian Representations in Heidegger and Nietzsche', in: Conditions Handsome and Unhandsome: The Constitution of Emersonian Perfectionism, Chicago: University of Chicago Press, pp. 33–63.

Conant, James, 2001, 'Nietzsche's Perfectionism: A Reading of Schopenhauer as Educator', in: Richard Schacht (ed.), Nietzsche's Postmoralism, Cambridge: Cambridge University Press, pp. 181–257.

Conway, Daniel, 1997, Nietzsche and the Political, London: Routledge.

Gerhardt, Volker, 1992, 'Selbstbegründung. Nietzsches Moral der Individualität', in: Nietzsche-Studien, 21, pp. 28–49.

Gerhardt, Volker, 1996, Vom Willen zur Macht. Anthropologie und Metaphysik der Macht am exemplarischen Fall Friedrich Nietzsches, Berlin: Walter de Gruyter.

Habermas, Jürgen, 1987, The Philosophical Discourse of Modernity: Twelve Lectures, trans. Frederick Lawrence, Cambridge: Polity Press.

Siemens, Herman, 2002, 'Agonal Communities of Taste: Law and Community in Nietzsche's Philosophy of Transvaluation', in: Journal of Nietzsche Studies, 24 (Special Issue on Nietzsche and the Agon), pp. 83–112.

Simmel, Georg, 1968, Das individuelle Gesetz. Ein Versuch über das Prinzip der Ethik (1913), in: Michael Landmann (ed.), Das individuelle Gesetz. Philosophische Exkurse, Frankfurt a. M.: Suhrkamp, pp. 174–230.

Translations

Untimely Meditations, ed. Daniel Breazeale, trans. R. J. Hollingdale, Cambridge: Cambridge University Press, 1997.

Beyond Good and Evil, trans. R. J. Hollingdale, introduction Michael Tanner, London: Penguin, 1990.

The Gay Science, trans. Walter Kaufmann, New York: Vintage Books, 1974.

---

for a source of orientation, a new 'goal', is coupled with the call for a unifying counter-force that cannot be met by individual self-legislation: for 'a new love'.

Part IV

Nietzsche's Contemporaries

Geschichte or *Historie*? Nietzsche's Second *Untimely Meditation* in the Context of Nineteenth-Century Philological Studies

Anthony K. Jensen

In the second *Untimely Meditation* 'On the Uses and Disadvantages of History for Life',[1] Nietzsche outlines three dominant trends in historiography: the 'critical', the 'antiquarian', and the 'monumental'. These designations were not wholly invented, but indicate a then significant debate in classical philology, which Nietzsche knew intimately.[2] Indeed, the dispute between the historical trends Nietzsche names in 'On the Uses and Disadvantages' 'was to become the most famous quarrel between philologists in the history of modern scholarship, and the exchange between Nietzsche and Wilamowitz furnished the concluding satyr play' (Müller 1990, p. 232). Throughout the nineteenth century, scholars such as Friedrich August Wolf, Gottfried Hermann, Karl Lachmann, and August Boeckh debated as to which of those trends should mark the proper impulse of the field they labelled *Altertumswissenschaft*, how philological pursuits should be carried out, and how their students might most properly be instructed. It was a feud seen rekindled by the politically unpopular actions of Nietzsche's teachers, Friedrich Ritschl and Otto Jahn. Nietzsche's meditations on that debate are manifest in a plain, if somewhat psychologistic and oversimplified, manner. His very selection of the borrowed academic term *Historie* over the more expected *Geschichte* in the title of his essay suggests that his aim was not simply to proffer a purely theoretical reflection on the nature or essence of history, but equally to expound on the current state and direction of the discipline.[3] Thus, to understand the complicated turns of argu-

1 For a sampling of the consulted secondary literature on this essay, see Campioni 1975; Zuckert 1976; Salaquarda 1984; Gerhardt 1988; and Stambaugh 1987.

2 See Nietzsche's 'Encyclopädie der klassischen Philologie' for a more closely philological statement of many of the issues I read into UM II. See especially, KGW II.3, p. 365, pp. 366–376.

3 I owe the recognition of this point to Glenn Most, who was generous enough to share with me a draft that visits many of these same themes, but does so with an

ment in the second *Untimely Meditation* requires an acquaintance with that historical debate and Nietzsche's place within it.

Similarly, the content of 'On the Uses and Disadvantages' should not be considered in isolation from the well-known circumstances that surrounded the publication of the *Birth of Tragedy*. Now famous is the somewhat over-blown struggle between Ulrich von Wilamowitz-Moellendorff and Nietzsche's friend Erwin Rohde. For his part Nietzsche never publicly commented on the *Zukunftsphilologie!* review, at least not in a philological fashion, but poured his labour into writing an essay on history and its value as a field of study within the German university. This essay will indicate how his remarks on the characteristic traits of historians was informed significantly by his lessons in classical studies and how the *Meditation* is itself partly an attempt to address the criticisms of Wilamowitz and to justify the manner of history Nietzsche believed himself to be doing in *The Birth of Tragedy*.[4] I wish to contend that the second *Untimely Meditation* is a book about history, foremost, but also a book about a particular set of historians. I do not wish to contend that this was Nietzsche's only motivation for writing—certainly it was not—nor that the few philologists I can name here were the only historians Nietzsche was characterizing; but the backdrop of Nietzsche's philological environment has been underestimated in connection with his *Meditation* and so it is my purpose here to highlight and emphasize those theoretical aspects which have a tangible root in Nietzsche's own biography.

An Overview of Classical Studies in the Nineteenth Century

A generation after the matriculation of Friedrich August Wolf into the University of Göttingen in 1777 as a *Studiosus Philologiae*[5] and his later lectures at Halle, which attracted the admiration of figures the likes of Goethe, von Humboldt, and the Schlegel brothers, the field known as *Al-*

---

eye more concentrated on the role Wilhelm von Humboldt played in the structuring of humanistic pedagogy.

4 This way of reading is given weight when we note how similar the language and themes between the two works are, and, moreover, the timing involved—the Wilamowitz pamphlet appeared 30 May 1872, while the manuscript of UM II was likely begun in early September 1873 (see Calder 1983, p. 228; also Salaquarda 1984, p. 5).

5 Of the day Wolf declared his ambition to study philology at Göttingen (8 April 1777), Nietzsche proclaims a 'birthday of philology' (Nachlaß March 1875, KSA 8, 3[2]).

tertumswissenschaft became divided along the widening breach between two pillars of the discipline. Gottfried Hermann and August Boeckh became heads of two rival schools, the first of which has sometimes been labelled 'positive', 'critical', or *Wort-Philologie*, approaching antiquity with the tools of textual emendation, codices, and literary criticism, while the second, being more concerned to effectively demonstrate the writ-large spirit of antiquity and to implant that ideal into the hearts and minds of their students, was variously named 'hermeneutical', 'antiquarian', 'humanistic', or *Sach-Philologie*.[6] The antiquarians sought to construct classical world-views, while the critical philologists tried to tear down their speculative fancies in the name of philological certainty and interpretive precision. Nietzsche, as we shall see, was on the front lines of this debate during his education at both Bonn and Leipzig.

Turning first to Johann Gottfried Jakob Hermann, we find the precursor of what might now be called a philological positivist. His primary interest was the critical study of classical languages. In his grammar, the *De emendanda ratione Graecae Grammaticae*, in his dissertation on the term *αὐτός*, and in his 'Four Books on the particle *ἄν*, Hermann insisted on the central importance of syntactical perfection as the prior condition of any knowledge of antiquity (Sandys 1908, vol. 3, p. 91). A hermeneutical rendering of classical texts without a grounded insight into the myriad uses and meanings of the words and grammar of those texts would prove empty. Not a fanciful construal of the ancient world—like something out of Hölderlin, Goethe, or Schiller[7]—but a certain, precise, and elemental philological method should be the aim of established scholars' research, as well as their sole pedagogical goal. For how could an author such as Goethe say anything about the 'spirit' of Iphigenia or Prometheus without an adequate knowledge of Aeschylus' grammar or the history of the emendations of Aeschylean texts? No, the aim of philology is, as Nietzsche quotes Hermann, *ut recte intellegantur scripta verterum* (BAW 4, p. 6). Among Hermann's many respected students at Leipzig, of particular note are Moritz Haupt and Theodor Bergk, on whom Nietzsche relied for his own 'On

6 I myself am inclined to believe that this division is too simplistic. However, this seems to be the picture Nietzsche inherited from his instructors and from the histories of philology composed at the time. So, whether or not the debate was as significant as I make it out to be here, my point is that Nietzsche himself thought it was—so much so that he spent several chapters of UM II discussing it, that he nearly completed a work that dealt explicitly with it ('Wir Philologen'), and that his own inauguration lectures at Basel were caught up with it.

7 Among the best studies on these influences are those of Politycki 1981 and 1989, Siemens 2004, and Ulfers/Cohen 2004.

the History of the Collection of the Theognideian Anthology' (KGW II.1, pp. 3–58), and also, interestingly enough, Friedrich Ritschl (Sandys 1908, p. 95).

August Boeckh exemplified a methodology antithetical to that of Hermann. The student of Wolf and Schleiermacher, Boeckh sought to explain from the broad scope of a comprehensive *Weltanschauung* what he considered the most pedagogically important aspects of antiquity, rather than focusing on the grammatical or technical mastery of any particular text. His lectures stand as a profound application of previous generations of scholarship to holistic branches of classical learning: methods of inscription or household management. The aim of studying the classics was eventually to emulate the classical models; and to accomplish this task, one must sense the overarching spirit of the classics, something which a single-mindedly technical focus on individual words and phrases was likely to retard. Unlike Hermann and his later followers, Boeckh viewed grammatical and technical scholarship as a mere tool towards the more interesting and more pedagogically valuable portrayal of antiquity as a whole. Among his prize students at Berlin is Nietzsche's professor Otto Jahn. And though Ritschl studied under Hermann, the great historian of philology John Edwin Sandys categorizes him 'among the warmest admirers of Boeckh' (1908, vol. 3, pp. 100–101).

The generation of classical scholars that followed was effectively polarized into either the camp of Boeckh or else the school of Hermann. In the antiquarian group, we find Gottfried Bernhardy, after whom Nietzsche patterned his own attempt at a *Geschichte der griechischen Literatur* (KGW II.5, pp. 7–353) and at an *Encyclopädie der klassischen Philologie* (KGW II.3, pp. 339–437). Following the critical school of Hermann was Karl Lachmann, who largely codified textual criticism into a strict methodological discipline—something he believed was not much respected by those poets, artists, and musicians more concerned with generalizations about the 'true majesty of antiquity' (Bursian 1883, vol. 2, p. 789). Along with his close friend Moritz Haupt, Lachmann maintained an almost guard-dog attitude toward the teachings of Hermann. Among the most important students of this pair—one who actually obtained his doctorate under Haupt—was none other than Ulrich von Wilamowitz-Moellendorff.

Although this historical sketch is all too general, it is a fair summary of the two major opposing trends of philological scholarship with which Nietzsche was certainly familiar. The stage is now set for us to discuss the most important philologists with respect to Nietzsche: Friedrich Ritschl and Otto Jahn. Once the student of Boeckh, Otto Jahn was for a time employed as an archaeologist alongside the philology department of Hermann and his disciple turned son-in-law Mortiz Haupt at Leipzig. Having developed a personal friendship with the pair, Jahn joined with Haupt in the 1849 political agitation for the maintenance of the imperial constitution, for which he was prosecuted for high treason and relieved of his professorship. Ritschl would later invite Jahn to join the 'humanistic' school at Bonn as the successor of F. G. Welcker, who had been an ardent critic of Hermann. All involved parties were apparently unaware of Jahn's growing distaste for antiquarianism, initiated by his friendship with the Hermannians at Leipzig.[8] Now, while Ritschl likely considered himself to be of the Hermannian 'critical' school—having written his dissertation under Hermann with the title *Schedae Criticae* and having even turned Hermann's failed initiatives on Plautus into his own life's work (Vogt 1990, p. 390)—his ties were becoming strained. Ritschl's student and biographer Otto Ribbeck believed this was due to increasing political tensions during the 1850s between Ritschl and his Hermannian counterparts at Berlin: Lachmann, and now Haupt (Ribbeck 1879–1881, vol. 2, pp. 332–381).[9] The first break likely occurred in 1839, when Ritschl secured the co-editorship of the *Rheinisches Museum für Philologie* along with F. G. Welcker. The long-time Bonn journal had been founded there by Boeckh in 1827. As a former student of Hermann, though, Ritschl's appointment to its helm would have raised some curious eyebrows, doubting Ritschl's loyalty to the principles of his teacher.[10] Now, when Ritschl split with Jahn,[11] he transferred the editorial offices of the journal to Leipzig. Rather than handing over the reigns of Boeckh's journal to his colleague and Boeckh's own student Jahn, Ritschl deprived his old university and one-

8 There was a further complication in the story. Jahn admits in his letters that he long aspired to be the successor to Welcker's chair at Bonn, and when Ritschl issued the call in 1854 he found himself positively delighted (see Petersen 1913, nos. 91–92). However, it seems that Ritschl never informed Welcker, who had been on temporary leave at the time. Jahn reasonably assumed that Welcker himself was supportive of the appointment. Welcker, however, was deeply and justifiably insulted at being replaced without consultation after decades of excellent and loyal service at Bonn, though to his credit he did not direct his indignation towards Jahn. Ritschl was made the culprit, and, since he was then the senior scholar, the tensions around the department heightened proportionately (see Müller 1990, p. 231).

9 Haupt was professor at both Leipzig and Berlin. He taught at Leipzig from 1837 to 1850. After his forced resignation, he was invited to Berlin by Lachmann, where he taught from 1853 to 1874.

10 Ernst Vogt (1979, pp. 103–121) suggests that the impetus underlying the division between Jahn and Ritschl was in fact due to their opposing sides in the Hermann and Boeckh conflict. This view, however, reduces the complexity of Ritschl's own professional standing.

11 There were several reasons for their break, which the author is presently exploring in a comprehensive treatment of the influence of Ritchl and Jahn on Nietzsche.

time friend of their most important publication. Moreover, Leipzig was the school of Hermann;[12] to have Boeckh's old journal published in the university of his great rival would have made something of a scene (Sandys 1908, vol. 3, p. 135). In short, Ritschl was a Hermannian with growing sympathies towards antiquarianism; Jahn was a student of Boeckh but became personally linked to the Hermannians. When Ritschl gave his commencement address at Leipzig in October 1865, with a young Nietzsche in tow, he entered the lecture hall as a conquering hero for the 'antiquarian' school. But in the eyes of Lachmann, Haupt, and now Jahn, Ritschl had conducted something of a coup; and any student who followed him from Bonn to Leipzig, much less one who was signalled out by Ritschl at his inaugural address, would have been cast in the same light.

What transpired politically affected the scholarly ideals held by all parties. Ritschl attempted but largely failed to embrace two traditions of scholarship that at the time were not to be commingled. Jahn, in the time he had left, moved closer to the Hermannians and impressed upon his students a definite distaste for the methodologies and personality of Ritschl—and among those students at Bonn was Wilamowitz. Nietzsche, whose motivations for following his teacher were more personal than philological,[13] was at first hastily regarded in the same light as Ritschl. But Nietzsche would not remain a disciple for very long (a fact Ritschl recognized and lamented) but would in his first two books, with a powerful new voice, reject both traditions on the way to positing a third way of his own.

Nietzsche's Untimely Response to Wilamowitz

What in Ulrich von Wilamowitz-Moellendorff's now infamous *Zukunftsphilologie!* review was labelled the *πρῶτον ψεῦδος* [first falsehood] is Nietzsche's presumption that the music of Richard Wagner could contribute to our understanding of antiquity.[14] As Hermann and Lachmann had argued a generation earlier, a comprehensive understanding of scholarly emendations and textual analyses was prerequisite for philological re-

12 Nietzsche understood well the significance of Hermann's influence at Leipzig. See Nietzsche to Hermann Mushacke, 30 August 1865, KSB 2, no. 478, p. 81.

13 See Nietzsche to Carl von Gersdorff, 25 May 1865, KSB 2, no. 467, p. 56.

14 'His [Nietzsche's] solution is to belittle the historical-critical method, to scold any aesthetic insight which deviates from his own, and to ascribe a 'complete misunderstanding of the study of antiquity' to the age in which philology in Germany, especially through the work of Gottfried Hermann and Karl Lachmann, was raised to an unprecedented height' (Wilamowitz-Moellendorff 2000, p. 4).

search—what could a musician know of this? Nietzsche, it is further claimed, did not engage in any sustained source criticism; he was more concerned with his 'intuitions'—*Anschauungen*—or 'glorious experiences' about the 'inner truths' of the tragic age. Completely unacceptable was his shamelessness, the arrogance of his self-proclaimed 'certainty of something directly apprehended [*unmittelbaren Sicherheit der Anschauung*]' (BT 1). Rather than a serious philological enquiry in the manner of Hermann, Nietzsche believed he could shortcut the scholarship for the sake of his 'direct intuitive faculty'—something more often propounded by a Romantic poet or novelist than a university professor.

We might defend Nietzsche against some of Wilamowitz's criticisms: for to say that Nietzsche did not engage in source criticism reveals that Wilamowitz could not have meant the works already published in *Rheinisches Museum*, one of which was even titled, *De Laertii Diogenis fontibus*, and a German version named, *Beiträge zur Quellenkunde und Kritik des Laertius Diogenes*. To say that Nietzsche was ignorant of the secondary scholarship completely overlooks the fact, which Wilamowitz surely knew, that Nietzsche himself was responsible for the creation of a 176-page catalogue indexing the first twenty-four volumes of the *Rheinisches Museum*.[15] Contra Wilamowitz, Nietzsche was thoroughly acquainted with the sort of 'critical scholarship' Wilamowitz found lacking in *The Birth of Tragedy*, and was, moreover, proficient in the critical techniques.[16] Yet this itself reveals that Nietzsche had consciously transferred his efforts *away* from a more critical philology in order to pursue a more magisterial pronouncement on the nature of tragedy—and it was especially this disengagement, from a chaired philologist no less, that attracted the

15 A commentary and publication history on this project can be found in Brobjer (2000, pp. 157–161).

16 Furthermore, the work of Hermann himself pervades Nietzsche's *philologica*. Nietzsche's 'Griechische Rhythmik' (KGW II.3, pp. 101–201), 'Aufzeichnungen zur Metrik und Rhythmik' (KGW II.3, pp. 205–261), 'Zur Theorie der quantitirenden Rhythmik' (KGW II.3, pp. 267–280), and 'Rhythmische Untersuchungen' (KGW II.3, pp. 285–338) were each heavily indebted to Gottfried Hermann's *De metris poetarum graecorum et latinorum* (Hermann 1796), *Elementa doctrinae metricae* (Hermann 1816), and *Epitome doctrinae metricae* (Hermann 1818). Hermann also edited most of the canonical editions of and commentaries on the works of Aeschylus and Sophocles that were available to Nietzsche at that time, and Nietzsche drew freely on their contents in his Basel lectures, 'Prolegomena zu den Choephoren des Aeschylus' (KGW II.2, pp. 1–30), and 'Kommentar zu den Choephoren' (KGW II.2, pp. 45–104).

young Wilamowitz's invective.[17] In his eyes, Nietzsche's transformation was even worse than Ritschl's; for whereas Ritschl had tried to embrace two traditions, Nietzsche seemed to ignore the only one Wilamowitz regarded as valid. Whereas Wilamowitz—who was the last disciple of the then recently departed Jahn and who thereafter obtained his doctorate under Haupt—fancied himself at this time to be the defender of the critical school,[18] Nietzsche—the favourite of the defector Ritschl—was *seen* as an apologist for the antiquarians. But for various reasons, the remaining antiquarian scholars did not embrace Nietzsche, nor would Nietzsche ever really embrace them. As a result, Nietzsche felt himself almost wholly shut out of both scholastic factions. Any adequate response to the 'conspiracy', I think Nietzsche believed, would require him to disclaim both sides and to effectively demarcate his own position on the proper aims of historiography.

While Rohde was induced into a published response, and while Richard Wagner eagerly sought to rebuff the 'Berlin bum' (Gründer 1969, pp. 57–65) for his 'literary Jewry',[19] Ritschl urged Nietzsche to write a 'strict scholarly response' to Wilamowitz.[20] This never came: Ritschl and Nietz-

17 'In any case, I want nothing to do with N, the metaphysician and apostle. Were he only this, I would not have bothered to appear as a "new Lycurgus" against this Dionysian prophet, because I would have then hardly encountered his revelations. Yet Mr. N is also Professor of Classical Philology. He engages some of the most important questions of Greek literature' (Wilamowitz-Moellendorff 2000, p. 4). For a more complete evaluation of Wilamowitz's objections, see Porter 2000, pp. 225–288.

18 Wilamowitz himself would later realize the arrogance of this attitude, writing in his autobiography, 'Have I taken up this struggle because I had perverted conceptions, crude errors, all in all philological sins to reproach [Nietzsche's theory] with? Or was it a tendency, as may perhaps be believed of me, to turn my efforts against *Anschauung* of art as a whole, against the method of science? No, there yawns an unbridgeable gulf here. To me, the highest idea is the unfolding of the world according to regular laws, full of life and reason. Gratefully do I look upon the great minds who, proceeding from level to level, have wrested out the world's secrets: with wonder do I seek to draw nearer the light of the eternally beautiful which art, in every different instance of its appearance, expresses in its special way; and in the science which fills my life, I strive to follow the path of those who free my judgement, because I have willingly given myself to their charge.' The letter is preserved in Gründer 1969, p. 134.

19 This was a common phrase shared between members of the Wagner circle. See, for example, Carl von Gersdorff to Nietzsche, 31 May 1872, KGB II 4, no. 326, p. 9.

20 See Ritschl to Nietzsche, 7 February 1872, KGB II 4, no. 335, p. 33.

sche remained silent in the journals.[21] Nevertheless, what I would like to suggest in what remains of this essay is that the opening chapters of 'On the Uses and Disadvantages' can at least in part be read as a specific type of response, that Nietzsche's criticisms of the critical and antiquarian schools, and his proposal of the perhaps fanciful 'monumental' mode of historicity, are to be viewed—again, at least in part—as a response to the indictment of Wilamowitz and to that whole tradition which he believed had spurned both himself and his teacher before him. Yet, if this is a response to specific scholarly trends, then it was intentionally not carried out as Ritschl had once advised—in a 'strict scholarly fashion'. Nietzsche never challenges Wilamowitz's many specific criticisms nor addresses any of his scholarly arguments. In fact, Wilamowitz is never once mentioned by name. Nietzsche's response is one that is both philosophical and psychological in one breath. It cuts more deeply because it targets the instincts and motivations of the historian, rather than simply those scholarly methods outlined here. It is in effect even more *ad hominem* than had been Wilamowitz's diatribe because it understands their methods as the practical extension of their character. For as Nietzsche believes, it was not just that Wilamowitz's criticism was misguided, but that the entire historical perspective from which it was issued was 'unhealthy', and the type of historian who was driven to it, 'degenerate'. This goes some ways towards explaining why Wilamowitz, Hermann, Boeckh, and even Ritschl and Jahn are not mentioned in the text of the second 'Untimely Meditation': Nietzsche feels compelled to reveal their advantages and disadvantages for life as psychological types rather than as scholars with the methodologies I have described here.

Even if these individuals are not singled out, they are present throughout Nietzsche's writing during the early 1870s. Turning to the critical historian, Nietzsche writes in 'On the Future of our Educational Institutions':

> Others, again, pass their lives in counting the number of verses written by Greek and Roman poets, and are delighted with the proportions 7:13 = 14:26. Finally, one of them brings forward his solution to a question, such as the Homeric poems considered from the standpoint of prepositions, and thinks he has drawn the truth from the bottom of the well with *ἀνά* and *κατά*. All of them, however, with the most widely separated aims in view, dig and burrow in Greek soil with a restlessness and a blundering awkwardness that must surely be painful to a true friend of antiquity. (FEI 3, KSA 1, p. 702)

21 At least part of the reason, which has been overlooked, is that shortly after Wilamowitz wrote his pamphlet he undertook a long-planned pilgrimage to Italy and Greece. He did not return until 1874, when the anger on both sides of the debate had somewhat cooled (see Fowler 1990, p. 492).

Now, he who is said to consider the Homeric poems from the perspective of its prepositions is Gottfried Hermann, who wrote on the Homeric hymns in 1806, whose dissertation was on the word *αὐτός*, and who wrote four entire volumes on the particle *ἄν*. But, 'What does the teaching of Greek particles have to do with the meaning of life?' (Nachlaß March 1875, KSA 8, 3[63]). As for the type that takes joy in discovering the hidden proportions of Greek and Roman verses, it was Karl Lachmann who counted among his greatest achievements the discovery that the total number of lines assigned to chorus and actors in tragedy was invariably divisible by seven.[22] Nietzsche labels them 'pedantic micrologists' (UM II 2, KSA 1, p. 258). Notice, however, that he does not quibble with any particular philological 'fact' here—he never disputes the numerical reductions or the applicability of *κατά*. It is ever only the spirit, drives, or intentions of these positivistic philologists that suffer his rancour: it comes down to their discipline's efficacy within educational institutions to shape the future of culture and society, to their discipline's value for life.

As Nietzsche says in the never-completed *Wir Philologen*, 'Those who say, "But certainly classical culture survives as an object of pure scholarship, even if all its educational aims are disavowed," deserve this reply: 'Where is pure scholarship here? Achievements and qualities have to be assessed, and the assessor has to stand above what he assesses. So your first concern must be to surpass antiquity. Until you do that, your scholarship isn't pure, but impure and limited' (Nachlaß Beginning of 1875–Spring 1876, KSA 8, 5[53]). Every type of scholarship must recognize its pedagogical dimension; what distinguishes them rests on a certain quality of character. These critical philologists tend to exhibit a lack of that grand and majestic taste required of the true philologist to create new, similarly grand idols to overcome, and are, Nietzsche thinks, thereby unable to assess the greatness of the Greek culture. The critical historian only tears down what others have built up: 'he does this by bringing it before the tribunal, scrupulously examining it and finally condemning it' (UM II 3, KSA 1, p. 269). Their destruction of the old antiquated world-views by means of source criticism and meticulous textual analysis is an advantage for life; their *Nachteil* is their 'wanton analytic drive' (UM II 6, KSA 1, p. 295) to reduce all philology to this destructive task. As Nietzsche writes in

22 Nietzsche certainly knew this, writing, 'So profoundly and frequently oppressive is the uncertainty in prediction that it now and then becomes a morbid passion for believing at any price and a desire to be certain: e.g., as concerns Aristotle, or in discovering numerical necessities—almost a disease in Lachmann' (Nachlaß Beginning of 1875–Spring 1876, KSA 8, 3[36]).

his *Encyclopädie*, 'Criticism in itself cannot be goal, but only the means to the *full understanding* [Kritik selbst kann nicht Ziel sein, sondern nur Mittel für das *Volle Verständniß*]' (KGW II.3, p. 375).

Unlike the critical type, who discredits the inherited constructions of antiquity for their lack of 'objective' critical analysis, the antiquarian scholar recognizes the interrelation of their personal world-views and their representations of historical topics. Their need or instinct to find and exposit a *Gesammtanschauung* is set in stark contrast to grammatical reductionism: theirs is an instinct towards artistic virtuosity, towards the production of a 'plastic apprehensible portrait' of the world. But reality, Nietzsche believes, especially the tangled web of history, does not allow representation of its comprehensive structuring without the intrusion of the artistic impulse of that active subject. More hermeneutically minded, they recognize the influence of their own ideals upon their historical presentations. 'The history of his city becomes for him the history of himself; he reads its walls, its towered gate, its rules and regulations, its holidays, like an illuminated diary of his youth and in all this he finds again himself, his force, his industry, his joy, his judgement, his folly and vices' (UM II 3, KSA 1, p. 265). This necessarily does 'violence' to what some call 'the facts', and affords us only a history that prohibits the designation 'objective' in the critical sense. But not only is this 'violence' a non-issue for Nietzsche, he endorses it with his stamp of 'necessity'. The individual has never been born, he thinks, who could represent the world in itself, unencumbered by an already determined cluster of epistemological and, more importantly, psychological categories. 'Thus man spins his web over the past and subdues it, thus he gives expression to his artistic drive—but not to his drive towards truth or justice' (UM II 6, KSA 1, p. 290).

In UM II, Nietzsche maintains that the 'antiquarian' serves life by adding value to what is inherently valueless, and in this respect he is better off than his critical counterpart. No aspect of the past has value in and of itself; value is only bestowed by the legislating activity of the historian. 'The small, truncated, decaying, and obsolete acquire their own dignity and inviolability through the fact that the preserving and revering soul of the antiquarian man has emigrated into them and made there a homely nest' (UM II 3, KSA 1, p. 265). These scholars consciously or unconsciously value the noble, the tranquil, and the scholarly; it is therefore no wonder that the image of antiquity they construct tends to highlight these aspects of the past. It is a prejudiced account, to be sure, but an honest sort of prejudice since they admit the intuitional status of their accounts: they recognize that their world-views are really *their* world-views. This creative activity is the healthy aspect of the illusion these historians have created for themselves.

But Wolf, Boeckh, and the rest of the 'antiquarian' school are not spared Nietzsche's venom either: they too have impulses that Nietzsche finds distasteful.[23] For the antiquarian type, present-day life stands in poor comparison with what he has elected to represent to himself of the past, and his turning back to some perceived 'good old days' (which, again, is the result of the philologist's creative intuition) carries the effect of turning him away from the present. Frustrated by his inability to render the present at all palatable and incapable of creating new idols for the future, he devotes his efforts to frantically preserving the glories of the past. The past and dead become the *only* sources of value, while what is to come can only ever be of lesser worth. His ideal of the classical reveals what, to Nietzsche, is a thoroughgoing 'mummification of life' (UM II 3, KSA 1, p. 268). It is no longer inspired by the fresh air of the present, much less the hope for the future. Nietzsche quips, 'For it knows only how to preserve life, not how to engender it; it always undervalues that which is becoming because it has no instinct for divining it—as does monumental history, for example' (UM II 3, KSA 1, p. 268).

Though again this is but summary, we turn now briefly to the third type of historian characterized in 'On the Uses and Disadvantages': the so-called monumental type. Note how Nietzsche, against the philological tradition, implicitly justifies the type of historicity already displayed in *The Birth of Tragedy*:

> If you are to venture to interpret the past you can do so only out of the fullest exertion of the vigour of the present: only when you put forth your noblest qualities in all their strength will you divine what is worth knowing and preserving in the past. Like to like! Otherwise, you will draw the past down to you. Do not believe historiography that does not spring from the head of the rarest minds. (UM II 6, KSA 1, pp. 293–294)

We see once again Nietzsche's tendency to regard 'types' of life rather than specific scholarly conclusions. Statements such as these rarely inform us as to how we should carry out the work, but do tell us what sort of scholar we should or should not be: the historian must be such a 'master', who from his own salutary conglomeration of instincts can intuit what is worth knowing and preserving in the past. What separates the real philologist from the mere philological labourer is not a degree of technical apti-

23 For example, Nietzsche writes about Wolf, 'Our terminology already indicates our tendency to misrepresent the ancients. For example, the exaggerated taste for literature —or Wolf, who, speaking of the "inner history of classical erudition", calls it "the history of learned enlightenment"' (Nachlaß Beginning of 1875–Spring 1876, KSA 8, 3[5]). Nietzsche's quotation is to F. A. Wolf, *Kleine Schriften* (Wolf 2003, vol. 1, p. 844).

tude; rather, it is that essential quality—elusive in the extreme—which makes one 'masterly' or 'great' or 'wise' (Pöschl 1979, pp. 141–155). This is the high perspective, the *Distanz* from which the monumental philologist does his work, devises his world-view. If antiquity is to be interpreted as a grand idol of the past which can and must be repeated another time in the present—something necessary if it is to be employed in the education of the young—then only the grandest souls of the present are capable of assessing it. In order to evaluate the classical, one must have *surpassed* the classical models themselves: the assessor must stand above what he assesses. '[H]istory is written by the experienced and superior man. He who has not experienced greater and more exalted things than others will not know how to interpret the great and exalted things of the past' (UM II 6, KSA 1, p. 294).

Just as the antiquarian scholar discovers in his antiquity the scholarly, the noble, and the tranquil, and is thereby trapped in that antiquity when all he sees around him is the worthlessness of the present, so too the 'monumental' historian artistically paints his own antiquity with a selective quality of judgement. 'The past itself suffers harm: whole segments of it are forgotten, destroyed, and flow away in an uninterrupted colourless flood, and only individually exaggerated facts rise out of it like islands' (UM II 2, KSA 1, p. 262). Like the antiquarian, too, the monumentalist finds nobility in the past, but only because he knows his evaluation was actually a legislation of that nobility. '[N]ow it would be right to say that only he who constructs the future has a right to judge the past. If you look ahead and set yourself a great goal, you at the same time restrain that wanton analytical drive' (UM II 6, KSA 1, pp. 294–295). For even this young Nietzsche, still heavily indebted to Schopenhauer's epistemology, the claim to represent 'antiquity in itself' is nonsense; facts of representation are never free from our evaluations. The past is only great 'in fact' because the great man has 'evaluated' it as such. Because he recognizes his interpreting as a creating, and that by interpreting the past as 'great' he has in fact created something 'great'—'a chain of moments in the struggle of the human individual which unites mankind across the millennia like a range of human mountain peaks' (UM II 2, KSA 1, p. 259)—the monumentalist becomes aware of the possibility of further creating such idols and exemplars for the future. He too discovers himself in his antiquity. The distinction drawn between the antiquarian and the monumentalist is thus not methodological but psychological, that is, not in their tendency to create holistic portraits of antiquity but in the manner in which they regard this picture. The present for the monumentalist leads not to the inescapable feeling of pessimism that characterizes the antiquarian whose optimism deserts him whenever he walks out of the library door, but to the cheerful recognition that the eternally becoming allows for the perpetual revaluation of what is to be considered classical.

Instead of 'mummifying' life, the monumentalist engenders it by acknowledging that something great can once again return to the present through his own activity. 'As long as the soul of historiography lies in the great stimuli that a man of power derives from it, as long as the past has to be described as worthy of imitation, as imitable and possible for a second time, it of course incurs the danger of becoming somewhat distorted, beautified, and coming close to free poetic invention' (UM II 2, KSA 1, p. 262), but despite the danger, this is where the pedagogical value of historiography lies—how *Historie* and not just *Geschichte* can be used for *das Leben*, how historians themselves can engender life. Such was the value of Nietzsche's own work in philology, a value misunderstood by both sides of the debate and by the then young Wilamowitz. As the last pages of the *The Birth of Tragedy* read, 'Let no one believe that the German spirit has lost its mythical home for ever, if it can still understand so clearly the voices of the birds which tell of its homeland. One day it will find itself awake ... then it will slay dragons, destroy the treacherous dwarfs, and awaken Brünnhilde—and not even Wotan's spear itself will be able to bar its path!' (BT 24, KSA 1, p. 154). Only with 'On the Uses and Disadvantages of History for Life' in mind can we rightly comprehend why a purportedly philological book about Greek tragedy should conclude with an exhortation to the German youth: this was a call to give rebirth to Nietzsche's classical ideal through the spirit of music. Not only is such a statement consistent with the monumental ideal of 1874, that ideal demands such a call to rebirth. And should we ask who the model for this 'reinvigorating' historical impulse is, we would be once again confronted with Nietzsche's other great 'master': Wagner—the same 'scholar' whom Wilamowitz jeered as the *πρῶτον ψεῦδος* of Nietzsche's philology, the same Wagner who responded to Wilamowitz's pamphlet with language all too familiar to that of the 'monumental' historian here. 'We [Wagner and Nietzsche] by ourselves look out from the mountain top over the wide plain without disturbance from the scuffling peasants in the tavern below us' (Gründer 1969, pp. 57–65). And so neither a Wolf, nor a Hermann, nor a Boeckh, nor even a Ritschl or a Jahn stand as the proper heirs to antiquity, but a Wagner, a Goethe, and ideally a Nietzsche himself—and these not for reasons of scholarly method nor styles of interpretation, but for psychological grounds, for the quality of character these are said to have, which are instinctually driven to value only the healthiest aspects of antiquity for the sake of reinvigorating culture.

There are both historical and philosophical problems with Nietzsche's conception of 'monumental historiography', to say the least, and partial solutions have now and again been proffered. I cannot address these issues here, and can only restate what I think is an important and overlooked aspect of Nietzsche's essay, namely, his engagement with the scholarly environment in which it was written. From what I have said, I hope it is clear that Nietzsche's account of the 'critical' and 'antiquarian' historians was not a purely theoretical construction, but was his attempt to outline a conflict with which he, as a prodigy scholar and philology professor, had up-close experience. Nietzsche never responded to the attack on his own philological work, but wrote an essay shortly after whose opening sections serve to critique both sides of the debate, not on philological grounds, but from the standpoint of his budding, psychology-laden philosophy. If true history is done by those who possess a certain greatness of character, as Nietzsche sought to demonstrate, then a Wilamowitz didn't qualify to contest a Nietzsche—or so at least he convinced himself.[24] By positing the rather fanciful ideal of the 'monumental' historian, I believe Nietzsche hoped to rise above the debate that ensnared his mentor, to thereby not only silence the grumblings of the petty 'conspiracy' against him, but to show both historians and philologists alike the sort of ends on which they should be focused. And if his ardour for these historical ideals waned with time, I think it is further evidence for my position to see that it did so in proportion to the growing distance that grew between him and his philological career. In the end, Nietzsche's resignation was plain: 'Wort- und Sach-Philologie—stupid quarrel!' (Nachlaß Spring–Summer 1875, KSA 8, 5[106]).

References

Brobjer, Thomas, 2000, 'Nietzsche's Forgotten Book. The Index to the Rheinisches Museum für Philologie', in: New Nietzsche Studies, 4/1, pp. 157–161.

Bursian, Conrad, 1883, Geschichte der klassischen Philologie in Deutschland von den Anfängen bis zur Gegenwart, 2 vols., Munich, Leipzig: Oldenbourg.

24 Notes from this period testify to this point. 'Contemporary philologists have proven themselves unworthy of being permitted to consider me and my book as one of their own. It is hardly necessary to affirm that, in this case as well, I leave it up to them whether they want to learn anything or not. But I still do not feel in the least inclined to meet them half way. May that which now calls itself "philology" [*Philologie*, written without its definite article] (and which I designate only neutrally on purpose) ignore my book this time as well. For this book has a manly temperament and is of no value for castrati. It is more seemly for them to be seated at the loom of conjecture' (Nachlaß Summer 1872–Beginning of 1873, KSA 7, 19[58]).

Calder III, William M., 1983, 'The Wilamowitz-Nietzsche Struggle. New Documents and a Reappraisal', in: Nietzsche-Studien, 12, pp. 214–254.

Campioni, Guiliano, 1975, 'Malattia dell'ideale e antistoricismo nel giovane Nietzsche', in: Annali della Scuola Normale Superiore Pisa, 5/4, pp. 1537–1566.

Fowler, Robert L., 1990, 'Ulrich von Wilamowitz-Moellendorff', in: Ward W. Briggs / William M. Calder III (eds.), Classical Scholarship: A Biographical Encyclopedia, New York: Garland, pp. 489–522.

Gerhardt, Volker, 1988, 'Leben und Geschichte. Menschliches Handeln und historischer Sinn in Nietzsches zweiter "Unzeitgemäßer Betrachtung"', in: Pathos und Distanz. Studien zur Philosophie Friedrich Nietzsches, Stuttgart: Reclam Verlag, pp. 133–162.

Gründer, Karlfried, 1969, Der Streit um Nietzsches 'Geburt der Tragödie': Die Schriften von E. Rohde, R. Wagner, U. v. Wilamowitz-Moellendorff, Hildesheim: Georg Olms Verlag.

Hermann, Gottfried, 1796, De metris poetarum graecorum et latinorum, 3 vols., Leipzig: Fleischer.

Hermann, Gottfried, 1816, Elementa doctrinae metricae, Leipzig: Fleischer.

Hermann, Gottfried, 1818, Epitome doctrinae metricae, Leipzig: Fleischer.

Müller, Carl Werner, 1990, 'Otto Jahn', in: Ward W. Briggs / William M. Calder III (eds.), Classical Scholarship: A Biographical Encyclopedia, New York: Garland, pp. 227–238.

Petersen, Eugen (ed.), 1913, Otto Jahn in seinen Briefen, Leipzig: Teubner.

Politycki, Matthias, 1981, Der frühe Nietzsche und die deutsche Klassik, Munich: Donau Verlag.

Politycki, Matthias, 1989, Umwertung aller Werte? Deutsche Literatur im Urteil Nietzsches, Berlin: Walter de Gruyter.

Porter, James I., 2000, Nietzsche and the Philology of the Future, Stanford: Stanford University Press.

Pöschl, Viktor, 1979, 'Nietzsche und die klassische Philologie', in: Hellmut Flashar, et al. (eds.), Philologie und Hermeneutik im 19. Jahrhundert. Zur Geschichte und Methodologie der Geisteswissenschaften, Göttingen: Vandenhoeck und Ruprecht, pp. 141–155.

Ribbeck, Otto, 1879–1881, Friedrich Wilhelm Ritschl. Ein Beitrag zur Geschichte der Philologie, 2 vols., Leipzig: Teubner, pp. 332–381.

Salaquarda, Jörg, 1984, 'Studien zur Zweiten Unzeitgemäßen Betrachtung', in: Nietzsche-Studien, 12, pp. 1–45.

Sandys, John Edwin, 1908, A History of Classical Scholarship, 3 vols., Cambridge: Cambridge University Press.

Siemens, Herman, 2004, 'Nietzsche and the "Classical". Traditional and Innovative Features of Nietzsche's Usage with Special Reference to Goethe', in: Paul Bishop (ed.), Nietzsche and Antiquity. His Reaction and Response to the Classical Tradition, Rochester, NY: Camden House, pp. 391–410.

Stambaugh, Joan, 1987, The Problem of Time in Nietzsche, trans. John F. Humphrey, Philadelphia: Bucknell University Press.

Ulfers, Friedrich / Cohen, Mark Daniel, 2004, 'Nietzsche's Ontological Roots in Goethe's Classicism', in: Paul Bishop (ed.), Nietzsche and Antiquity. His Reaction and Response to the Classical Tradition, Rochester, NY: Camden House, pp. 425–440.

Vogt, Ernst, 1979, 'Der Methodenstreit zwischen Hermann und Böckh und seine Bedeutung für die Geschichte der Philologie', in: Hellmut Flashar, et al. (eds.), Philologie und Hermeneutik im 19. Jahrhundert. Zur Geschichte und Methodologie der Geisteswissenschaften, Göttingen: Vandenhoeck und Ruprecht, pp. 103–121.

Vogt, Ernst, 1990, 'Friedrich Ritschl', in: Ward W. Briggs / William M. Calder III (eds.), Classical Scholarship: A Biographical Encyclopedia, New York: Garland, pp. 389–394.

Wilamowitz-Moellendorff, Ulrich von, 2000, 'Future Philology! A Reply to Friedrich Nietzsche's "The Birth of Tragedy"', trans. Gertrude Postl, in: New Nietzsche Studies, 4/1, pp. 1–32.

Wolf, Friedrich August, 2003, Kleine Schriften in lateinischer und deutscher Sprache, reprint of the Halle 1869 edn, ed. Gottfried Bernhardy, 2 vols., Hildesheim: Georg Olms Verlag.

Zuckert, Catherine, 1976, 'Nature, History, and the Self. Friedrich Nietzsche's Untimely Considerations', in: Nietzsche-Studien, 5, pp. 55–82.

'An Uncanny Re-Awakening': Nietzsche's Renascence of the Renaissance out of the Spirit of Jacob Burckhardt

Martin A. Ruehl

In the fall of 1895, the Catholic bi-weekly *Historisch-Politische Blätter* published a long article on 'Friedrich Nietzsche's Intellectual Development and Philosophy'. At a time when Nietzsche's writings were rapidly gaining in popularity and some of his concepts were becoming catchwords of various counter-cultural currents in fin-de-siècle Germany, the (anonymous) author of the article tried to enumerate the various trends in nineteenth-century thought that had shaped the ideas of the self-proclaimed 'untimely' philosopher. One of the intellectual influences he singled out was the Swiss cultural historian and one-time colleague of Nietzsche at the university of Basel, Jacob Burckhardt. In particular, he suggested that it was Burckhardt's colourful description of the ruthless, neo-pagan despots of early modern Italy in his 1860 book *The Civilization of the Renaissance in Italy* that had inspired Nietzsche's vision of a future race of 'violent men' (*Gewaltmenschen*) beyond the good and evil of Christian morality ('Friedrich Nietzsche's Intellectual Development' 1895, p. 871). Shortly after the appearance of the article, the Catholic historian Ludwig Pastor, who had recently begun to correspond with Burckhardt, told the latter of his dismay over these 'unreasonable suppositions' and proposed a rectification in the *Blätter* (Pastor 1950, pp. 289–290).[1] In his response of 13 January 1896, Burckhardt politely declined Pastor's offer. In view of his advanced age and poor health, he wrote, he preferred to keep his peace 'with all the world' and would refrain from a correction. His communications with Nietzsche had been 'serious and peaceful', yet infrequent, and about the *Gewaltmenschen* they had never actually discoursed. At any rate, he, Burckhardt, had never been 'an admirer of the violent men and outlaws [*Gewaltmenschen und Out-laws*] in history' and rather considered them to

1 On Burckhardt's late but warm acquaintance with Pastor see Kaegi 1982, vol. 7, pp. 165–182.

be 'flagella Dei' (scourges of God) whose psychological construction he 'gladly' left to others (Burckhardt 1986, vol. 10, p. 263).[2]

The letter to Pastor has become a *locus classicus* in the extensive literature on what might be called the Burckhardt–Nietzsche problem, that is, the long-standing scholarly debates about the nature of Nietzsche's relationship with Burckhardt during his Basel years (1869–1879); the question whether there existed a genuine 'congruence', as Nietzsche believed,[3] between their aesthetic and political convictions; and, finally, the extent to which the Swiss historian's reflections on the interplay of 'force and freedom' in civilizations past and present affected the philosophizing of his young German colleague.[4] In these debates, Burckhardt's denunciation of the *Gewaltmenschen* in his reply to Pastor almost invariably serves as a piece of textual evidence for commentators anxious to stress the general intellectual distance between his 'classical', liberal, and humanist worldview from Nietzsche's neo-Romantic, anti-bourgeois reflections on transgression and excess, as well as the more specific dissimilarity between their respective interpretations of the Renaissance.[5]

This essay, by contrast, takes its cue from the suspicion voiced in the *Historisch-Politische Blätter* that Burckhardt shaped Nietzsche's intellectual development to a considerable extent and that his conception of the Renaissance in particular had a profound impact on Nietzsche's philosophy. Its aim is to show that Nietzsche's understanding not just of the early modern period, but of history and time as such, drew on the idea of the Renaissance as formulated by Burckhardt.[6] Burckhardt's reading of the Renaissance, which will be examined below, conditioned Nietzsche's thinking on recurrence and change, the possibility of cultural renewal, and the sociopolitical parameters for a future overcoming of Christian 'slave morality'. Nietzsche's turn to the Renaissance, as a historical reference

2 Unless otherwise indicated, all translations from the German are my own.

3 See his letter to Erwin Rohde of 29 May 1869 (KSB 3, p. 13).

4 The most notable contributions are Andler 1926, Barth 1943, von Martin 1947a, Salin 1948, Rossi 1987, and Heller 1988. For a survey of the literature see Ruhstaller 1988.

5 See, e.g., von Martin 1947a, p. 139, Gay 1967, p. 198, Janssen 1970, pp. 221–222, Kaegi 1982, vol. VII, pp. 69–70, and Gossman 1999, p. 904. Gossman 2000, pp. 432–434, while pointing up shared concerns about the 'social question' as it posed itself in Basel since the 1860s, similarly stresses the fundamental differences between Burckhardt's *altliberal* humanism and Nietzsche's radically anti-democratic rejection of the masses as well as his 'excessiveness' and immoralism.

6 Burckhardt's influence on the development of Nietzsche's historical thought has been largely ignored in the relevant literature: see, e.g., Schlechta 1958, Brose 1973, Pletsch 1977, Maurer 1993, Brose 1994, Gilbert 2000, Brobjer 2007.

point and cultural ideal,[7] in the 1870s allowed him to question a set of values and notions that had determined his early thought: the Protestant inheritance from Röcken and Naumburg; the philhellenist belief, instilled in him at Schulpforta, Bonn, and Leipzig, in the absolute and exclusive model character of Greek antiquity; Schopenhauer's radically anti-historical philosophy of the will; and, most importantly perhaps, the medievalizing, neo-Romantic nationalism of Richard Wagner. However, the Renaissance also became a crystallization point, especially in the 1880s, for Nietzsche's most radically anti-humanist, anti-liberal ideas about tyranny and individuality, war and culture, violence and health. Burckhardt had good reasons to dissociate himself from the *Gewaltmenschen* glorified in Nietzsche's later writings—but the latter nonetheless bore a striking family resemblance to the tyrants and *condottieri* described in his book on the Renaissance.

1. The Dark Cradle of Modernity: Tyrants and Transgressors in Burckhardt's *Civilization of the Renaissance*

If Burckhardt's book provided a template for Nietzsche's reflections on the meaning of history and the possibilities for cultural revival, its depiction of the Italian Renaissance as a distinctive period of Western civilization, characterized by a secular individualism and neo-classical zest for beauty, was itself determined by previous attempts, both historiographical and fictional, to uncover the origins of modern subjectivity.[8] These attempts began with one of the foundational texts of the German *Sturm und Drang* movement, Wilhelm Heinse's popular multi-volume novel *Ardinghello* of

7 It is one of the contentions of this essay that the Renaissance represented not just an aesthetic concept for Nietzsche and that he conceived of early modern rulers like Frederick II and Cesare Borgia, whose image he culled largely from the first chapter of Burckhardt's book, as distinctively historical figures, *pace* Nehamas 1985, pp. 225–227, who interprets them as purely 'literary' characters. According to Nehamas, Cesare was little more than a fictional construct in Nietzsche's œuvre, without a genuine historical identity—and thus should not be misread as a model or type of the superman. As will be argued here, Cesare in particular and the Renaissance in general possessed a very definite historical significance for Nietzsche.

8 On the changing interpretations of the Renaissance in Germany before Burckhardt see Ferguson 1948, pp. 78–179, Stierle 1987 and Körner 1980. On the historiographical associations of the Renaissance with the birth of modern individualism see Baldwin 2001, esp. pp. 341–345.

1787.[9] Heinse's paean to the sensual, morally uninhibited life of artists and aristocrats in cinquecento Italy fundamentally shaped the idea of the Renaissance in the German literary imagination throughout the long nineteenth century.[10] In the decades following its publication, Romantic authors such as Ludwig Tieck glorified the unfettered egoism and 'aesthetic immoralism' (W. Brecht) of demonic Renaissance princes like the Duke of Bracciano.[11] In contradistinction to Heinse and the Romantics,[12] Goethe and Schiller projected an image of the Renaissance that stressed the 'responsibility of power' (G. Craig) and a classical, harmonious *Humanitätsideal* which Goethe saw realized in the works of Raphael, Mantegna, and even Cellini.[13]

While the ruthless, overreaching Renaissance despots of Romantic fiction often seemed modelled on Napoleon, the historiographical approaches to the Renaissance in the early nineteenth century were fundamentally indebted to the liberal, republican ideals of the Enlightenment and the French Revolution. Thus in the 1820s and 1830s Heinrich Leo and Carl Friedrich von Rumohr traced the notion of civic liberty back to the fourteenth-century Italian city-states which had proudly defended their independence against the encroaching Holy Roman Empire.[14] In the decades of

9 See Heinse 1998. On Heinse's conception of the Renaissance see Rehm 1924, pp. 61–78, and Ferguson 1948, pp. 128–131.

10 See the brief 'reception history' of the book in Heinse 1998, pp. 560–600. A number of literary critics around 1900 read Heinse's hero Ardinghello as an eighteenth-century precursor of the Renaissance *Herrenmensch* idealized by Nietzsche; see Heinse 1998, pp. 596, 598, 607, 610, 612.

11 See Brecht 1911, Rehm 1924, pp. 159–181, and Weibel 1925, pp. 44–54, 121–127.

12 See Jacobs 1998, p. 900: 'With its extreme idealization of the sensuous-ecstatic life, Heinse's *Ardinghello* was diametrically opposed to the striving for harmony and artistic autonomy that informed Goethe's image of the Renaissance.' See also Baeumer's comments in Heinse 1998, pp. 643–648, and Borcherdt 1949, pp. 149–166, who distinguishes between Heinse's 'Dionysian' and Goethe's 'Apollonian' conception of the Renaissance (p. 159).

13 See Craig 1967, pp. 125–144. On Goethe's idea of the Renaissance see Jacobs 1996 and Jacobs 1997. Baron 1960, p. 211 comments: 'In studying [the cinquecento artist Benvenuto] Cellini, Goethe had formed the idea of an age which had brought forth men of rare passions ... but also [of] higher yearnings: an honest respect for religious and ethical values ... and for noble enterprises.' According to Baron, Goethe's *Cellini* was an important source for *The Civilization of the Renaissance*, and the force of its psychological interpretation of the artist's self-formation 'is felt throughout Burckhardt's analysis of the development of the individual'. See also Janssen 1970, pp. 217–219.

14 See Leo 1832, vol. 3, pp. 378–387, vol. 4, pp. 1–36; von Rumohr 1920, pp. 126–222; and Ferguson 1948, pp. 127, 145–146.

political reaction following the Congress of Vienna (1814–1815), the educated bourgeois elites of Germany (*Bildungsbürgertum*) readily interpreted Leo's *History of the Italian States* (1829–1832) as a genealogy of their own emancipatory hopes.[15] The art historian Rumohr, in his *Italian Researches* (1827–1831), depicted the struggle for independence of the thirteenth-century communes as the backdrop both to the genesis of modern constitutional political theory and the great artistic revival of the Italian Renaissance.[16] The acclaimed *History of the Italian Republics in the Middle Ages* (1809–1818) by the Swiss-born political economist and historian Simonde de Sismondi, similarly, related the *énergie de liberté* sparked by participatory politics to the great outburst of cultural activity especially in the trecento. With the ascendancy of the tyrants and despots in northern Italy at the end of the fourteenth century, this *énergie*, according to Sismondi, began to wane and with it the flowering of Italian civilization.[17] The cultural vitality of the Renaissance cities, for both Sismondi and Rumohr, was intimately connected with their republican liberty.

Burckhardt's *Civilization of the Renaissance* marked a profoundly ambivalent intervention in these representations of early modern Italy. On the one hand, it contributed to and reinforced the liberal idealizations of the Renaissance typical of the first half of the nineteenth century. Like so many bourgeois intellectuals of his age, Burckhardt viewed the Italian quattrocento, which he famously labelled the 'mother and home of modern man' ('Mutter und Heimat des modernen Menschen'),[18] as a dress rehearsal for the civil society of contemporary Europe: an energetically meritocratic world of atomized individuals competing with one another on an equal basis, without regard for traditional religious, social, and moral constraints. In that respect, Burckhardt's Renaissance Men bore a striking resemblance to the early capitalists described in Marx's *Communist Manifesto*, published twelve years earlier: both were secularizers, rationalizers,

15 See Leo 1832, vol. 4, pp. 138–420.

16 See von Rumohr 1920, pp. 178–222.

17 See Simonde de Sismondi 1826, vol. 7, pp. 351–395, and vol. 8, pp. 2–6. On Sismondi's interpretation of Renaissance culture see Ferguson 1948, pp. 165–168, and Bullen 1994, pp. 38–59.

18 This is how Burckhardt described the Renaissance in his letter of May 1858 to King Maximilian II of Bavaria, see Kaegi 1982, vol. 3, p. 664. Burckhardt 1988, p. 3, calls the Renaissance the 'mother' of modern civilization. Gilbert 1990, p. 61, counts thirty passages in which Burckhardt 'identifies the Italy of the Renaissance with the modern age'. Baron 1960, p. 213, remarks that '[n]o other *leitmotif* occurs as often in the text [i.e., *The Civilization of the Renaissance in Italy*] as the contention that the Italian of the Renaissance "was the firstborn among the sons of modern Europe"'.

and demystifiers, pioneering self-made men, as efficient as they were ruthless.[19]

On the other hand, *The Civilization of the Renaissance* was a forceful critique of the emancipatory, progressivist interpretations of the Renaissance formulated by Sismondi, Leo, and Rumohr.[20] For Burckhardt, the sociopolitical corollaries of modernization—initiated and epitomized, in his eyes, by the French Revolution—had had a fateful effect on contemporary civilization. Writing in the aftermath of 1848, he believed that the revolutionary unrest of the Fourth Estate posed as much of a threat to the cultural legacy of 'old Europe' (*Alteuropa*) as the bourgeoisie's complacent desire for 'security' (*Sekurität*).[21] He tried to imagine the Renaissance as the beginning of an alternative modernity, one quite distinct from his own mundane, unheroic 'modern age', which in his eyes was defined by the crass materialism, timid acquiescence, and soulless scientism of the middle class, on the one hand, and the proletariat's increasingly vociferous demands for political participation and social justice on the other. As a consequence, Burckhardt, in contrast to previous liberal historians of the early modern period, associated the new secular, individualist spirit of the Renaissance not so much with merchants, scholars, or artists, but with military leaders and despots like the Sforza and Visconti, whose complete immoralism both fascinated and disturbed him.

By opening his book with a series of vivid vignettes recounting the cold-blooded machinations of despotic rulers (*Gewaltherrscher*) from Ezzelino da Romano to Cesare Borgia, Burckhardt made the tyrant the embodiment of what he regarded as the two essential features of Renaissance civilization: the 'objective judgement and treatment of ... all the things in the world' (*objektive Betrachtungsweise und Behandlung ... der sämtlichen Dinge dieser Welt*) and the 'development of an autonomous personality' (*Entwicklung der auf sich selbst gestellten Persönlichkeit*), freed from

19 Their new, problematic sense of self also resembles that of the early Puritans described in Max Weber's *Protestant Ethic and the Spirit of Capitalism* (1904/1905): see Hardtwig 1996, pp. 170–180.

20 See Kahan 2003. Kaegi 1962, pp. 133–134, argues that refuting Sismondi's 'republican' interpretation of the Renaissance was one of the 'principal aims' of Burckhardt's book: 'Burckhardt saw the flowering of the Renaissance not in the context of the Italian city-states struggling for liberty, but against the dark backdrop of the demonic concentration of power in the *signorie*, of liberty lost.'

21 For his critique of bourgeois *Sekurität* see Burckhardt 1982, pp. 282–283, 236–237. On Burckhardt's almost pathological fear of a proletarian revolution see Wenzel 1967, pp. 25–32. That Burckhardt wrote *The Civilization of the Renaissance in Italy* in response to what he perceived as a time of sociopolitical unrest and cultural crisis is powerfully argued in Gossman 1994, pp. 409–427.

medieval corporatism and religious paternalism (Burckhardt 1988, pp. 5, 99). Although he also examined the emergence of this new 'realist' politics and individualist ethos in the context of republican city-states like Florence and Venice, Burckhardt evidently considered the tyrants to be the first and foremost incarnations of the new 'thisworldly' mind-set of Renaissance Italy. Thus he remarked of the Hohenstaufen emperor Frederick II (1194–1250), whom he called a 'model' (*Vorbild*) for the later despots, that he had accustomed himself early in his life to an 'utterly objective view of world', which made him 'the first modern man on the throne'. The foreign policy of Maria Galeazzo and Lodovico il Moro he described in similar terms, as a 'completely objective treatment of international affairs, free of prejudices and moral qualms' (Burckhardt 1988, pp. 4, 5, 67). What Burckhardt identified as the distinctive aspect of Renaissance politics was not so much the brutality as the strictly *realpolitisch* approach of the *condottieri* and princes, their uninhibited, yet carefully calculated, deployment of force. This was *ragione di stato*—Machiavelli, significantly, featured as Burckhardt's chief theoretical witness for the chapters on tyrannical rule—a perfectly pragmatic form of politics fundamentally at odds with the abstract, 'artificial' power structures of feudal northern Europe (Burckhardt 1988, p. 72).[22]

The new brand of politics practised in the petty despotic states of Italy set the stage for what Burckhardt considered to be the second great contribution of Renaissance civilization to the genesis of modernity: the emergence of the autonomous individual. Exploding the medieval system of rank and inheritance, the tyrants, according to Burckhardt, found themselves in an unusually volatile predicament. Unable to rely on what Max Weber would later call 'traditional authority',[23] princely bastards like Ferrante of Aragon and military leaders of humble social origins like Francesco Sforza had to depend entirely on their own talents in their bid for political power. They created their state just as they created their own identity: as if *ex nihilo*, as a 'work of art' (*Kunstwerk*). The illegitimacy of their rule and the radical 'insecurity' (*Garantielosigkeit*) of their existence forced them to develop virtuoso personalities.[24] But in the shadow of their

22 See also Burckhardt 1988, pp. 12–13. But cf. Hale 1973, p. 63, who argues that Burckhardt greatly overestimated the degree of rationality and centralization in Renaissance politics.

23 See Weber 1978, pp. 226–241.

24 In his lectures 'On the Study of History', which Nietzsche attended in the winter semester of 1870/1871, Burckhardt described Napoleon in a strikingly similar fashion as 'unpredictability incarnate' (*die Garantielosigkeit in Person*): Burckhardt 1982, p. 397. For Nietzsche's idealization of Napoleon as a *Gewaltmensch*

rule, the 'individualization' of their courtiers and vassals also received a powerful 'stimulus' (Burckhardt 1988, pp. 4, 7–8, 101). The 'sovereign subject' of modernity thus emerged, paradoxically, against the backdrop of political unfreedom.

Whereas Sismondi and Rumohr had stressed the connection between 'liberty and letters' and praised the republican ethos of the city-states as the necessary political context for the 'revival of the arts', Burckhardt argued that the productivity and excellence of the Renaissance artists was not only compatible with, but actually enhanced by, the violent politics of the tyrants. Raphael's early paintings of St George (c. 1504) and St Michael (c. 1505), he remarked, could have been inspired by the bloody street fighting between the Baglione and their enemies in Perugia, where the young artist had worked as an apprentice in the 1490s (Burckhardt 1988, p. 24).[25] For Burckhardt, the tyrants were more than just great patrons of Renaissance art and science. There existed a genuine elective affinity, he believed, between the despots, with their virtuoso personalities and plastic political skills, and 'those who also thrived by dint of their own talent: the scholars, poets, musicians and artists'. Leonardo da Vinci's extended stay at the court of Lodovico il Moro was evidence that 'a higher element was alive' in the tyrant. That Leonardo subsequently served Cesare Borgia, similarly, suggested his appreciation of the latter's 'extraordinary nature' (Burckhardt 1988, p. 33).

Insofar as they tore away the medieval 'veil' of Christian beliefs and feudal power structures and helped to launch the laical, scientific 'discovery of the world and of man' (Burckhardt 1988, pp. 99, 203),[26] Burckhardt's tyrants seem to anticipate the secular, emancipatory ideals of the nineteenth-century *Bürger*.[27] The values they represent—rationality, pragmatism, individual talent, competitive struggle—belong to the catechism of what Adorno termed the 'bourgeois religion of success' (Adorno 1970, p.

---

and *Herrenmensch* in the tradition of the Hohenstaufen emperor Frederick II see von Martin 1947a, pp. 148–160.

25 According to Burckhardt 1988, p. 24, the figure of the heavenly horseman in Raphael's *Expulsion of Heliodorus from the Temple* (1511–1512) was modelled on Astorre Baglione. For a differing interpretation of the painting cf. Traeger 1971, pp. 31–34.

26 See also Burckhardt 1988, pp. 16 and 72, where the meritocratic individualism of the Italian princes is favourably contrasted with the feudal 'class prejudice' (*Kastenhochmut*) of Northern Europe. Burckhardt borrowed the expression 'the discovery of the world and of man' (*die Entdeckung der Welt und des Menschen*) from Jules Michelet's *Histoire de France*, see Michelet 1883, vol. 3, p. 3.

27 See Gay 1967, p. 184, who describes Burckhardt's Renaissance men as 'human types that might be walking through nineteenth-century cities'.

53). In that respect, they were indeed the 'first-born among the sons of contemporary Europe'. Yet Burckhardt also constructed them as anti-types of the modern bourgeois.[28] His lively, detailed descriptions of their 'colossal crimes' and 'endless atrocities' established a stark contrast between the violent, immoral universe of the Italian Renaissance and the *Biedermeier* propriety of nineteenth-century Central Europe. With an almost Gothic literary sensibility, he evoked the realm of the tyrants' courts as one of constant deception, danger, and dread. Theirs was a 'monstrous' (*ungeheuer*) and 'uncanny' (*unheimlich*) world far removed from the comfortable, orderly life and utilitarian concerns of civil society (Burckhardt 1988, pp. 6, 99, 11, 10).[29]

Like the authors of the *Sturm and Drang* and the Romantic period, Burckhardt was fascinated by the transgressive elements of Renaissance civilization.[30] His tyrants embodied more than just a 'worldly' individualism: they were Faustian overreachers, 'godless' (*gottverlassen*), at times demonic characters, full of 'daring profanity' (*Frevelmut*) and diabolical genius. Of Cesare Borgia, whose inhumanity ultimately seems to have repelled him, he wrote that his cruelty took on a 'completely satanic character' (Burckhardt 1988, pp. 10, 26, 38). Yet Burckhardt largely refused to apportion moral blame even to the most blasphemous tyrannical deeds. The apostate son of Basel's chief Protestant minister, whose 'anti-Christian sentiment' was notorious in his hometown,[31] he more or less suspended judgement on the despots and related their crimes with the same cool objectivity for which he praised Machiavelli. These crimes, he argued, were

28 Kaegi 1932, p. xxx, calls them 'bogeymen of the bourgeoisie' (*Bürgerschrecke*).

29 For more instances of the Gothic in his depiction of the Renaissance tyrannies see Burckhardt 1988, pp. 11, 20, 26, 28, 33. See also Kaegi 1982, vol. 3, pp. 710–711.

30 See Janssen 1970, pp. 11–15, 217–223. In this respect, *The Civilization of the Renaissance* echoes not just Heinse's *Ardinghello*, but also works of Romantic fiction like Hoffmann's *Elixiere des Teufels* (1815) and Tieck's *Vittoria Accorombona* (1830); see Rehm 1960, pp. 19, 54–64, and Rehm 1924, p. 69.

31 Gelzer 1907, p. 340, reports that the young Burckhardt was imbued with an 'almost fanatical anti-Christian animus'. It should be noted, however, that in the 1870s, Burckhardt's attitude to Christianity, and to Catholicism in particular, changed, partly, it seems, in response to the experience of Bismarck's *Kulturkampf*; see his letter to Max Alioth of 12 May 1889 (Burckhardt 1986, vol. 9, p. 185). Stadelmann 1930, p. 504, argues that under the impact of the *Kulturkampf*, Burckhardt came to 'appreciate Catholicism as a harbour of liberty' for all things intellectual that were threatened by the 'brutality of state power'. On Burckhardt's changing attitude towards Christianity see Ernst 1948, Zeeden 1954, and Howard 2000, pp. 110–170; for a somewhat different view see von Martin 1947a, pp. 131–133 and von Martin 1947b, esp. pp. 18–22, 39–53, 155–216.

the expression of an essentially 'naive' amoralism that was, in the last instance, beyond 'ethical judgement' (Burckhardt 1988, pp. 15, 40). The development of the new political and individual entities of Renaissance Italy, for him, was ultimately an aesthetic phenomenon and had to be valuated accordingly.

Burckhardt's Renaissance Man, thus, was something quite different from the harmonious, classically *gebildet* individual idealized by Goethe and Schiller. In contrast to the Weimar classicists and a number of influential nineteenth-century historians like Georg Voigt,[32] he did not consider the 'revival of classical antiquity' to be the defining feature of the Italian Renaissance. According to Burckhardt, humanism played but a secondary, or indeed tertiary, role in the making of modern subjectivity.[33] The process of individualization, for him, was shaped not so much by neo-Platonism or the Ciceronian ideal of *humanitas*, but altogether more worldly factors, chief among them, as we have seen, the violent politics of the north Italian tyrannies. Burckhardt's identification of the Renaissance with the birth of a new 'autonomous' personality, consequently, was both a contribution to the 'bourgeois religion of success' and a forceful attack on the neo-humanist notions of selfhood so dear to the German-speaking *Bildungsbürger*. His idea of 'tyrannical self-fashioning', similarly, challenged the traditional liberal associations between individualism and political participation, autonomy and security, self-cultivation and the private sphere—just as his depiction of the despots as congenial patrons and catalysts of the Renaissance artists called into question the 'civic humanist' association of 'liberty and letters'.[34]

---

32 See Voigt/Lehnerdt 1893. On Voigt's conception of the Renaissance see Ferguson 1948, pp. 159–163, and now Todte 2004.

33 It is telling that Burckhardt turns to the revival of arts and letters at the hand of the humanists only in the third part of his Renaissance book, that is, after the long opening section on 'The State as a Work of Art' and the treatment of 'The Development of the Individual' in section 2. He begins his survey of Italian humanism—see Burckhardt 1988, p. 171—with a categorical qualification of the significance hitherto attached, in histories of the Renaissance, to the revival of antiquity, insisting that all the major cultural and intellectual transformations in early modern Italy would have taken place 'without it'.

34 This interpretation of the Italian Renaissance, which goes back to Adam Ferguson's *Essay on the History of Civil Society* (1767), is now generally associated with the work of Hans Baron, who developed his notion of 'civic humanism' or *Bürgerhumanismus* partly in response to Burckhardt's claim that Renaissance culture had first flourished at the tyrannical courts. For Baron, it was born out of the politically engaged, republican spirit of certain Florentine humanists in the period around 1400, when the city-state struggled against the Visconti of Milan: 'The

The Civilization of the Renaissance, while continuing earlier master-narratives of emancipation and secularization, hence marked an intervention not just in the interpretations of the Renaissance since Heinse and Goethe, but also in topical, post-revolutionary debates about the nexus between *Geist* (intellect) and *Macht* (political power), morality and greatness, culture and violence. In the age of 'blood and iron' and especially after the foundation of the Second German Empire in 1871, this intervention would take on a new significance for bourgeois intellectuals wondering about the fate of the German *Kulturnation* in Bismarck's recently established *Nationalstaat*.

2. 'The Golden Age of the Millennium': Nietzsche's Uses of the Renaissance contra Wagner and Luther

One of these intellectuals was Friedrich Nietzsche. While Nietzsche's bitter attacks on the culture of Bismarck's *Reich* have received much scholarly attention,[35] relatively little is known about the impact that Burckhardt's conception of the Renaissance had on his *Kulturkritik*.[36] And yet throughout the 1870s and 1880s, the 'idea of the Renaissance', which he selectively adopted, as we shall see, from Burckhardt's book, proved to be a powerful inspiration for Nietzsche's denunciation of the liberal, *kulturprotestantisch* ethos of the Second Empire which he regarded as one of the main reasons for the decadence of contemporary German culture.[37] But Burckhardt's Renaissance also prompted him to challenge the Germanic ideology underlying Wagner's music dramas, to overcome Schopenhauer's ahistorical pessimism and to rethink the sociopolitical framework within which the future transvaluation of values and the revival of European civilization could come about.

---

places which held cultural predominance in the first decades of the *Quattrocento* were not as yet the seats of the tyrants, later to become famous, but rather the remaining city-state republics led by Florence': Baron 1966, p. xxv. On the genesis of Baron's concept of 'civic humanism' see Fubini 1992, Hankins 1995, Hankins 2000, and Ladwig 2004, pp. 278–360.

35 See, e.g., Schieder 1963, Kaufmann 1974, pp. 121–178, Bergmann 1987, pp. 81–107, and Gossman 2000, pp. 413–439.

36 Even an otherwise astute analysis of the various factors contributing to the formation of Nietzsche's anti-modern *Geschichtsbild* in Basel like Cancik 1995, pp. 23–34, largely ignores Burckhardt's role in this process. But see the perceptive comments in Ross 1980, pp. 312–319, Hofmann 1971, and Large 2000.

37 On Burckhardt as a critic of the Second Empire and its culture see Ressing 1951, Zeeden 1963, and Gossman 2000, pp. 439–442, 445–447.

That Nietzsche was familiar with Burckhardt's conception of the Renaissance is beyond doubt. His personal library contained two copies of *The Civilization of the Renaissance*, both second editions of 1869, the year he arrived in Basel to take up the chair in classical philology at the city's distinguished university.[38] One of these copies must have been a present from Burckhardt himself: its title page bears a short but amicable inscription 'to Prof. Dr. Nietzsche' by the author.[39] Both are heavily marked in Nietzsche's hand, especially the first three sections, entitled 'The State as a Work of Art', 'The Development of the Individual', and 'The Revival of Antiquity'. An entry in the diaries of Cosima Wagner reveals that Nietzsche sent the Wagners a copy of the book as early as December 1870.[40] In the summer semester of 1871, at any rate, he presented a series of lectures on 'The History of Classical Philology' (*Enzyclopaedie der klassischen Philologie*), the first of which, entitled 'The Discovery of Antiquity in Italy', drew heavily on Burckhardt's *Civilization of the Renaissance.* The lecture manuscript shows that Nietzsche adopted—often verbatim—some of Burckhardt's central arguments, for instance the notion that it was the 'Italian national genius' (*italienischer Volksgeist*) that had corroded the feudal ties of the Middle Ages or that there existed 'innermost affinities'

38 See Brobjer 1997, pp. 691–692.

39 The copies are preserved in Nietzsche's private library, which is now part of the Herzogin Anna Amalia Bibliothek in Weimar. They are listed as items C482a and C482b, respectively. The former bears the following inscription on the title page: 'Herrn Prof. Dr. Nietzsche in Verehrung dargebracht vom Verf.[asser]'. For Nietzsche's markings see C482a, esp. pp. 106–110, 112, 421; and C482b, pp. 136–139, 141, 147, 149, 154–155, 163, 171, 174, 197–198, 212–215. Nietzsche's library contains a number of other works on the Italian Renaissance, most notably Émile Gebhart's *Études méridionales. La Renaissance italienne et la philosophie de l'histoire*, Paris 1887, and Albert Trolle's *Das italienische Volkstum und seine Abhängigkeit von den Naturbedingungen*, Leipzig 1885. That his conception of the Renaissance was nonetheless indebted primarily to *The Civilization of the Renaissance* is suggested not just by the much more expansive markings and marginalia in the latter, but also by the fact that Gebhart's book itself drew heavily on Burckhardt. Brobjer 1995, p. 81, n. 37 argues that Nietzsche's later transvaluation of virtue as Machiavellian *virtù* or 'virtue free of moralistic acid' (*moralinfreie Tugend*, A 2) was inspired by Gebhart, not Burckhardt. In a footnote to the first section of his book, however, Burckhardt describes Machiavelli's notion of *virtù* in a way that is perfectly congruous with Nietzsche's subsequent use of the concept, namely, as a 'synthesis of force and talent' that is 'compatible with *sceleratezza*': Burckhardt 1988, p. 409.

40 See Wagner 1977, vol. 1, p. 320 (4 December 1870): 'Prof. Nietzsche sends Burckhardt's book on the Renaissance.' See also Borchmeyer/Salaquarda 1994, vol. 1, p. 109.

(*die innersten Beziehungen*) between the despots and the scholars residing at their courts (KGW II.3, pp. 348, 350).[41]

Although Nietzsche was thus evidently well acquainted with Burckhardt's interpretation of the Renaissance in the summer of 1871, he concealed this knowledge very skilfully in his first major philosophical work. *The Birth of Tragedy*, in fact, presented an image of the Renaissance that was decidedly at odds with Burckhardt's. It was indebted almost entirely to Richard Wagner, who in the early 1870s exerted a strong influence on Nietzsche's ideas about ancient Greece and its cultural legacy. For Wagner, Renaissance Italy was a 'corrupt' world, imbued with a superficial aestheticism whose dissemination into the north proved 'detrimental' to the development of a genuine German *Kultur*.[42] The Renaissance humanists' attempt to revive classical antiquity had been an abject failure, according to Wagner, because they lacked a true understanding of the tragic nature of ancient Greek civilization and their thinking was perverted by the villainous rulers they served.[43] In *The Birth of Tragedy*, Nietzsche denounced the Renaissance in very similar terms, dismissing quattrocento humanism as a shallow 'theoretical' imitation of antiquity and Renaissance civilization in general as a 'false idyll' constructed by 'Socratic' men (BT 19).[44] It may

41 On Nietzsche's liberal borrowings from *The Civilization of the Renaissance* in his lecture manuscript see Campioni 1998, pp. 96–102, and Volpi 1999.

42 See Wagner 1977, vol. 1, p. 506 (2 April 1872): 'At table, he [i.e. Richard Wagner] rails against the Renaissance, saying that it did enormous damage to the Germanic development; this age showed as little appreciation of antiquity as of Christianity; men of prodigious talent placed themselves in the service of a power that corrupted everything; and as always, the naïve Germans let themselves be so impressed by a foreign civilization that their own feeling nearly perished'. See also Wagner 1977, vol. 2, p. 617 (3 November 1880), where Cosima mentions further 'invectives against the Renaissance', and vol. 2, pp. 836–837 (2 December 1881), where she reports Wagner's 'disgusted' reaction to the 'pernicious' eagerness of Renaissance artists to 'make everything look beautiful' and to 'avoid harshness' (*das Herbe*). On Wagner's repudiation of the Renaissance see Campioni 1998, pp. 88–91.

43 See Wagner 1977, vol. 2, p. 287 (10 January 1879): 'A modern man like Machiavelli ... cuts a poor figure in comparison [to the ancient Greeks]; what a corrupt world formed the background to his being!'

44 'Imitative' and 'decorative' are typical terms of abuse in Wagner's diatribes against Renaissance art and civilization; see, e.g., Wagner 1977, vol. 1, p. 1002, and vol. 2, pp. 621, 682, 867, 933. That Nietzsche was aware of Wagner's distaste for the Renaissance is evidenced by his notes for the fourth *Untimely Meditation* (Nachlaß Beginning of 1874–Spring 1874, KSA 7, 32 [58]), in which he meditates on the composer's 'ambition' to measure himself against great figures of the past like Goethe, Beethoven, Luther, and the Greek tragedians: 'only to the Renaissance could he not relate' (*nur zur Renaissance fand er kein Verhältnis*).

have been out of consideration for Wagner, who closely followed the composition of the book, that Nietzsche refrained from using the Latinate term 'Renaissance', referring instead to the imminent revival of Greek culture through Wagner's music as a *Wiedergeburt* or 'rebirth' (BT 16).[45] In accordance with Wagner's *nationalprotestantisch* and profoundly anti-Roman views, Nietzsche identified the great moment of spiritual emancipation in European history not with the Renaissance, but with the Reformation. It was out of Luther's choral, he remarked, that the music of Bach, Beethoven, and Wagner was born (BT 23).

These critical remarks contrast with Nietzsche's altogether more positive, 'Burckhardtean' assessment of the Renaissance in the lectures on classical philology. The fact that the latter were delivered just as he was completing the manuscript of *The Birth of Tragedy* suggests a certain tension in his view of early modern Italy. While officially paying tribute to Tribschen, Nietzsche had already obtained a different perspective on the quattrocento, thanks to his new Basel associate. Basel, which had been a 'focal point for contact between German intellectuals and Italian ideas' in the early sixteenth century (Tracy 1968, p. 282), when the city hosted numerous renowned Renaissance scholars drawn to the circle around Erasmus and the humanist publisher Johann Froben, was still a vibrant centre of cultural exchange between northern and southern Europe in the 1870s. For Nietzsche, it provided an alternative *geistige Lebensform* (mode of intellectual existence),[46] a corrective to the heady mix of Nordic myths, Romantic medievalism, and patriotic pathos in Wagner's operas, which had cast a powerful spell on the young German classicist since he first heard the prelude to the *Mastersingers* in the fall of 1868.[47] Other resident Italophiles like Johann Jakob Bachofen also played a part in this emancipatory process,[48] but the major impulse came from Burckhardt, a sharp-

45 Nietzsche consistently speaks of a *Wiedergeburt der Tragödie*, *Wiedergeburt des griechischen Alterthums*, *Wiedergeburt der hellenischen Welt*, and so on; see Campioni 1998, p. 93. Gerhardt 1995, p. 153, remarks: 'Already in his first philosophical work [i.e., *The Birth of Tragedy*] Nietzsche expressed hope in a rebirth of tragedy out of the German spirit of music and thus—even though he refrained from using the term, out of respect for Wagner—a renaissance.' See also Hinz 1989.

46 The phrase is borrowed from Thomas Mann's 1926 lecture 'Lübeck als geistige Lebensform'; see Mann 1953.

47 On Nietzsche's early 'Wagnerianism' see Love 1963, Ross 1980, pp. 168–177, Janz 1978, vol. 1, pp. 246–252, and Borchmeyer/Salaquarda 1994, vol. 2, pp. 1278–1295. Kaegi 1982, vol. 7, p. 52, maintains that Burckhardt represented 'the most dangerous ferment' of Nietzsche's discontent with Wagner in the middle of the 1870s.

48 On Nietzsche's relation to Bachofen see Cesana 1994 and Cancik 1995, pp. 25–26.

tongued critic of German chauvinism (at least since the Wars of Unification) and Wagnerian music,[49] who quickly became a revered colleague, mentor, and *ersatz* master for Nietzsche.[50] Throughout the 1870s, while transforming himself into a proselyte of the *philosophes* and a cosmopolitan 'free spirit' (HA II 87),[51] Nietzsche used Burckhardt's Renaissance as a compass and signpost on his gradual retreat from Bayreuth. Wagner was aware of the role that Burckhardt had played in the apostasy of his former disciple.[52] 'People like Nietzsche', he remarked to Cosima in 1881, 'via the Renaissance man Burckhardt' (*durch den Renaissance-Mann Burckhardt*), had revealed their true colours when they identified themselves with 'odious' figures such as Erasmus and Petrarch (Wagner 1977, vol. 2, p. 837).[53]

49 Burckhardt reveals his dislike of Wagner in a letter to Max Alioth (24 July 1875), which alludes to the composer's 'lurid' (*grell*) and 'formless' (*herrenlose*) fantasy: Burckhardt 1986, vol. 6, pp. 42–43. Burckhardt 1986, vol. 5, pp. 43 and 183 and vol. 6, pp. 48–49, 81, 151–152, 192, denounces Wagner's music as a 'romantic swindle' and describes its oppressive, domineering effects on the listener, anticipating some of Nietzsche's later arguments contra Wagner, e.g., in HA II 3. Bergmann 1987, p. 95, misreads Burckhardt's letter to Friedrich von Preen of 31 December 1872 (see Burckhardt 1986, vol. 5, p. 183) as an expression of support for Wagner's Bayreuth project. For a more accurate assessment of his attitude towards Wagner and Wagnerianism see Kaegi 1982, vol. 7, pp. 40, 54, von Martin 1947a, pp. 44–45, 212–213, and Salin 1948, p. 54.

50 Even Elisabeth Förster-Nietzsche, a one-time associate of Bayreuth and one of the most influential propagators of the 'German' Nietzsche in the first third of the twentieth century, acknowledged the moderating impact that Burckhardt's francophile, cosmopolitan outlook had on her brother in the early 1870s (Förster-Nietzsche 1928, p. 38): 'Jakob [sic] Burckhardt surely exerted a great influence on my brother who always considered him [i.e. Burckhardt] a representative of Latin culture. Especially during the time of the [Franco-Prussian] War, when intellectual arrogance prompted many Germans to put down their victories ... to their "Bildung", Burckhardt was an excellent counter-weight [and allowed my brother] to view the world-historical events with a certain detachment, beyond German sensibilities. My brother had always embraced such supra-national views, but found it hard to hold on to them in those days, when even Richard Wagner (who at the time was his greatest and closest friend) got so carried away by the incredible euphoria in the wake of the proud victory... that he [i.e. Wagner] spoke out with bitterness and condescension against Latin civilization'.

51 See Campioni 1976.

52 See Ross 1980, p. 316: 'they [i.e., Richard and Cosima Wagner] knew who their opponent and rival was in Basel'.

53 See also the Wagners' objection to the 'arrogant, coldly critical tone' of Burckhardt's art-historical judgements in the *Cicerone* (à propos the *duomo* in Florence), in which they discerned 'traces' of his 'influence on Nietzsche': Wagner 1977, vol. 2, p. 589 (30 August 1880). A little less than a year later, Wagner decried 'the admirers of the Renaissance' as 'Jew lovers' (*Juden-Freunde*)—a curious charge,

Nietzsche's little essay on 'The Greek State', composed in 1871 and originally intended for inclusion in *The Birth of Tragedy*,[54] already hinted at his imminent departure from the Wagnerian evaluation of the Renaissance as a falsely optimistic, 'idyllic' revival of Greek antiquity. In this early piece, Nietzsche put the 'men of the Renaissance in Italy' on a par with the ancient Hellenes as 'political men par excellence' (CV 3, KSA 1, p. 771), imbued with violent, agonal instincts which he presented—this, too, was an implicit challenge to Wagner's more Winckelmannian, 'civic humanist' conception of ancient Greece—as necessary preconditions for the establishment of a great culture.[55] A few years later, in the second *Untimely Meditation* (1874), he went a step further and held up the men of the Renaissance as the exact opposite of cerebral Socratism. Far from being theoretical men, they were a powerful elite of practical individuals who had lifted the culture of early modern Italy on their strong 'shoulders' (UM II 2). In a little aside, he acknowledged the scholar who had drawn his attention to this first successful rebirth of the ancient world. The Renaissance, he remarked, had awakened 'once again the ancient Italian genius', thus producing 'a "wondrous echo of the ancient string-instruments", as Jacob Burckhardt puts it' (UM II 3).[56]

With the publication of *Human, All Too Human* (1878–1880), Nietzsche made his break with Wagner explicit. In this work, significantly, he offered his first sustained commentary on the historical and, to some extent, the personal import of the Renaissance. Redefining his own philosophy as a continuation and execution of the Enlightenment project, Nietzsche presented the Enlightenment as an extension of the Renaissance. The Renaissance, which he now hailed, quite unambiguously, as 'the golden age of the millennium', had comprised 'all the positive forces to which we

and one quite misplaced at least with regard to Burckhardt who was a convinced (if conventional) anti-Semite: Wagner 1977, vol. 2, p. 763 (16 July 1881). Burckhardt's hostility towards Jews and Jewish emancipation, which was intimately connected with his anti-modern anxieties, is evident in Burckhardt 1986, vol. 3, p. 69, vol. 6, p. 214, vol. 7, pp. 190, 204, vol. 8, p. 228, vol. 9, p. 90, vol. 10, pp. 26, 251. On Burckhardt's anti-Semitism see Mattioli 1999.

54 See von Reibnitz 1992, pp. 43–46, and Ruehl 2003, pp. 67–69.

55 On the anti-Wagnerian force of this little essay see Kaegi 1982, vol. 7, p. 47, and Ruehl 2003. That Wagner also had a more 'civic humanist' perspective on the politics and culture of late medieval and early modern Italy is suggested by his great appreciation of Sismondi, see Wagner 1977, vol. 1, pp. 1005–1009, 1011–1012, 1019.

56 Nietzsche is quoting Burckhardt 1988, p. 183: 'a partial reawakening of the ancient Italian soul ... a wondrous echo of ancient string-instruments' (*ein wundersames Weiterklingen eines uralten Saitenspiels*). See Kaegi 1982, vol. 7, p. 51.

owe modern culture'. Nietzsche's description of these forces reads like a keyword synopsis of Burckhardt's book: 'the liberation of thought, disregard for authority, the triumph of education over the presumption of lineage, a passion for science ... the unchaining of the individual ... a disdain for appearances and mere effect' (HA I 237).[57] Just as he himself set out to revive the secular world-view of the *philosophes* and the ethical scepticism of the *moralistes*, most notably Montaigne and La Rochefoucauld,[58] the latter had 'carried on' the secularizing and rationalizing 'task' of the Renaissance. In doing so, they had revived not just the ideas, but the actual psychological disposition of classical antiquity to which Nietzsche now, significantly, apportioned Rome:

> Reading Montaigne, La Rochefoucauld, La Bruyère ... one is closer to antiquity than with any other group of authors ... —together, they form an important link in the great continuing chain of the Renaissance ... With their resurrection of the great stoic world of ancient Rome, the French have continued the task of the Renaissance in a most honourable fashion ... —they began with the creative imitation of ancient forms and ended up having splendid success in recreating ancient characters. (HA II 214, 216)

If these remarks suggest a new attitude towards rationalism and 'Latin' culture, they also reflect a changed conception of history. The interpretation of Greek tragedy in *The Birth of Tragedy* had been informed, to a large extent, by Schopenhauer's irrational metaphysics of the will and his synchronic-pessimistic vision of the world as aimless suffering and striving. Nietzsche's reference now to a developmental 'chain' of enlightenment, from the ancients, via the Renaissance humanists and the *philosophes*, to present-day sceptics like himself, indicates a more diachronic, optimistic perception of the past. For Schopenhauer, the notion of history as meaningful change or indeed progress had been one of the most pernicious effects of what he called the 'stultifying Hegelian *Afterphilosophie*'. As he commented in the second volume of *The World as Will and Representation*:

> The Hegelians, who regard the philosophy of history as the aim of all philosophy, ought to be taught some Plato, who untiringly repeats that the object of philosophy lies in the unchangeable and in what lasts, and not in the things which are now like this, and now like that. All those who make such claims about the world in motion, or as they call it, history, have not grasped the

57 See Andler 1926, pp. 151–152. On Nietzsche's early 'uses' of Burckhardt's idea of the Renaissance see Farulli 1990, pp. 54–58.

58 See Molner 1993, Vivarelli 1993, Vivarelli 1998, and Donnellan 1979. Kaegi 1982, vol. 3, p. 540, reminds us that Burckhardt greatly contributed to Nietzsche's discovery of La Rochefoucauld's *Maximes et pensées* in the 1870s. See also Ross 1980, pp. 316–317.

> fundamental truth of philosophy: that, philosophically speaking, what really is is the same at all times. (Schopenhauer 1988, vol. II, pp. 514–515)[59]

While in *The Birth of Tragedy*, Nietzsche had still largely embraced this 'fundamental truth', in the second *Untimely Meditation* and, even more emphatically, in *Human, All Too Human*, he discarded most of Schopenhauer's ahistorical metaphysics. Though staunchly opposed to Hegel's progressivist philosophy of history (in this respect at least, he remained a faithful disciple of his former 'educator'), he began to think more historically. Ironically, it was Jacob Burckhardt, a self-professed Schopenhauerian,[60] who more than anyone else effected this historical turn.[61] *The Civilization of the Renaissance* seems to have played a dual role in this process. On the one hand, its glowing depiction of the autonomous personalities emerging from and shaping, in turn, the culture of the quattrocento furnished Nietzsche with a counter-weight to Schopenhauer's philosophical deconstruction of the *principium individuationis* and led him to rethink the significance of individual agency in history. It thus provided a historical precedent and an inspiration for one of the central ideas underlying his *Kulturkritik* in the 1870s: the belief that the great task of cultural renewal could be carried out by a small group of superior human beings. On the other hand, Burckhardt's account of the Renaissance gave Nietzsche a broader and more complex understanding of European cultural history and made him question earlier absolutes, most notably the belief in the inimitable greatness and timeless model character of ancient Greece. It contained ample evidence that Western civilization had not lain dormant since the fifth century BC, waiting to be awoken from its Socratic slumber by the kiss of Richard Wagner, the 'most German man' (Wagner 1975, p. 86).[62]

59 On Schopenhauer's critique of Hegel's philosophy of history (and its influence on Nietzsche) see Gottfried 1975, esp. pp. 337–338.

60 His allegiance to Schopenhauer's philosophy is expressed in Burckhardt 1986, vol. 5, pp. 105, 112, 119, 129, 139; vol. 6, pp. 30, 55, 134, 276; and vol. 7, p. 83. See also von Martin 1947a, pp. 27–28, Kaegi 1982, vol. 5, pp. 280–283, 491–497, vol. 6, pp. 109–113, and Jung 1991. But cf. Joël 1918, pp. 62, 245–246, who argues that Burckhardt was not 'an orthodox Schopenhauerian'.

61 Nietzsche's exposure to *The Civilization of the Renaissance* was as important, in this context, as his attendance of Burckhardt's lectures 'On the Study of History' in the winter semester of 1870/1871. Burckhardt's influence on Nietzsche's conception of history in the second *Untimely Meditation* is discussed in Bauer 2001, pp. 213–222.

62 For Wagner, by contrast, the 'excavation' of classical antiquity by the Renaissance humanists was a mere 'misfortune'; unlike his own *Gesamtkunstwerk*, Renaissance art was 'not destined to be redemptive': Wagner 1977, vol. 2, p. 1041 (7 November 1882).

The 'Italian genius', as Burckhardt put it, had already achieved a first revival—and not just of the (cultural) glory that was Greece, but also of the noble, martial values that Nietzsche would later identify with the label 'Rome' (see, e.g., GM I 16). For those contemporary Europeans seeking a second revival, there were important lessons to be learnt from the Renaissance, with regard both to its enabling factors and the reasons for its ultimate failure. In his reflections on the latter, Nietzsche soon came to single out Martin Luther.

Nietzsche's critical reassessment of Luther and the Reformation since the mid-1870s went hand in hand with his ideological emancipation from Wagner, who had recently taken a Protestant turn.[63] Wagner's new-found religiosity—'Incredible! Wagner had become pious', as Nietzsche put it, many years later, in *Ecce Homo* (EH III HA 5)—found expression in his last opera, *Parsifal*, the libretto of which was completed in the spring of 1877.[64] *The Birth of Tragedy*, as we have seen, had posited a close connection between the composer and the reformer, hailing Wagner as a product of that 'glorious, internally healthy, primordial force' of the German 'essence' (*Wesen*), which had also manifested itself in the Reformation (BT 23). In his second lecture 'On the Future of Our Educational Institutions', delivered in Basel on 6 February 1872, he still expressed his commitment to this German essence which had inspired the 'German Reformation [and] German music' (EI 2, KSA 1, p. 691). The preparations for the fourth and the (unfinished) fifth *Untimely Meditation*, written in the first half of 1875, however, already betrayed a more ambivalent relation to Luther. On the one hand, Nietzsche, evidently with an eye on his Wagnerian friends, applauded the Reformation as a 'protest against the decorative culture of the Renaissance'; on the other hand, he conceded that it had 'separated us from antiquity' (Nachlaß Spring 1875–Summer 1875, KSA 8, 5[28]). In another fragment from 1875, he remarked that the Renaissance showed 'an awakening of truthfulness in the south, as did the Reformation in the north', but added that the anti-Christian approach to classical antiquity

63 On Nietzsche's changing attitude towards Luther in those years see Hirsch 1998, Bluhm 1950, Bluhm 1953, Bluhm 1956, and Orsucci 1996, pp. 352–364. For a different reading see Bertram 1921, pp. 42–63.

64 See Gregor-Dellin 1980, pp. 739–740. On Wagner's increased interest in and admiration for Luther during the 1870s see Wagner 1977, vol. 1, pp. 741, 744, 748–753, 756, 775–777, 805, 1014, and vol. 2, pp. 206–210. See also Gregor-Dellin 1980, pp. 763–764, and Ross 1980, p. 519: 'In the meantime [i.e., the mid-1870s], she [i.e., Cosima Wagner] had become a good Protestant, the *Kulturkampf* was in full swing, and Wagner considered himself a descendant of Luther.'

taken by the Italian humanists had been 'purer' than that of the German reformers (Nachlaß Spring 1875–Summer 1875, KSA 8, 5[107]).

As he moved further away from Bayreuth in the second half of the 1870s, Nietzsche began to concentrate on the negative effects of the Reformation. In the first volume of *Human, All Too Human* (1878), a book that he retrospectively stylized as his response to *Parsifal* (see EH III HA 5),[65] he made Luther responsible for delaying the 'full awakening and supremacy of the sciences' and for preventing the 'complete synthesis [*In-Eins-Verwachsen*] of the ancient and the modern spirit' attempted in the Italian Renaissance. Insofar as it caused the Counter-Reformation, the Reformation, which he now decried as 'a vociferous protest of reactionary minds who had not yet had their fill of the medieval world-view', helped to re-establish a 'self-defensive Catholic Christianity' (HA I 237). It was Luther, Nietzsche contended in *The Gay Science* (1882), who had launched the fateful 'peasants' revolt of the north' against the 'noble' (*vornehm*) values and institutions of the south, a revolt that brought the 'common', 'plebeian' instincts back to the fore, 'emaciated' German culture and 'flattened' the European mind for centuries to come (GS 358).

Nietzsche's polemical juxtapositions of the Renaissance and the Reformation became more pronounced in his so-called 'transvaluative' writings. They reached a climax in *The Antichrist* (completed in September 1888), which included a lengthy counterfactual speculation about what might have ensued had Cesare Borgia ascended the papal throne in the early 1500s. Taking his cue from Burckhardt's redolent conjectures about the imminent decline of the papacy and the possible 'secularization' of the Papal States in Cesare's hands (see Burckhardt 1988, pp. 85, 87), Nietzsche mused that such an attack on the Church 'from within' would have brought about the realization of the Renaissance project, which he identified squarely with the 'transvaluation of Christian values'. What undermined this project, in the end, was not so much Cesare's premature death in 1507 as the intervention of a certain 'German monk':

> To attack at the decisive place, at the seat of Christianity itself, and there to enthrone the noble values ... I see before me the possibility of a ... heavenly ... spectacle ... Cesare Borgia as pope! ... Am I understood? ... Well then, that would have been the only sort of victory that I desire today: with that, Christianity would have been abolished!—What happened? A German monk, Luther, came to Rome. This monk, with all the vengeful instincts of a

65 On the completion of Nietzsche's break with Wagner in 1878 and the significance of *Human, All Too Human* in this context see Borchmeyer/Salaquarda 1994, vol. 2, pp. 1316–1333.

> failed priest, rebelled against the Renaissance in Rome ... Luther saw the depravity of the papacy when in fact the exact opposite was becoming apparent: ... Christianity no longer occupied the papal chair! Instead there was life! The triumph of life! The great yea to all lofty, beautiful and reckless things! ... And Luther restored the church: he attacked it... The Renaissance—an event without meaning, a great 'in vain' [*ein großes Umsonst*]. (A 61)[66]

That Burckhardt had influenced the revision of Nietzsche's formerly uncritical Protestant view of history is suggested by a letter to Heinrich Köselitz from October 1879, in which Nietzsche confessed that 'for a long time', he had been 'incapable of saying anything respectful' about Luther.[67] He put this down to the recent perusal of 'a huge collection of material' to which Jacob Burckhardt had drawn his attention. 'Here, for once,' he commented, 'we don't get the falsified Protestant version of history we have been taught to believe in' (KSB 5, p. 451). The 'material' in question was the second volume of the *History of the German People since the End of the Middle Ages* (published a little earlier in 1879) by the Catholic historian Johannes Janssen, which offered a fiercely partisan account of the confessional struggles in sixteenth-century Central Europe.[68] While Janssen's *History* evidently contributed to his reassessment of Luther in the 1880s,[69] Nietzsche's particular conception of the Reformation as a fateful interruption of and lasting impediment to the secularization and rationalization of the Western world was shaped more directly by his reading of *The Civilization of the Renaissance.* In the section on 'Morality and Religion', Burckhardt had speculated that the Renaissance would have 'swiftly done away with' outdated Christian institutions like the Mendicant orders, 'if the German Reformation and the Counter-Reformation had not interfered'. To

66 For a brilliant analysis of this passage see Sommer 2000, pp. 627–646. Nietzsche repeats his critique of the Reformation as a tragic interruption of the secularization process begun in the Renaissance in EH III CW 2.

67 On the Protestant values that determined Nietzsche's education in Röcken and Naumburg see Bohley 1987, Bohley 1989, and Pernet 1989.

68 On Burckhardt's deep respect for Janssen's scholarship see Kaegi 1982, vol. 5, pp. 56–58. Ludwig Pastor, who was Janssen's pupil and friend, reports that Burckhardt called Janssen's *History of the German People* 'essential' for the understanding of the 'end of the fifteenth and the beginning of the sixteenth century', because it 'finally told us the truth about the so-called Reformation'; 'up to now, we have only had uplifting stories [*Erbauungsgeschichten*] by Protestant pastors': Pastor 1950, p. 276.

69 See Hirsch 1998, pp. 175–179, and Orsucci 1996, pp. 353–364; but cf. Benz 1956, pp. 73–79, who rightly points out (p. 75) that, unlike Nietzsche, Janssen viewed Renaissance humanism as an ally of Protestantism.

the secular eyes of his Renaissance men, these orders appeared 'either comical or disgusting'. 'And who knows', he remarked ambiguously, 'what would have been in store for the papacy then, if the Reformation had not saved it' (Burckhardt 1988, pp. 337–338).[70]

3. The Makings of Renaissance Man: Individualism without Humanism

If Burckhardt prompted Nietzsche to reflect critically on the reasons for the failure of the Renaissance, he also made him consider the 'causes and conditions' for the 'superiority of Renaissance Man' (Nachlaß November 1887–March 1888, KSA 13, 11[133]) as well as the possibilities for a second cultural renewal. The first one, as he learnt from Burckhardt, was brought about by a new breed of men, who actively dismantled the values and institutions of the Middle Ages. Burckhardt's insistence that the makers and shapers of Renaissance culture were not humanist scholars like Coluccio Salutati or Lorenzo Valla left an impression on the young German philologist, who soon came to despise the pedagogical and methodological assumptions of his profession.[71] The humanists' revival of learning—this was made abundantly clear in the first three sections of Burckhardt's book—was little more than an epiphenomenon, an effect, rather than the cause or the essence, of the Renaissance (see Burckhardt 1988, pp. 127–128). Nietzsche adopted this reinterpretation of the Renaissance for his own speculations about the coming rebirth of antiquity, a task, he believed, that required a certain practical, activist disposition rather than classical training and scholarly erudition. The agonal, aristocratic spirit of ancient Greece and Rome would be restored not by men who had studied the ancient texts, but by men who embodied the values of the ancients. He already suggested as much in the second *Untimely Meditation* (1874) à propos the uses of monumental history:

70 Other remarks about the Reformation—see, e.g., Burckhardt 1988, pp. 93–94—reiterate the same idea, viz. that Luther prevented the secularization of the Church 'from within' (*von innen heraus*). They seem to confirm David Norbrook's contention that Burckhardt shared Nietzsche's later 'unease with Protestantism' and his conviction that 'Renaissance Italy was fortunate to maintain an aristocratic freedom from this extreme form of slave religion' (Norbrook 1989, p. 109). Benz 1956, p. 77, claims that more than anyone else, Burckhardt conditioned Nietzsche's anti-Protestant turn in the 1870s. But cf. Janssen 1970, p. 61, who argues that Burckhardt's comments on the Reformation in *The Civilization of the Renaissance* 'are, on the whole, appreciatory'.

71 See, e.g., his scathing remark on the 'castrated', 'philistine' empiricism of German classical philology (Nachlaß Spring–Summer 1875, KSA 8, 5[109]).

> Let us assume that somebody believes it would take no more than a hundred productive men, effective people brought up in a new spirit, to put an end to the superficial culture [*Gebildetheit*] that has become fashionable in Germany right now, how must it strengthen him to see that the culture of the Renaissance raised itself on the shoulders of such a group of a hundred men. (UM II 2)

The Civilization of the Renaissance, as we shall see, provided a blueprint for Nietzsche's reflections on the psychological make-up of this new breed of 'productive' men and the sociopolitical conditions that enabled the growth of their personalities.[72]

Like Burckhardt, Nietzsche regarded the development of a secular individualism as a defining characteristic of the Renaissance, which he described as a 'return to a heathen and profoundly personal ethos' (*Anlauf in's Heidnisch-stark-Persönliche zurück*) (Nachlaß Summer–Autumn 1873, KSA 7, 29[132]). Like his senior colleague, who belonged to one of the oldest families of the Basel patriciate, he viewed this new personal ethos of the Renaissance as the privilege of a new elite. In one of his drafts for the unfinished fifth *Untimely Meditation*, to be entitled 'We Philologists' (*Wir Philologen*), he cited Burckhardt's remark about the 'sophisticated' nature (*das Unvolksthümliche*) of Renaissance civilization (Nachlaß Spring–Summer 1875, KSA 8, 5[108]).[73] He had already elaborated this thought in an earlier draft, arguing that the sovereignty of the individual in Renaissance Italy produced an aristocratic culture which no longer drew on the forces of the people: 'The new education [*Bildung*] of the Renaissance ... also sought a corresponding art form ... The soil of the new art is no longer the people ... The individual dominates, that is, he

72 On Nietzsche's reliance on Burckhardt in his psychological reconstruction of Renaissance Man see Farulli 1990b, esp. pp. 42–49.

73 It should be emphasized, however, that while Nietzsche considered the elitism of Renaissance civilization a 'terrible fact' (*eine furchtbare Thatsache*) (Nachlaß Spring–Summer 1875, KSA 8, 5[108]), Burckhardt 1988, p. 128, coolly accepted the 'separation of the cultivated from the uncultivated' (*Scheidung von Gebildeten und Ungebildeten*) brought about by the new humanistic education as a 'necessary' and indeed immutable aspect of cultural evolution. The young Nietzsche, arguably under the influence of Wagner's democratic ideal of a new 'music for the masses', still assessed this development more sceptically. See also his remarks in the lectures on the 'History of Classical Philology' (1871), KGW II.3, p. 348: 'This [i.e., the pedagogical reforms of the Renaissance humanists] immediately transposed the central division of medieval culture, that between priest and layman, into the new education which became elitist [*unvolksthümlich*] and thus produced a rift from which all of us suffer today: from now on, there are cultivated and uncultivated men [*Gebildete und Ungebildete*] in Europe.'

contains within himself the forces that previously lay dormant in great masses. The individual as the extract of the people: withering away for the sake of one blossom' (Nachlaß 1871, KSA 7, 9[107]).

Implicit in these observations on the aristocratic individualism of the Renaissance was the condemnation of what Nietzsche regarded as a levelling of education and culture in contemporary European society. Both Nietzsche and Burckhardt constructed an image of early modern Italy that could be held up as a mirror to present-day Northern Europe whose schools and universities invoked the legacy of Renaissance learning and yet failed miserably to produce the kind of individuals that inhabited quattrocento Italy. Renaissance individualism, for both, was diametrically opposed to the bourgeois, liberal individualism that informed the pedagogical as well as the political ideals of Germany and Switzerland in the second half of the nineteenth century.[74] Both men believed that these ideals—the Rousseauean faith in the natural goodness of man, universal rights, equality of opportunity, the promotion of general welfare, and so on—would open the door to various forms of 'massification' and eventually usher in a 'great rabble- and slave-rebellion' (Nachlaß Summer–Autumn 1884, KSA 11, 26[324]) that was bound to destroy the last remnants of individual autonomy and genuine *Bildung*.[75] Both, accordingly, rejected universal suffrage, the shortening of working hours—in Basel from twelve to eleven hours per day—the abolition of child labour, and the broadening of humanistic education, in particular the establishment of 'educational associations' (*Bildungsvereine*) for workers.[76] As Nietzsche observed in the notes for his lectures 'On the Future of Our Educational Institutions', delivered in 1872 to a packed auditorium in Basel's *Aula*,'universal education' was a 'preliminary stage of communism ... the precondition for communism' (Nach-

74 On the modernization, in particular the expansion and 'democratization', of the German educational system in the second half of the nineteenth century see Jeismann/Lundgreen 1987, pp. 71–250, 317–362, and Berg 1991, pp. 147–371, 411–473. On educational reform in nineteenth-century Basel see Gossman 2000, pp. 69–77 .

75 See Burckhardt's letter to Heinrich von Geymüller of 27 December 1874 (Burckhardt 1986, vol. 5, pp. 261–262): 'Since the Paris Commune, anything is possible anywhere in Europe, mainly because there are well-meaning, splendid liberal people everywhere who do not rightly know where justice ends and injustice begins ... They are the ones opening the gates and paving the way for the dreadful masses everywhere.' On Nietzsche's fear of the masses see Marti 1993.

76 On Nietzsche's attitude to these contemporary sociopolitical issues see Naake 1985, esp. pp. 61, 86, 89, and Cancik 1995, pp. 23–24, 27–31; on Burckhardt's standpoint see Bächtold 1939, esp. pp. 286–299, and Bauer 2001, pp. 87–101.

laß Winter 1870/1871–Autumn 1872, KSA 7, 8[57]).[77] Like Burckhardt, Nietzsche hailed the Renaissance as the harbinger of a modern sense of self, while dismissing all modern attempts to universalize and democratize its humanist legacy of self-formation.

Nietzsche's conception of Renaissance individualism, however, was more radically anti-democratic, anti-liberal, and anti-humanist than Burckhardt's.[78] The latter, after all, though he acknowledged the new division into 'educated' and 'uneducated' people brought about by humanism, nonetheless stressed the competitive, meritocratic aspects of Renaissance civilization, the foundation of which, he believed, was a 'universal society' (*allgemeine Gesellschaft*) characterized by an 'equality of estates' (Burckhardt 1988, p. 106).[79] Nietzsche, by contrast, exclusively dwelt on the 'noble', aristocratic elements of the Renaissance. While Burckhardt conceded that, alongside the tyrannical courts, republican city-states like Florence also allowed for the growth of 'individuality' and cultural productivity (Burckhardt 1988, p. 10),[80] Nietzsche chose to ignore this republican alternative. The tyrants, he observed in *The Gay Science* (1882), were not just the 'first-born' of the new individuals (*Erstlinge der Individuen*), they were the only *raison d'être* for a people. True self-fashioning, for him, was only possible in the radically insecure, violent sphere of tyranny (GS 23). It

77 On the anti-modern animus and elitist ethos of Nietzsche's lectures see Gossman 2000, pp. 423–424, 427–430. Cf. Burckhardt 1982, p. 182: 'The latest thing in our world: the demand for culture [*Cultur*] as a human right, which is a veiled desire for a life of luxury [*Wohlleben*].' On Burckhardt's anti-democratic conception of *Bildung* in particular see Schmidt 1976, pp. 18–22. Wagner, by contrast, enthusiastically welcomed the 1880 Schulreform in Basel city, which made secondary school education free of charge and thus (at least in principle) accessible to the lower orders; see Wagner 1977, vol. 2, p. 570.

78 The anti-democratic and anti-humanist *Weltanschauung* underlying Nietzsche's vision of a reborn humanity is almost completely overlooked in Boeschenstein 1982.

79 See also Burckhardt 1988, p. 262, which contains a brief eulogy on the new social mobility and general disregard for lineage in early modern Italy. All of this, Burckhardt observed, 'gave the impression' that the Renaissance ushered in an 'age of equality' (*Zeitalter der Gleichheit*). Janssen 1970, pp. 202–203, rightly points out that Burckhardt hailed Renaissance society as meritocratic and homogenous insofar as it exploded the feudal, hierarchical structures of the Middle Ages, while highlighting those new forms of (cultural) stratification and elitism that had unfortunately been eroded, he believed, in the 'mass societies' of modern Europe.

80 It is nonetheless significant that Burckhardt's discussion of the political context of Renaissance individualism is devoted first and foremost to the tyrannies: only twenty of the roughly one hundred pages that make up the first section of *The Civilization of the Renaissance* are dedicated to the republican city-states.

is significant that in his otherwise very warm response to the *Gay Science*, a complimentary copy of which had been sent to him by the author immediately after publication,[81] Burckhardt expressed mild concern over Nietzsche's 'possible propensity towards tyranny' (*Anlage zu eventueller Tyrannei*) which he thought was revealed in aphorism 325 of the book (Burckhardt 1986, vol. 8, p. 87). Entitled 'What Belongs to Greatness',[82] the aphorism in question reads: 'Who is going to achieve great things if he does not feel within himself the force and the will to cause great pain? The ability to suffer is the least ... But not to perish by dint of inner distress and uncertainty when one inflicts great suffering and hears the cry of this suffering—that is great, that belongs to greatness' (GS 325).

But tyrannical self-fashioning, according to Nietzsche, did not just produce great individuals with stony hearts and Machiavellian minds; it also aided the growth of culture. Under a tyranny, he argued, 'the individual is usually most mature and "culture", consequently, most developed and fertile'. The tyrant was a catalyst for the creation of 'bold', 'transgressive' individuals as well as artists (GS 23). Again and again, Nietzsche returned to this juncture between oppression and individualization, destruction and cultural production, the 'mysterious connection', as he called it in his essay on the 'Greek State', 'between political greed and artistic creation, battlefield and work of art' (CV 3, KSA 1, p. 772). Even if the 'aristocratic radicalism' that informed his later writings went far beyond Burckhardt's more conservative 'cultural pessimism',[83] there can be little

81 Nietzsche sent copies of all his books to Burckhardt, whom he considered his most discerning reader, and continued to do so long after the latter had stopped even acknowledging receipt of the shipments from his young friend and admirer (that is, after the publication of *The Genealogy of Morality* in 1887); see KSB 8, pp. 80, 187, 205, 489, 547.

82 Note that Burckhardt's lecture series 'On the Study of History' contained a long segment on the nature of 'historical greatness', in which Burckhardt proffered a—qualified—'dispensation' of the 'great man' from the 'normal moral law' (*Dispensation von dem gewöhnlichen Sittengesetz*): Burckhardt 1982, pp. 401–402. His observation (p. 396) that the 'first task of the great man is to assert and to increase his power' and the categorical statement (p. 401) that 'power has never been established without crime; yet the most important material and spiritual possessions of a nation can develop only when they are protected by power' suggest that Nietzsche's 'tyrannical' definition of greatness in *The Gay Science* was not complete anathema to him. Kaegi 1982, vol. 8, p. 63, remarks that Burckhardt 'knew his Machiavelli well enough not to be too perturbed' by the aphorism in question.

83 The Danish critic Georg Brandes first coined the expression 'aristocratic radicalism' to describe the strange mixture of revolutionary and elitist elements in the thought of Nietzsche, who emphatically embraced it; see his letter to Brandes of 2 December 1887 (KSB 8, p. 206). See also Detwiler 1990. On Burckhardt's more

doubt that the author of *The Civilization of the Renaissance*, who had detected an aesthetic quality in the steely state-building of the tyrants and *condottieri*, drew his attention to this connection in the first place. Nietzsche's slightly obscure observation, in a note of August 1881, that the Pitti Palace in Florence represented a renunciation of everything that was 'pretty and pleasing' and expressed the sublime 'contempt for the world' typical of a *Gewaltmensch* (Nachlaß Spring 1881–Autumn 1881, KSA 9, 11[197]) was a quotation from Burckhardt, who had reverently described the creators of the palace as 'superhuman beings' (*übermenschliche Wesen*).[84]

4. Radicalizing the Renaissance: the Borgia versus the *Bürger*

Following Burckhardt, the later Nietzsche glorified Renaissance Man, even more than the ancient Greeks, as a synthesis of the will to power and the will to form, the incarnation of an entirely amoral plastic instinct for self-creation and self-assertion. If in 1878 he had still insisted that the new 'artistic natures' of the Renaissance possessed 'the highest moral purity' and that the quattrocento was a golden age 'despite its flaws and vices' (HA I 237), he successively inverted this judgement over the next ten years, extolling the very 'flaws and vices' of early modern Italy as signs of a new pagan master-morality. Beginning with *Beyond Good and Evil* (1886), he embarked on a fundamental transvaluation of various Renaissance figures who had been traditionally decried as demonic, corrupt, or blasphemous. He held up these figures in a decidedly 'monumental' fashion: as ideals and incentives for the 'new ruling caste' of a post-Christian

---

conservative brand of *Kulturkritik* see Mommsen 1986. For a comparative assessment of their positions see Löwith 1966, pp. 31–34, and Sautet 1981, pp. 138–142.

84 Nietzsche is paraphrasing Burckhardt's *Cicerone, or Guide to the Enjoyment of the Artworks of Italy* (1855), a book that greatly shaped his own experience and assessment of Renaissance art; see Burckhardt 2001a, p. 151. It is not clear whether Burckhardt meant to attach the label *Gewaltmensch* to the patron of the palace, Luca Pitti, or its architect, Luca Fancelli (or indeed Filippo Brunelleschi, whom Burckhardt, following Vasari, erroneously credited with the original plans for the building). Note that in *The Civilization of the Renaissance*, he reserved the epithet 'Gewaltmensch' for the *uomo universale* Leon Battista Alberti whose fame rested on his extraordinary talents in art and architecture (as well as poetry and philosophy), not on any ruthless political actions; see Burckhardt 1988, p. 104. That Nietzsche was familiar with Burckhardt's portrait of the artist as *Gewaltmensch* is suggested by the marginalia in his copy of *The Civilization of the Renaissance*; see C482a, p. 110.

Europe, exempla all the more inspiring because of their concrete historical identity (Nachlaß May–July 1885, KSA 11, 35[73]). Nietzsche invoked a number of early modern characters in this context, including the Hohenstaufen emperor Frederick II, Michelangelo, and Machiavelli,[85] but his favourite Renaissance 'monument' by far was Cesare Borgia.

Burckhardt's assessment of Cesare, as we have seen, was ambivalent. On the one hand, he depicted him, not without some appreciation, as the most ruthless of the new breed of quattrocento tyrants. His callousness, Burckhardt wrote, was as extreme as his 'talent' and his attempts to centralize the Papal States had 'great prospects'. Like Leonardo, Burckhardt evidently saw something 'extraordinary' in his character. On the other hand, he recoiled from the extremity of Cesare's crimes, observing, quite unambiguously, that the monstrosity of the means outstripped 'the actual as well as the imaginable ends' of his actions—and thus ceased to be comprehensible, even within a purely 'objective', Machiavellian frame of re-

85 Nietzsche mentions Frederick II on a number of occasions, generally as a great antagonist of the medieval papacy and an early European free spirit; see, e.g., EH III Z 4, where he calls Frederick 'an atheist and enemy of the church comme il faut ... one of my closest relatives'. That Nietzsche's understanding of Frederick II was conditioned by Burckhardt's *Civilization of the Renaissance* is argued in Hampe 1925, p. 51. See also Janssen 1970, pp. 104–109. Nietzsche's portrait of Michelangelo (see Nachlaß April–June 1885, KSA 11, 34[149]) as the revolutionary creator of new artistic norms and forms seems equally indebted to Burckhardt and the marginalia in his copy of the *Cicerone*, also preserved at the Herzogin Anna Amalia Bibliothek (see C483 pp. 667, 669), evidence his familiarity with Burckhardt's views on Michelangelo. However, by explicitly placing Michelangelo above the 'Christian' Raphael, who 'faithfully' adhered to the classic standards of the ancients, Nietzsche went against one of Burckhardt's most fundamental verdicts on early modern art. In his art-historical writings as well as his correspondence, Burckhardt emphatically and repeatedly exalted Raphael's 'classicism' over Michelangelo's 'Titanism'. What Nietzsche hailed as Michelangelo's 'yearning instincts' for a new art and new artistic 'values', Burckhardt denounced as 'demonic', 'arbitrary', and 'reckless'; see Burckhardt 2001a, pp. 267–269, 273, 276, and Burckhardt 1986, vol. 8, p. 192. But Burckhardt's judgement of Michelangelo was not entirely negative. His comments on Michelangelo's paintings, for instance, notably those in the Sistine Chapel, betray a certain reluctant admiration for the Promethean aspects of his art, which he repeatedly describes as 'superhuman' (Burckhardt 2001b, pp. 124–129). Wölfflin 1934, pp. xxiv–xxv, and von Martin 1947a, pp. 140–142, overstate his traditionalism and classicism in this respect. Gossman 1999, pp. 904–905, and Kaegi 1982, vol. 3, pp. 510–513, offer a more balanced account. In their assessment of Machiavelli, at any rate, both men were in broad agreement again: like Burckhardt, Nietzsche praised Machiavelli primarily on account of his political 'realism'; see, e.g., TI 'What I Owe to the Ancients' 2. On Nietzsche's 'Machiavellianism' see Dombowsky 2004, pp. 131–168, and Vacano 2007.

ference (Burckhardt 1988, pp. 84–85).[86] Nietzsche, by contrast, unconditionally hailed Cesare as the incarnation of the wholly secular, 'noble way of evaluating all things' (*der vornehmen Werthungsweise aller Dinge*) which he regarded as typical of Renaissance civilization in general (GM I 16). When Nietzsche defended Cesare's aggressive and transgressive traits against modern detractors who had judged them depraved and degenerate, he also challenged Burckhardt's judgement:

> One altogether misunderstands the beast of prey and the man of prey (Cesare Borgia for example), one misunderstands 'nature', as long as one looks for something 'sick' at the bottom of these healthiest of all tropical monsters. (BGE 197)[87]

In *The Civilization of the Renaissance*, Burckhardt had related the malefactions of Cesare and his father Alexander VI in the style of Gothic horror stories, as manifestations of a darkly demonic force in the otherwise translucent world of Renaissance *ragione di stato*. Nietzsche listed Cesare amongst the 'great virtuosi of Life' (Nachlaß November 1887–March 1888, KSA 13, 11[153]) and the main representatives of the 'brilliant-uncanny re-awakening of the classical ideal' (*glanzvoll-unheimliches Wiederaufwachen des klassischen Ideals*) in early modern Italy (GM I 16). For Burckhardt, whose conception of antiquity differed in many important ways from that of Winckelmann, Goethe, and Schiller, this 'classical ideal' nonetheless implied *humanitas*, harmony and *Bildung* as defined by 'Weimar classicism'.[88] For Nietzsche, it meant the 'noble', pagan values of ancient Rome, a pre-Christian 'master morality' that imbued the ancient elites with warlike ardour and a 'pathos of distance' towards lesser beings (BGE 257). After having lain dormant in the Middle Ages, these values, he believed, had come to life again in Cesare Borgia. Within the historical framework underlying *The Genealogy of Morality*, Cesare thus occupied a central place as the originator and driving force of the first 'transvaluation of Christian values'—and although eventually stopped short by the Reformation, his transvaluation bore the promise of a second, more complete reversal. His strategic significance for the entire transvaluative project is highlighted at the very end of *The Antichrist* where Nietzsche identified him with the Renaissance war against Christianity, a war, he contended, that had to be continued at all costs:

---

86 See also Burckhardt 1988, p. 331.

87 See also A 46, where Nietzsche approvingly cites Domenico Boccaccio's assessment of Cesare —'è tutto festo'—which he translates as 'immortally healthy, immortally cheerful [*heiter*] and well-turned out [*wohlgerathen*].'

88 See von Martin 1947a, pp. 40–43, 139–141.

> So far there has been only this one great war, so far there has been no more decisive question than that of the Renaissance—my question is its question. (A 61)

The extent to which Nietzsche equated Cesare's historical 'task' with his own is evidenced by a late letter to Georg Brandes, dated 20 November 1888, in which he compares the anti-Christian polemics of *Ecce Homo* with Cesare's 'overcoming' of Christianity by dint of his superior 'vital instincts' (KSB 8, pp. 482–483). These remarks suggest that in the late 1880s at least, Nietzsche viewed himself as a kind of *Cesare redivivus*, a continuator of his work and the harbinger of a new, more complete attack on Christendom.[89]

This new attack, of course, would roll back not just the religious institutions of Christianity, but Christian 'slave morality' in all its laical nineteenth-century permutations. Cesare's noble values were *Gegenwerthe* (counter-values) to the 'life-denying' ascetic doctrines of the Christian Church as much as to the universalist humanitarian ideals underlying contemporary ideologies like liberalism and socialism. In *The Gay Science*, Nietzsche had already suggested that these latter ideologies were born out of the 'plebeian', egalitarian doctrines formulated in the Reformation. He returned to this narrative of decline in *The Twilight of the Idols* (completed in September 1888), where he condemned the modern demands for 'humanity' (*Humanität*) as signs of cultural as well as physical decay, contrasting them sharply with the vital, agonistic instincts of Cesare's Italy:

> We moderns, very delicate, very vulnerable ... really have the conceit that our tender humanity, our unanimous consensus to be merciful, helpful, and trusting, is a positive advance, that with this we have gone far beyond the men of the Renaissance ... What is certain is that we may not place ourselves in Renaissance conditions, not even by an act of the imagination: our nerves would not endure it, let alone our muscles. But such incapacity is not a sign of progress ... Let us not doubt that we moderns, with our thickly padded humanity [*mit unsrer dick wattirten Humanität*] ... would have provided Cesare Borgia's contemporaries with a comedy at the sight of which they would have laughed themselves to death. (TI 'Reconnaissance Raids' 37)

Whereas in earlier works like *Human, All Too Human*, Nietzsche had posited a trajectory from the Renaissance to the rationalist, scientific world-view of modern Europe, the later, transvaluative writings largely capped these positive links. In the 1880s, Nietzsche dwelt almost exclusively on those elements of Renaissance civilization that stood in diametri-

89 See Gerhardt 1989, p. 109, and Gerhardt 1988.

cal opposition to the modern bourgeois world and its 'de-vitalizing' ethics of weakness:

> Strong ages, *noble* cultures all consider pity, 'neighbourly love', and the lack of self and self-assurance as something contemptible. Ages are to be measured by their *positive strength*—and if we apply this yardstick, the lavish, fateful age of the Renaissance [*jene so verschwenderische und verhängnissreiche Zeit der Renaissance*] emerges as the last *great* age. We moderns, by contrast ... with our virtues of ... modesty, legality, and scientism [*Wissenschaftlichkeit*] ... emerge as a *weak* age. (TI 'Reconnaissance Raids' 37)

Like Burckhardt, Nietzsche criticized the bourgeois *satisfait*, with his ideal of legal *Sekurität* and his utilitarian concerns, by holding up the Renaissance as an era of ruthlessness, violence, and 'dangerous living'.[90] However, where Burckhardt suspended judgement, Nietzsche offered explicit praise.[91] His Renaissance Men, most notably Frederick II and Cesare, were unscrupulous, immoral beings, splendid 'criminals' (Nachlaß Fall 1887, KSA 12, 10[50]),[92] who thought and acted in blissful disdain for the moral precepts of the Christian slave religion. They possessed a new, superior type of moral fibre, 'virtue in the Renaissance style, *virtù*, that is: virtue free of moralistic acid' (*moralinfreie Tugend*), which allowed them to increase their 'power' (A 2).[93] For Burckhardt, as we have seen, Cesare's crimes transcended the calculus of cruelty that characterized the Machiavellian politics of early modern Italy. For Nietzsche, such an excess of cruelty was in complete accordance with the amoral *virtù* of those 'higher

90 See Burckhardt's letter to Nietzsche of 26 September 1886 (Burckhardt 1986, vol. 9, pp. 50–51), which indicates his sympathetic interest in Nietzsche's reflections on the 'antithesis between the great security and comfort of well-being [*Assecuranz des Wohlbefindens*] and the desirable education through danger'. Even Kaegi, who generally emphasizes their ideological differences, concedes that Burckhardt shared Nietzsche's critical perspective on 'contemporary European man'; see Kaegi 1982, vol. 7, p. 67.

91 See Janssen 1970, pp. 32–33, 153–156, 215–216.

92 In his *History of the Popes* (1834–1836), Leopold von Ranke had called Cesare a 'virtuoso of crime' (*Virtuos des Verbrechens*): Ranke 1890, p. 34.

93 The lines from A 2: 'What is happiness?—The feeling that power increases, that a resistance is being overcome. Not contentment, but more power; not peacefulness, but war; not virtue [*Tugend*], but efficiency [*Tüchtigkeit*] (virtue in the Renaissance style, *virtù*, virtue free of moralistic acid' echo the following observation in the Nachlaß (Autumn 1887, KSA 12, 10[50]): 'in the age of the Renaissance, the criminal flourished and acquired his own type of virtue — virtue in the Renaissance style, of course, *virtù*, that is: virtue free of moralistic acid'. On Nietzsche's neologism *moralinfrei* and its anti-Semitic connotations see Sommer 2000, pp. 98–99. There is some evidence that Nietzsche adopted the notion of 'criminal virtue' from Burckhardt; see the marginalia in C482a, p. 12, n. 3.

men' destined to destroy the Christian idols and to proclaim, once again, the 'great yea to all lofty, beautiful and reckless things' (A 61). Burckhardt regarded Cesare as an extreme, Nietzsche as an exemplary embodiment of the features that, in the eyes of both, defined the civilization of the Renaissance in Italy: secular individualism, remorseless *Realpolitik*, and a 'premoral intensity' that sought its 'most permanent expression in art' (Norbrook 1989, p. 109). Within the dramatic-allegorical subtext of Burckhardt's Renaissance book, Cesare played the role of transgressor and overreacher, but ultimately he remained a historical figure, a deeply contradictory product of the new, morally ambiguous world that was the cradle of modern man. Nietzsche, by contrast, used Cesare primarily as a symbol and type: the iconic negation of all the sickly instincts, the 'thickly cushioned humanity' and 'herd animal morality' of those 'last men' populating contemporary Europe. This does not mean, however, that his Cesare was an arbitrary, 'literary' construct, as some recent critics have claimed (see Nehamas 1985, pp. 225–227). When Nietzsche referred to Cesare as a model of the superman and, indeed, 'a kind of superman' (TI 'Reconnaissance Raids' 37),[94] he had in mind a very specific moment in the history of Western civilization, a turning point that, ultimately, failed to turn, but that nonetheless held rich promise for the future.[95] Given the centrality of Cesare in particular and of early modern Italy in general to Nietzsche's vision of a transformation of European culture, one might justifiably describe the essence of this vision as a renascence of the Renaissance.

5. Legacies: A Language of Blood and Beauty

The notion of Cesare Borgia as a neo-pagan superman became one of the most prominent emblems of the first wave of German Nietzscheanism in the 1890s,[96] when hosts of eager new disciples celebrated the Renaissance as an age of unrestrained subjectivity and a Dionysian zest for life.[97] Look-

94 See also Nietzsche's letter to Malwida von Meysenbug of 20 October 1888 (KSB 8, p. 458), in which he explains that a 'figure like Cesare Borgia' comes a 'hundred times' closer to his idea of the 'superman' than 'the figure of Christ'.

95 Sommer 2000, pp. 630–631, points out that it is the very failure of the Renaissance that necessitates and legitimizes Nietzsche's own transvaluative efforts.

96 This important feature of the early 'Nietzsche legacy' in Wilhelmine Germany is curiously overlooked in Aschheim 1992.

97 See Rehm 1929, O'Pecko 1976, Ritter-Santini 1974, and especially Uekermann 1985, who demonstrates (pp. 55–67) the profound influence that Nietzsche's ideas had on fin-de-siècle *Renaissancismus*.

ing back at this time in 1918, Thomas Mann listed the 'modish mass effects' of Nietzsche's philosophy, beginning with *Renaissancismus*, 'the cult of the Superman', 'Cesare Borgia aestheticism' (*Cesare-Borgia-Ästhetizismus*) and 'the loudmouthed language of blood and beauty'.[98] Like Mann, Burckhardt recoiled from this language and it may have been the 'loudmouthed' glorifications of Renaissance evildoers by fin-de-siècle playwrights such as Rudolf Lothar and Oscar Panizza that prompted his explicit dissociation from Nietzsche's *Gewaltmenschen* in the letter to von Pastor.[99]

To be sure, Nietzsche—and the dramatists of the fin de siècle after him—radicalized the secular, transgressive elements of Renaissance culture to such an extent that their representations of early modern Italy seemed like a ghastly distortion of Burckhardt's. At the same time, there can be little doubt that their 'language of blood and beauty' drew its vocabulary from Burckhardt's evocative descriptions of all the Sforza, Malatesta, Visconti, and their Machiavellian machinations.[100] Whoever penned the 1895 article for the *Historisch-Politische Blätter* was not the only contemporary observer to discern a connection between the first section of *The Civilization of the Renaissance* and the aestheticization of violence in Nietzsche's later writings, which shaped the Renaissance cult of ruthlessness around 1900.[101]

98 Mann 1918, pp. 553–560. On Mann's ambivalent relation to *Renaissancismus* see Ruehl 2004.

99 See Lothar 1893, Panizza 1894, and Berthold Weiß' drama *Caesar Borgia*, which premiered in Zurich in 1893. On the scandal provoked by Panizza's *Liebeskonzil* in 1894 see Jelavich 1985, pp. 54–74.

100 Uekermann 1985, pp. 42–55, shows that the major playwrights of *Renaissancismus* did not just borrow their *dramatis personae* and plot-lines from Burckhardt's book, but adopted its very terminology, for instance the concept of the state as a 'work of art'.

101 Writing in 1917, the literary historian Franz Baumgarten called Burckhardt the 'historian' and Nietzsche the 'prophet' of *Renaissancismus*; see Baumgarten 1917, p. 5. The cultural historian Aby Warburg, who was better acquainted than most with the literature on early modern Italy, similarly believed that it was Burckhardt who had triggered the fin-de-siècle craze for the ruthless Renaissance hero; see Roeck 1991, p. 66. Fubini 1992, p. 563, remarks that Hans Baron's emphasis on the civic, urban, and 'proto-liberal' elements of Renaissance political thought in the early 1920s was an attempt to 'suppress' a line of interpretation that, 'through Burckhardt and Nietzsche, hailed the individualism of the Renaissance as the forerunner of the antibourgeois radical currents, both on the Left and the Right, that ran through Germany at the time'. See also Reinhardt 2002.

Pace Pastor, von Martin, and a host of later commentators,[102] it is not unreasonable to establish such a connection. *The Civilization of the Renaissance in Italy*, as we have seen, had a profound and lasting impact on Nietzsche's 'intellectual development and philosophy'. It fundamentally altered his view of European history and made him question the Protestant, philhellenist ideals that underwrote his early hopes for a German cultural renewal out of the spirit of Wagner's *Gesamtkunstwerk*. The central ideas that Nietzsche adopted from Burckhardt's book—about secularization and individualization, the redemptive power of science and moral scepticism, southern form and clarity, the 'Italian genius'—were important discursive weapons in his struggle for liberation from Bayreuth in the 1870s. But these weapons were double-edged.[103] Inspired by Burckhardt, Nietzsche reinvented the Renaissance as the cradle of tyrannical self-fashioning and dangerous living as well as the model for a future rebirth of European civilization under the sign of the Antichrist. As a *kulturkritisch* construct and historical reference point, Burckhardt's Renaissance contributed to his discovery of the *moralistes*, his reassessment of the Enlightenment and 'Latin' culture as well as his self-invention as a 'good European'. At the same time, it paved the way for his radically anti-humanist reflections on culture, politics, and the self in the 1880s. Burckhardt's book was thus not just a benign antidote to the Wagnerian enthusiasm of Nietzsche's youth. It also provided new stimulants that would fuel some of his most transgressive thoughts.

References

Adorno, Theodor W., 1970, 'Glosse über Persönlichkeit', in: Theodor W. Adorno, Stichworte: Kritische Modelle, Frankfurt a. M.: Suhrkamp, pp. 51–57.

---

102 See, e.g., Baron 1960, pp. 207–209, who highlights the 'difference in spirit' between Burckhardt's more Goethean, conservative conception of Renaissance individualism and Nietzsche's 'reinterpretation' which according to Baron helped to create a 'false image of the unscrupulous, ruthless, and lusty "superman" of the Renaissance'.

103 Ross 1980, pp. 313, 317, captures this ambivalence with characteristic acumen in his description of Burckhardt's struggle with Wagner over Nietzsche's allegiance as a metaphysical tug-of-war: 'Wagner was tugging at Nietzsche in order to drag him over to his side, just as in the Middle Ages demons were [believed to be] tugging at the disembodied souls of the recently deceased. Jacob Burckhardt was the guardian angel pulling [Nietzsche] in the other direction ... But one can also interpret the fight over Nietzsche's soul in a different way: with Wagner as the noble Germanic angel and Burckhardt as the sceptical Mephisto.'

Andler, Charles, 1926, Nietzsche und Jakob Burckhardt, Basel: Rhein-Verlag.
Aschheim, Steven E., 1992, The Nietzsche Legacy in Germany, 1890–1990, Berkeley: University of California Press.
Bächtold, Hermann, 1939, 'Jacob Burckhardt und das öffentliche Wesen seiner Zeit', in: H. Bächtold, Gesammelte Schriften, ed. Eduard Vischer, Aarau: Sauerländer, pp. 266–300.
Baldwin, Geoff, 2001, 'Individual and Self in the Late Renaissance', in: The Historical Journal, 44/2, pp. 341–364.
Baron, Hans, 1960, 'Burckhardt's "Civilization of the Renaissance" a Century after its Publication', in: Renaissance News, 13/3, pp. 207–222.
Baron, Hans, 1966, The Crisis of the Early Italian Renaissance: Civic Humanism and Republican Liberty in an Age of Classicism and Tyranny, 2 vols. [1955], 2nd edn, Princeton: Princeton University Press.
Barth, Hans, 1943, 'Der konservative Bürger und der revolutionäre Dynamiker (Burckhardt und Nietzsche)', in: Hans Barth, Fluten und Dämme: Der philosophische Gedanke in der Politik, Zurich: Fretz und Wasmuth, pp. 125–152.
Bauer, Stefan, 2001, Polisbild und Demokratieverständnis in Jacob Burckhardts 'Griechischer Kulturgeschichte', Basel: Schwabe.
Baumgarten, Franz F., 1917, Das Werk Conrad Ferdinand Meyers. Renaissance-Empfinden und Stilkunst, Munich: O. Beck.
Benz, Ernst, 1956, Nietzsches Ideen zur Geschichte des Christentums und der Kirche, 2nd edn, Leiden: E. J. Brill (= Beihefte der Zeitschrift für Religions- und Geistesgeschichte no. 3).
Berg, Christa (ed.),1991, Handbuch der deutschen Bildungsgeschichte, vol. 4, Munich: C. H. Beck.
Bergmann, Peter, 1987, Nietzsche, 'The Last Antipolitical German', Bloomington: Indiana University Press.
Bertram, Ernst, 1921, Nietzsche: Versuch einer Mythologie, 5th edn, Berlin: Georg Bondi.
Bluhm, Heinz, 1950, 'Nietzsche's Idea of Luther in *Menschliches, Allzumenschliches*', in: Proceedings of the Modern Language Association, 65/6, pp. 1053–1068.
Bluhm, Heinz, 1953, 'Nietzsche's View of Luther and the Reformation in *Morgenröthe* and *Die Fröhliche Wissenschaft*', in: Proceedings of the Modern Language Association, 68/1, pp. 111–127.
Bluhm, Heinz, 1956, 'Nietzsche's Final View of Luther and the Reformation', in: Proceedings of the Modern Language Association, 71/1, pp. 75–83.
Boeschenstein, Hermann, 1982, 'Nietzsche und das Problem der Humanität', in: Modern Language Notes, 97/3, pp. 636–655.
Bohley, Reiner, 1987, 'Nietzsches christliche Erziehung', in: Nietzsche-Studien, 16, pp. 164–196.
Bohley, Reiner, 1989, 'Nietzsches christliche Erziehung', in: Nietzsche-Studien, 18, pp. 377–395.
Borcherdt, Hans Heinrich, 1949, Der Roman der Goethezeit, Urach: Port.
Borchmeyer, Dieter, and Jörg Salaquarda (eds.), 1994, Nietzsche und Wagner. Stationen einer epochalen Begegnung, Frankfurt a. M.: Insel.

Brecht, Walther, 1911, Heinse und der ästhetische Immoralismus: Zur Geschichte der italienischen Renaissance in Deutschland. Nebst Mitteilungen aus Heinses Nachlaß, Berlin: Weidmann.

Brobjer, Thomas H., 1995, Nietzsche's Ethics of Character: A Study of Nietzsche's Ethics and its Place in the History of Moral Thinking, Uppsala: Uppsala University Press.

Brobjer, Thomas H., 1997, 'Nietzsche's Reading and Private Library', in: Journal of the History of Ideas, 58/4, pp. 663–693.

Brobjer, Thomas H., 2007, 'Nietzsche's Relation to Historical Methods and Nineteenth-Century German Historiography', in: History and Theory, 46/2, pp. 155–179.

Brose, Karl, 1973, Geschichtsphilosophische Strukturen im Werk Nietzsches, Bern: Peter Lang.

Brose, Karl, 1994, Nietzsche: Geschichtsphilosoph, Politiker und Soziologe, Essen: Verlag Die blaue Eule.

Bullen, Barrie, 1994, The Myth of the Renaissance in Nineteenth-Century Writing, Oxford: Clarendon Press.

Burckhardt, Jacob, 1982, Über das Studium der Geschichte. Der Text der 'Weltgeschichtlichen Betrachtungen' nach den Handschriften, ed. Peter Ganz, Munich: C. H. Beck.

Burckhardt, Jacob, 1986, Briefe. Vollständige und kritisch bearbeitete Ausgabe, ed. Max Burckhardt, 10 vols. (1949–1986), Basel: Schwabe.

Burckhardt, Jacob, 1988, Die Kultur der Renaissance in Italien. Ein Versuch [1860], 11th edn, ed. Konrad Hoffmann, Stuttgart: Kröner.

Burckhardt, Jacob, 2001a, Der Cicerone. Eine Anleitung zum Genuss der Kunstwerke Italiens I. Architektur und Sculptur [1855] = Jacob Burckhardt, Werke (Kritische Gesamtausgabe), 29 vols., ed. Andreas Cesana et al., vol. 2, ed. Bernd Roeck et al., Munich: C. H. Beck.

Burckhardt, Jacob, 2001b, Der Cicerone. Eine Anleitung zum Genuss der Kunstwerke Italiens II. Malerei [1855] = Jacob Burckhardt, Werke (Kritische Gesamtausgabe), 29 vols., ed. Andreas Cesana et al., vol. 3, ed. Bernd Roeck et al., Munich: C. H. Beck.

Campioni, Giuliano, 1976, 'Von der Auflösung der Gemeinschaft zur Bejahung des "Freigeistes"', in: Nietzsche-Studien, 5, pp. 83–112.

Campioni, Giuliano, 1998, 'Il Rinascimento in Wagner e nel giovane Nietzsche', in: Rinascimento, 38, pp. 81–121.

Cancik, Hubert, 1995, Nietzsches Antike: Vorlesung, Stuttgart: J. B. Metzler.

Cesana, Andreas, 1994, 'Bachofen und Nietzsche', in: David M. Hoffmann (ed.), Nietzsche und die Schweiz. Katalog der Ausstellung im Strauhof Zürich, Zurich 1994, pp. 55–64.

Craig, Gordon, 1967, 'Friedrich Schiller and the Problem of Power', in: Leonard Krieger / Fritz Stern (eds.), The Responsibility of Power: Historical Essays in Honour of Hajo Holborn, New York: Doubleday & Co., pp. 125–144.

Detwiler, Bruce, 1990, Nietzsche and the Politics of Aristocratic Radicalism, Chicago: University of Chicago Press.

Dombowsky, Don, 2004, Nietzsche's Machiavellian Politics, Basingstoke: Palgrave Macmillan.

Donnellan, Brendan, 1979, 'Nietzsche and La Rochefoucauld', in: The German Quarterly, 52/3, pp. 303–318.

Ernst, Joachim, 1948, 'Die Rolle der Religion bei Jacob Burckhardt', Zeitschrift für Religions- und Geistesgeschichte, 1, pp. 335–344.

Farulli, Luca, 1990a, 'Nietzsche und die Renaissance. Die Reflexion über "Grenze" und "Grenzüberschreitung"', in: August Buck (ed.), Renaissance und Renaissancismus von Jacob Burckhardt bis Thomas Mann, Tübingen: Max Niemeyer, pp. 54–70.

Farulli, Luca, 1990b, 'Nietzsche e l'uomo italiano della Rinascenza', in: Franca Janowski / Ennio Bispuri (eds.), Nietzsche und Italien. Ein Weg vom Logos zum Mythos? Tübingen: Stauffenburg-Verlag, pp. 41–59.

Ferguson, Wallace K., 1948, The Renaissance in Historical Thought: Five Centuries of Interpretation, Boston: Houghton Mifflin.

Förster-Nietzsche, Elisabeth and Henri Lichtenberger, 1928, Nietzsche und sein Werk, Dresden: Carl Reissner Verlag.

'Fried.[rich] Nietzsche's Geistesentwicklung und Geistesphilosophie', 1895, in: Historisch-Politische Blätter für das katholische Deutschland, 116, pp. 823–832; 865–873.

Fubini, Riccardo, 1992, 'Renaissance Historian: The Career of Hans Baron', in: Journal of Modern History, 64, pp. 541–574.

Gay, Peter, 1967, 'Burckhardt's Renaissance: Between Responsibility and Power', in: Leonard Krieger / Fritz Stern (eds.), The Responsibility of Power: Historical Essays in Honour of Hajo Holborn, New York: Doubleday & Co.

Gelzer, Heinrich, 1907, Ausgewählte kleine Schriften, Leipzig: Teubner.

Gerhardt, Volker, 1988, 'Die Rache des Phantoms. Nietzsche und die Renaissance', in: Frankfurter Allgemeine Zeitung (7 December 1988).

Gerhardt, Volker, 1989, Die Renaissance im Denken Nietzsches, in: August Buck / Cesare Vasoli (eds.), Die Renaissance im 19. Jahrhundert in Italien und Deutschland, Bologna: Il Mulino, pp. 93–116.

Gerhardt, Volker, 1995, Friedrich Nietzsche, 2nd edn, Munich: C. H. Beck.

Gilbert, Felix, 1990, History: Politics or Culture? Reflections on Ranke and Burckhardt, Princeton: Princeton University Press.

Gilbert, Merlio (ed.), 2000, Nietzsche et l'histoire: contributions présentées au colloque scientifique international sur la 'Seconde Intempestive' de Nietzsche = Études Germaniques, 55/2.

Gossman, Lionel, 1994, 'Cultural History and Crisis: Burckhardt's "Civilization of the Renaissance in Italy"', in: Michael S. Roth (ed.), Rediscovering History: Culture, Politics and the Psyche, Stanford: Stanford University Press, pp. 404–427.

Gossman, Lionel, 1999, 'The *Existenzbild* in Burckhardt's Art Historical Writing', in: Modern Language Notes, 114/5 (Comparative Literature Issue), pp. 879–928.

Gossman, Lionel, 2000, Basel in the Age of Burckhardt: A Study in Unseasonable Ideas, Chicago: University of Chicago Press.

Gottfried, Paul, 1975, 'Arthur Schopenhauer as a Critic of History', in: Journal of the History of Ideas, 36/2, pp. 331–338.

Gregor-Dellin, Martin, 1980, Richard Wagner. Sein Leben—Sein Werk—Sein Jahrhundert, Munich: Piper.

Hale, John R., 1973, Renaissance Europe, 1480–1520, London: Fontana.

Hampe, Karl, 1925, Kaiser Friedrich II. in der Auffassung der Nachwelt, Stuttgart: Deutsche Verlagsanstalt.

Hankins, James, 1995, 'The "Baron Thesis" after Forty Years and some Recent Studies of Leonardo Bruni', in: Journal of the History of Ideas, 56/2, pp. 309–338.

Hankins, James (ed.), 2000, Renaissance Civic Humanism: Reappraisals and Reflections, Cambridge: Cambridge University Press.

Hardtwig, Wolfgang, 1996, Jacob Burckhardt and Max Weber: Two Conceptions of the Origin of the Modern World, in: Reginald Lilly (ed.), The Ancients and the Moderns, Bloomington: Indiana University Press, pp. 170–180.

Heinse, Wilhelm, 1998, Ardinghello und die glückseligen Inseln. Eine italiänische Geschichte aus dem sechszehnten Jahrhundert [1787] (Kritische Studienausgabe), ed. Max L. Baeumer, Stuttgart: Reclam.

Heller, Erich, 1988, 'Burckhardt and Nietzsche', in: Erich Heller, The Importance of Nietzsche, Chicago: University of Chicago Press, pp. 39–55.

Hinz, Manfred, 1989, 'Nascita, rinascita e ripetizione nella "Nascita della tragedia"', in: August Buck / Cesare Vasoli (eds.), Die Renaissance im 19. Jahrhundert in Italien und Deutschland, Bologna: Il Mulino, pp. 117–131.

Hirsch, Emanuel, 1998, 'Nietzsche und Luther' [1921], in: Emanuel Hirsch, Lutherstudien, vol. 2 = Emanuel Hirsch, Gesammelte Werke, 48 vols. (1998ff.), ed. Hans Martin Müller, Waltrop: Spenner, vol. 2, pp. 168–206.

Hofmann, Hasso, 1971, 'Jacob Burckhardt und Friedrich Nietzsche als Kritiker des Bismarckreiches', in: Der Staat. Zeitschrift für Staatslehre, öffentliches Recht und Verfassungsgeschichte, 10, pp. 433–453.

Howard, Thomas A., 2000, Religion and the Rise of Historicism: W. M. L. de Wette, Jacob Burckhardt, and the Theological Origins of Nineteenth-Century Historical Consciousness, Cambridge: Cambridge University Press.

Jacobs, Angelika, 1996, 'Frühe Formen des Historismus in Goethes Renaissance-Rezeption. Zur ästhetisch-historischen Genese moderner Subjektivität', in: Harald Tausch (ed.), Historismus und Moderne, Würzburg: Ergon, pp. 81–97.

Jacobs, Angelika, 1997, Goethe und die Renaissance. Studien zum Konnex von historischem Bewußtsein und ästhetischer Identitätskonstruktion, Munich: Fink.

Jacobs, Angelika, 1998, 'Renaissance', in: Bernd Witte / Theo Buck et al. (eds.), Goethe Handbuch, 5 vols., vol. 4/2, Stuttgart: J. B. Metzler, pp. 898–905.

Janssen, Evert Maarten, 1970, Jacob Burckhardt und die Renaissance, Assen: van Gorcum.

Janz, Curt P., 1978, Friedrich Nietzsche. Biographie, 3 vols., Munich: Hanser.

Jeismann, Karl-Ernst, and Peter Lundgreen (eds.), 1987, Handbuch der deutschen Bildungsgeschichte, vol. 3, Munich: C. H. Beck.

Jelavich, Peter, 1985, Munich and Theatrical Modernism: Politics, Playwriting, and Performance, 1890–1914, Cambridge, MA: Harvard University Press.

Joël, Karl, 1918, Jacob Burckhardt als Geschichtsphilosoph, Basel: Helbing und Lichtenhahn.

Jung, Matthias, 1991, 'Das Unbehagen an der Geschichte. Arthur Schopenhauer und Jacob Burckhardt', in: Matthias Jung / Klaus-Jürgen Grün (eds.), Idee, Natur und Geschichte. Alfred Schmidt zum 60. Geburtstag, Hildesheim: Georg Olms, pp. 171–192.

Kaegi, Werner, 1932, 'Über die Renaissanceforschung Ernst Walsers', in: Ernst Walser, Gesammelte Studien zur Geistesgeschichte der Renaissance, Basel: Schwabe, pp. xi–lx.

Kaegi, Werner, 1962, 'Niederländische Blütezeit und italienische Klassik', in: Werner Kaegi, Europäische Horizonte im Denken Jacob Burckhardts, Basel: Schwabe, pp. 115–151.

Kaegi, Werner, 1982, Jacob Burckhardt. Eine Biographie, 7 vols. (1947–1982), Basel: Schwabe.

Kahan, Alan, 2003, 'The Burckhardt–Sismondi Debate over the Meaning of the Italian Renaissance', in: Yannick Portebois / Nicholas Terpstra (eds.), The Renaissance in the Nineteenth Century/Le XIXe siècle renaissant, Toronto: Publications of the Centre for Reformation and Renaissance Studies Series, pp. 159–171.

Kaufmann, Walter A., 1974, Nietzsche: Philosopher, Psychologist, Antichrist [1950], 4th edn, Princeton: Princeton University Press.

Körner, Elisabeth, 1980, 'Das Renaissancebild der Aufklärung', in: Richard Toellner (ed.), Aufklärung und Humanismus, Heidelberg: Niemeyer, pp. 23–34.

Ladwig, Perdita, 2004, Das Renaissancebild deutscher Historiker 1898–1933, Frankfurt a. M.: Campus Verlag.

Large, Duncan, 2000, '"Our Greatest Teacher": Nietzsche, Burckhardt, and the Concept of Culture', in: International Studies in Philosophy, 32/3, pp. 3–23.

Leo, Heinrich, 1832, Geschichte der italienischen Staaten, 5 vols. (1829–1832), Hamburg: Perthes.

Lothar, Rudolph, 1893, Cäsar Borgias Ende. Trauerspiel in einem Akt, Dresden: E. Pierson.

Love, Frederick R., 1963, Young Nietzsche and the Wagnerian Experience, Chapel Hill: University of North Carolina Press.

Löwith, Karl, 1966, Jacob Burckhardt. Der Mensch inmitten der Geschichte [1936], Stuttgart: Kohlhammer.

Mann, Thomas, 1918, Betrachtungen eines Unpolitischen, Berlin: S. Fischer.

Mann, Thomas, 1953, Lübeck als geistige Lebensform [1926], in: Thomas Mann, Altes und Neues. Kleine Prosa aus fünf Jahrzehnten, Frankfurt a. M.: S. Fischer, pp. 290–314.

Marti, Urs, 1993, 'Der große Pöbel- und Sklavenaufstand'. Nietzsches Auseinandersetzung mit Revolution und Demokratie, Stuttgart: J. B. Metzler.

Martin, Alfred von, 1947a, Nietzsche und Burckhardt. Zwei Repräsentanten einer Epoche, 4th edn, Munich: Erasmus-Verlag.

Martin, Alfred von, 1947b, Die Religion in Jacob Burckhardts Leben und Denken. Eine Studie zum Thema Humanismus und Christentum, 2nd edn, Munich: Erasmus-Verlag.

Mattioli, Aram, 1999, '"Odiose Kerle, Judenpack". Der schöngeredete Antisemitismus des Jacob Burckhardt', in: Die Zeit, 40 (30 September 1999), p. 82.

Maurer, Reinhart, 1993, 'Nietzsche und das Ende der Geschichte', in: Thomas Nipperdey / Anselm Doering-Manteuffel / Hans-Ulrich Thamer (eds.), Weltbürgerkrieg der Ideologien. Antworten an Ernst Nolte. Festschrift zum 70. Geburtstag, Berlin: Propyläen, pp. 421–439.

Michelet, Jules, 1883, Histoire de France [1833–1862], 5 vols., Paris: Hetzel.

Molner, David, 1993, 'The Influence of Montaigne on Nietzsche: A Raison d'Être in the Sun', in: Nietzsche-Studien, 22, pp. 80–93.

Mommsen, Wolfgang J., 1986, 'Jacob Burckhardt (1818–1897): Defender of Culture and Prophet of Doom', in: John A. Hall (ed.), Rediscoveries: Some Neglected Modern European Political Thinkers, Oxford: Oxford University Press, pp. 47–63.

Naake, Erhard, 1985, 'Friedrich Nietzsches Verhältnis zu wichtigen sozialen und politischen Bewegungen seiner Zeit', Diss. Jena.

Nehamas, Alexander, 1985, Nietzsche: Life as Literature, Cambridge, MA: Harvard University Press.

Norbrook, David, 1989, 'Life and Death of Renaissance Man', in: Raritan, 8/4, pp. 89–110.

O'Pecko, Michael T., 1976, 'Renaissancism and the German Drama, 1890–1910', Diss. Baltimore.

Orsucci, Andrea, 1996, Orient-Okzident. Nietzsches Versuch einer Loslösung vom europäischen Weltbild, Berlin: Walter de Gruyter.

Panizza, Oskar, 1894, Das Liebeskonzil. Eine Himmels-Tragödie in fünf Aufzügen. Zurich: J. Schabelitz.

Pastor, Ludwig, 1950, Tagebücher, Briefe, Erinnerungen, ed. Wilhelm Wühr, Heidelberg: F. H. Kerle.

Pernet, Martin, 1989, Das Christentum im Leben des jungen Nietzsche, Opladen: Westdeutscher Verlag.

Pletsch, Carl, 1977, 'History and Friedrich Nietzsche's Philosophy of Time', in: History and Theory, 16/1, pp. 30–39.

Ranke, Leopold von, 1890, Die römischen Päpste in den letzten vier Jahrhunderten, vol. 1 = Leopold von Ranke, Sämtliche Werke, 54 vols., Leipzig: Duncker & Humblot (1867–1890), vol. 37.

Rehm, Walther, 1924, Das Werden des Renaissancebildes in der deutschen Dichtung vom Rationalismus bis zum Realismus, Munich: C. H. Beck.

Rehm, Walther, 1929, 'Der Renaissancekult um 1900 und seine Überwindung', in: Zeitschrift für deutsche Philologie, 54, pp. 296–332.

Rehm, Walther, 1960, Jacob Burckhardt und Eichendorff, Freiburg im Breisgau: E. Albert.

Reibnitz, Barbara von, 1992, Ein Kommentar zu Friedrich Nietzsche 'Die Geburt der Tragödie aus dem Geiste der Musik' (Kapitel 1–12), Stuttgart: J. B. Metzler.

Reinhardt, Volker, 2002, Jacob Burckhardt und die Erfindung der Renaissance. Ein Mythos und seine Geschichte, Bern: Schweizerische Akademie der Geistes- und Sozialwissenschaften.

Ressing, Gerd, 1951, 'Jacob Burckhardt und Bismarck. Der Reichsgründer im zeitkritischen Bilde des Historikers', Diss. Cologne.

Ritter-Santini, Lea, 1974, 'Maniera Grande. Über italienische Renaissance und deutsche Jahrhunderwende', in: Roger Bauer et al. (eds.), Fin de Siècle. Zu Literatur und Kunst der Jahrhundertwende, Frankfurt a. M.: Klostermann, pp. 170–205.

Roeck, Bernd, 1991, 'Aby Warburgs Seminarübungen über Jacob Burckhardt im Sommersemester 1927', in: Idea (Jahrbuch der Hamburger Kunsthalle), 10, pp. 65–89.

Ross, Werner, 1980, Der ängstliche Adler. Friedrich Nietzsches Leben, Munich: Deutscher Taschenbuch Verlag.

Rossi, Roberto, 1987, Nietzsche e Burckhardt, Genoa: Tilgher.

Ruehl, Martin, 2003, 'Politeia 1871: Nietzsche *contra* Wagner on the Greek State', in: Ingo Gildenhard / Martin Ruehl (eds.), Out of Arcadia: Classics and Politics in Germany in the Age of Burckhardt, Nietzsche and Wilamowitz, London: Institute of Classical Studies, pp. 61–86.

Ruehl, Martin, 2004, 'Death in Florence: Thomas Mann and the Ideologies of Renaissancismus at the Fin de Siècle', in: Suzanne Marchand / David Lindenfeld (eds.), Germany at the Fin de Siècle: Culture, Politics, and Ideas, Baton Rouge: Louisiana State University Press, pp. 186–227.

Ruhstaller, Peter, 1988, 'Burckhardt und Nietzsche: Deutungen einer vieldeutigen Beziehung', Diss. Zurich.

Rumohr, Carl Friedrich von, 1920, Italienische Forschungen [1827–1831], ed. Julius Schlosser, Frankfurt: Frankfurter Verlagsanstalt.

Salin, Edgar, 1948, Jakob Burckhardt und Nietzsche [1938], 2nd edn, Heidelberg: L. Schneider.

Sautet, Marc, 1981, Nietzsche et la Commune, Paris: Le Sycomore.

Schieder, Theodor, 1963, 'Nietzsche und Bismarck', in: Historische Zeitschrift, 196, pp. 320–342.

Schlechta, Karl, 1958, 'Nietzsches Verhältnis zur Historie', in: Karl Schlechta, Der Fall Nietzsche, Munich: Hanser, pp. 44–72.

Schmidt, Martin, 1976, 'Der Liberalismus als Problem für die Kirche und Kirchengeschichte im 19. Jahrhundert', in: Martin Schmidt / Georg Schwaiger (eds.), Kirchen und Liberalismus im 19. Jahrhundert, Göttingen: Vandenhoeck und Ruprecht, pp. 9–32.

Schopenhauer, Arthur, 1988, Werke in fünf Bänden, ed. Ludger Lütkehaus, 6 vols., Zurich: Haffmans.

Simonde de Sismondi, Jean-Charles Léonard, 1826, Histoire des républiques italiennes du moyen age, 16 vols. (1809–1818), Paris: Treuttel & Würtz.

Sommer, Andreas Urs, 2000, Friedrich Nietzsches 'Der Antichrist'. Ein philosophisch-historischer Kommentar, Basel: Schwabe.

Stadelmann, Rudolf, 1930, 'Jacob Burckhardt und das Mittelalter', Historische Zeitschrift, 142, pp. 457–515.

Stierle, Karlheinz, 1987, 'Renaissance. Entstehung eines Epochenbegriffs', in: Reinhart Herzog / Reinhart Koselleck (eds.), Epochenschwelle und Epochenbewußtsein, Munich: Fink, pp. 453–492.

Todte, Mario, 2004, Georg Voigt (1827–1891). Pionier der historischen Humanismusforschung, Leipzig: Leipziger Universitätsverlag.

Tracy, James D., 1968, 'Erasmus Becomes a German', in: Renaissance Quarterly, 21/3, pp. 281–288.

Traeger, Jörg, 1971, 'Raffaels Stanza d'Eliodoro und ihr Bildprogramm', in: Römisches Jahrbuch für Kunstgeschichte, 13, pp. 29–99.

Uekermann, Gerd, 1985, Renaissancismus und Fin de Siècle. Die italienische Renaissance in der deutschen Dramatik der letzten Jahrhundertwende, Berlin: Walter de Gruyter.

Vacano, Diego A. von, 2007, The Art of Power: Machiavelli, Nietzsche and the Making of Aesthetic Political Theory, Lanham, MD: Lexington.

Vivarelli, Vivetta, 1993, 'Montaigne und der "Freie Geist"', in: Nietzsche-Studien, 23, pp. 79–101.

Vivarelli, Vivetta, 1998, Nietzsche und die Masken des freien Geistes. Montaigne, Pascal und Sterne, Würzburg: Königshausen und Neumann.

Voigt, Georg, and Maximilian Lehnerdt, 1893, Die Wiederbelebung des classischen Alterthums oder das erste Jahrhundert des Humanismus [1859], 2 vols., 3rd edn, Berlin: G. Reimer.

Volpi, Franco, 1999, 'Nietzsche ora si scopre che plagiava Burckhardt', in: La Repubblica (17 June 1999), pp. 40–41.

Wagner, Cosima, 1977, Die Tagebücher, 1869–1883, ed. Martin Gregor-Dellin / Dietrich Mack, 2 vols. (1976–1977), Munich: Piper.

Weber, Max, 1978, Economy and Society: An Outline of Interpretive Sociology, 2 vols., ed. Guenther Roth / Claus Wittich, vol. 1, Berkeley: University of California Press.

Weibel, Oskar, 1925, Tiecks Renaissancedichtung in ihrem Verhältnis zu Heinse und C. F. Meyer, Bern: Paul Haupt.

Wenzel, Johannes, 1967, Jacob Burckhardt in der Krise seiner Zeit, Berlin: Deutscher Verlag der Wissenschaften VEB.

Wölfflin, Heinrich, 1934, 'Einleitung', in: Jacob Burckhardt-Gesamtausgabe, ed. Emil Dürr, Werner Kaegi / Albert Oeri / Hans Trog / Felix Staehelin / Heinrich Wölfflin, 14 vols. (1929–1934), Stuttgart: Deutsche Verlagsanstalt, vol. 3, pp. i–xxv.

Zeeden, Ernst W., 1954, 'Die Auseinandersetzung des jungen Jacob Burckhardt mit Glaube und Christentum', in: Historische Zeitschrift, 178, pp. 493–514.

Zeeden, Ernst W., 1963, 'Zeitkritik und Gegenwartsverständnis in Jacob Burckhardts Briefen aus den Jahren der Reichsgründung (1859/1872)', in: Hans Rothfels / Waldemar Besson / Friedrich Hiller von Gaertringen (eds.), Geschichte und Gegenwartsbewußtsein. Historische Betrachtungen und Untersuchungen. Festschrift für H. Rothfels, Göttingen: Vandenhoeck und Ruprecht, pp. 86–105.

Part V

Tragic and Musical Time

Metaphysical and Historical Claims in *The Birth of Tragedy*

Katherine Harloe

What is *The Birth of Tragedy* about? From a contemporary critical perspective, the very attempt to pose this question may appear hopelessly naive. Even if the furthest reaches of the complex and varied history of the reception of Nietzsche's first book are ignored, debates among scholars over its coherence, content, and significance within Nietzsche's thought have shown no signs of abating, and Montinari's comment twenty-five years ago that 'the entire problem of interpretation of Nietzsche's philosophical firstling is still wide open' appears equally apposite today (Montinari 1980, p. 5). In this essay I wish to question an assumption which I believe has come increasingly to guide interpretation of *The Birth of Tragedy*, and which is common to many who hold very different substantive views of its content. This is the idea that it should be read primarily as a contribution to what was, admittedly, one of the major debates of German philosophy after Kant: that of the possibility of metaphysics. If this assumption is granted, the most important question to ask about Nietzsche's first book becomes whether or not he there asserts or denies the possibility of 'transcendent' knowledge of the ultimate nature of the world. Regardless of the substantive differences between the answers scholars have given to this question over the past few decades, agreement that *The Birth of Tragedy* is essentially an exercise in metaphysics has informed many influential readings.

Sometimes the assumption is very much a background presence in a discussion which focuses on different themes. Consider, for example, Alexander Nehamas' views as put forward in his book *Nietzsche: Life as Literature* (Nehamas 1985). His interpretation of Nietzsche places the doctrine of perspectivism centre-stage and is primarily concerned with the writings of the 1880s rather than the 1870s. *The Birth of Tragedy* is, however, mentioned in order to support the following observation:

> Nietzsche seems to have believed that there are some ultimate facts, some non-interpretive truths, concerning the real nature of the world ... he denied that these facts could ever be correctly stated through reason, language, and science. Yet he also believed (and here the influence of Schopenhauer became dominant) that tragedy, primarily through the musically inspired, 'Dionysian' chorus, can intimate the final truth that the ultimate nature of the world is to

> have no orderly structure: in itself the world is chaos, with no laws, no reason, and no purpose. (Nehamas 1985, pp. 42–43)

The Birth of Tragedy is here invoked as a document of Nietzsche's early faith in the possibility of metaphysics, and is thereby distinguished from the later writings, in which 'Nietzsche comes to deny the very contrast between things-in-themselves and appearance which was presupposed by his discussion of tragedy' (Nehamas 1985, p. 43). The assumption does rather more work in motivating the influential, deconstructive readings of *The Birth of Tragedy* offered by Paul de Man and Philippe Lacoue-Labarthe (de Man 1979, pp. 79–102; Lacoue-Labarthe 1971). As Henry Staten has convincingly shown, it is because de Man interprets *The Birth of Tragedy* as an attempt to depict an 'ontological hierarchy', according to which the Dionysian is genetically prior to the Apollonian, that his verdict on it as a text that is logocentric—and his consequent deconstruction—can operate (de Man 1979, pp. 83, 85; Staten 1990, pp. 187–216). More recently, James I. Porter has argued against the view that any metaphysical thesis is asserted in *The Birth of Tragedy* and in favour of reading it as an attempt 'to mimic and challenge—through a mixture of parody, irony, implausibility, and logical circularity—the metaphysical banalities that the work superficially conveys' (Porter 2000a, p. 87). While his reconstruction of the content of Nietzsche's argument could not be more opposed to that of Nehamas or de Man, his reinterpretation of Nietzsche as an anti-metaphysician nevertheless leaves the question of metaphysics in the foreground.

This first interpretative question is usually thought to be bound up closely with a second contested issue: the Schopenhauerianism of Nietzsche's first book. The connection seems straightforward enough: *The Birth of Tragedy*'s elaboration of the Apollonian-Dionysian polarity conspicuously deploys Schopenhauerian language, and Schopenhauer's magnum opus, *The World as Will and Representation* (Schopenhauer 1969 [English]; 1949a and b [German])[1], offers a systematic metaphysics in the traditional sense of a set of interconnected claims about the ultimate nature of the world. We might, therefore, take *The Birth of Tragedy*'s Schopenhauerianism as an indicator of its metaphysical commitment: insofar as Nietzsche's position there may justly be characterized as Schopenhauerian, he is defending a metaphysical thesis. It is my contention that this apparently plausible inference is in fact mistaken, and rests upon an oversimplification of what 'Schopenhauer' could have represented for Nietzsche at the time of

1 Translations from Schopenhauer's and Nietzsche's Nachlaß and letters are my own.

writing *The Birth of Tragedy*. Appreciating this leads us to recognize that *The Birth of Tragedy* may be 'Schopenhauerian' yet not 'metaphysical' in any straightforward sense.

My argument to this effect will proceed by means of a critique of one of the most recent attempts to give an overarching interpretation of *The Birth of Tragedy*: the aforementioned reading of Porter. Porter's discussion is important as it exposes of some of the puzzles and difficulties that arise when the interpretative question with which I began is answered in the affirmative. He is correct to insist that certain aspects of *The Birth of Tragedy*'s 'narrative structure'—its language, imagery, and train of argument—call into question the notion that its author is 'uncritically enthralled to a metaphysics that ... [he] later abandoned' (Porter 2000a, p. 20). In reinterpreting *The Birth of Tragedy* as an 'attack on metaphysics' (ibid., p. 28), however, and equating this with an attack on Schopenhauer, Porter repeats what I suggest are a mistaken interpretative assumption and attendant oversimplification. By responding to his arguments, then, I hope to be able to indicate why both ways of answering the question of metaphysics in relation to *The Birth of Tragedy* miss what is really at issue.

It would, of course, be impossible to provide an adequate response to Porter in the course of this essay. This is not just because his reading of *The Birth of Tragedy* is based on an detailed and broad-ranging consideration of Nietzsche's notebooks and early philological writings,[2] but also because he attributes to Nietzsche a deliberate strategy of what Quentin Skinner has termed 'oblique reference' (Skinner 1969, pp. 32–35). Put crudely, this is the writing of something one does not believe in order to disguise as well as to set out what one means to say. As Skinner points out, oblique strategies pose particular problems of interpretation, assessment of which requires close attention to the possible linguistic (textual) contexts of a particular work in order to decide whether its author is subverting or sustaining the ideas, generic conventions, topoi and so on, of his predecessors and contemporaries. Porter interprets *The Birth of Tragedy* as a subversive text; the immediate target of its critique is Wagnerian and Schopenhauerian metaphysics. My comments here are intended to draw attention to some aspects of the Schopenhauerian and Wagnerian linguistic

2 In addition to Porter 2000a, the focus of my discussion here, this reading is extended in Porter 2000b.

context, overlooked by Porter, which I believe support a different interpretation.

One of the cornerstones of Porter's reading is his interpretation of the Dionysian as a 'pleat in the texture of appearance' (2000a, p. 49; see pp. 33–50, passim) and hence of metaphysics as something 'generated from within' appearance itself. Repeatedly in *The Birth of Tragedy*, Nietzsche seems to claim that the Dionysian is ontologically prior to the Apollonian—'the eternal and original artistic power that first calls the whole world of phenomena into existence' (BT 25, quoted in Porter 2000a, p. 36). This message is, however, undermined by passages such as the allusion to Lucretius in *The Birth of Tragedy* 1, which implies that all divinities, Dionysus included, are the product of dreams. Just as much as Apollo, then, who is explicitly associated with dreams and deception, the opening section of *The Birth of Tragedy* provides a hefty hint that Dionysus is illusory: an aspect of human psychology rather than a constituent of the deeper reality behind appearances. Nevertheless—and this is the flip-side of Porter's reading—Nietzsche tells us that such illusions cannot simply be done away with. They are the product of the deep-seated human need to project some higher meaning onto existence.

These are noteworthy observations, but do they, as Porter thinks, amount to a decisive move away from a Schopenhauerian or Wagnerian position? Let us consider the crucial passage where Nietzsche states that 'As Lucretius envisages it, it was in dream that the magnificent figures of the gods first appeared before the souls of men' (BT 1, KSA 1, p. 26). The sentence continues by quoting Wagner:

> In dream the great image-maker saw the delightfully proportioned bodies of superhuman beings; and the Hellenic poet, if asked about the secrets of poetic procreation, would likewise have reminded us of dream and would have given an account much like that given by Hans Sachs in the *Meistersinger*:
>
> My friend, it is the poet's task
> To mark his dreams, their meaning ask.
> Trust me, the truest phantom man doth know
> Hath meaning only dreams may show:
> The arts of verse and poetry
> Tell nought but dreaming's prophecy. (ibid.)

In the following paragraph, which continues the theme of dreaming, we are referred to Schopenhauer:

> Philosophical natures even have a presentiment that hidden beneath the reality in which we live and have our being there also lies a second, quite different reality; in other words, this reality too is a semblance, and Schopenhauer actually states that the mark of a person's capacity for philosophy is the gift of

> feeling occasionally as if people and all things were mere phantoms or dream-images. (ibid.)

Porter is, I think, correct to interpret these passages as implicating the Dionysian and the supposedly higher reality it symbolizes in 'appearances', but how are we to read the specific allusions to Wagner and Schopenhauer in this context?

The immediate Schopenhauerian allusion is to a passage from his Nachlaß,[3] but the theme is treated at greater length in volume 2 of *The World as Will and Representation*, in a chapter tellingly titled 'On Man's Need for Metaphysics' (Schopenhauer 1969, vol. 2, pp. 160–187/1949b, pp. 175–209). There Schopenhauer talks of man as an *animal metaphysicum*, permanently afflicted by the desire for metaphysical knowledge. In the face of the evident suffering and misery of life, humans are compelled to wonder why the world exists. The desperate need for an answer to this question is, Schopenhauer says, the origin of all 'metaphysical' thought, both religious and philosophical:

> Temples and churches, pagodas and mosques, in all countries and ages, in their splendour and spaciousness, testify to man's need for metaphysics, a need strong and ineradicable, which follows close on the physical. (1969, vol. 2, p. 162/1949b, p. 177)

The difference between religion and philosophy does not consist in the claim, common to both, to embody a truth beyond appearances, but rather in their mode of presentation. Religions provide a 'popular metaphysics' resting upon revelation, and can be true solely *sensu allegorico*. Philosophy, by contrast, appeals to thought and conviction and claims to be true *sensu proprio* (1969, vol. 2, pp. 166–168/1949b, pp. 183, 185). Nevertheless, both arise from humans' need, faced with the misery of life, to make 'metaphysical assumptions' about the existence of another world whose real character is separated by 'a deep gulf, a radical difference' from anything of which they can conceive (1969, vol. 2, p. 178/1949b, pp. 197, 198). Belief in metaphysical doctrines is, then, a human cognitive response to misery and helplessness in the face of existence, and both religion and philosophy, as forms of metaphysics, gain their content by a projection of the antithesis of the world of 'appearances' into an assumed beyond. In this

3 'He who does not feel occasionally as if people and all things were mere phantoms or dream-images has no gift for philosophy. For it arises out of the contrast of individual things with the Idea of which they are the appearance' (Schopenhauer 1864, p. 295). An annotated copy of this work survives among Nietzsche's personal effects, although the date at which he purchased it is unknown (see Oehler 1942, p. 21).

chapter of *The World as Will and Representation*, Schopenhauer presents a view of the origins of metaphysical thought which is surprisingly similar to Porter's interpretation of the hidden message of *The Birth of Tragedy* 1.

The heavily annotated copy of *The World as Will and Representation* which survives among Nietzsche's personal possessions is part of the *Collected Works* edited by Julius Frauenstädt and published in 1873–1874, after the appearance of *The Birth of Tragedy* (Oehler 1942, p. 21). While there can be no doubt that Nietzsche read *The World as Will and Representation* extensively in the years 1865–1872, it is impossible to prove which chapters he studied most attentively. It is, however, extremely likely that he was familiar with the chapter discussed above, as it contains Schopenhauer's problematic and much-commented-upon claim that, unlike the systems of his predecessors, his metaphysics is *not* transcendent:

> And although no one can recognize the thing-in-itself through the veil of the forms of perception, on the other hand everyone carries this within himself, in fact he himself is it; hence in self-consciousness it must be in some way accessible to him, although still only conditionally. Thus the bridge on which metaphysics passes beyond experience is nothing but just that analysis of experience into phenomenon and thing-in-itself in which I have placed Kant's greatest merit. For it contains the proof of a kernel of the phenomenon different from the phenomenon itself. It is true that this kernel can never be entirely separated from the phenomenon, and be regarded by itself as an *ens extramundanum*; but it is known always only in its relations and references to the phenomenon itself. The interpretation and explanation of the phenomenon, however, in relation to its inner kernel can give us information about it which does not otherwise come into consciousness. Therefore in this sense metaphysics goes beyond the phenomenon, i.e., nature, to what is concealed in or behind it (*τὸ μετὰ τὸ φυσικόν*), yet always regarding it only as that which appears in the phenomenon, not independently of all phenomenon. Metaphysics thus remains immanent, and does not become transcendent; for it never tears itself entirely from experience, but remains the mere interpretation and explanation thereof, as it never speaks of the thing-in-itself otherwise than in its relation to the phenomenon. This, at any rate, is the sense in which I have attempted to solve the problem of metaphysics, taking into general consideration the limits of human knowledge which have been demonstrated by Kant. (Schopenhauer 1969, vol. 2, pp. 182–183/1949b, pp. 203–204)

This claim was interrogated by Rudolf Haym in his 1864 essay on Schopenhauer, which Nietzsche read in 1866.[4]

4 Nietzsche to Hermann Mushacke, 27 April 1866, KGB I 2, pp. 126–129; Nietzsche to Carl von Gersdorff, end-August 1866, KGB I 2, pp. 156–161 (see Barbera 1994).

The likely linguistic contexts of the appeal of *The Birth of Tragedy* 1 to Wagner complement this picture of congruence between Nietzsche's arguments and Schopenhauerian themes. The passage Nietzsche quotes centres around the paradoxical notion of the 'truest phantom' or 'illusion' (*wahrster Wahn*), and is taken from Act III of *Die Meistersinger*, in which *Wahn* is a prominent theme.[5] It is therefore relevant to consider Wagner's letter to Ludwig II of Bavaria (Wagner 1911 [German]/1995 [English]), published in 1873 under the title *Über Staat und Religion*, which Nietzsche read in manuscript in 1869.[6] In this letter, Wagner combines a Schopenhauerian metaphysical standpoint with a lengthy analysis of political and religious ideas as forms of *Wahn*, necessary illusions:

> Blindness is the world's true essence, and not Knowledge prompts its movements, but merely a headlong impulse, a blind impetus of unique weight and violence, which procures itself just so much light and knowledge as will suffice to still the pressing need experienced at the moment. So we recognize that nothing really happens but what has issued from this not far-seeing Will that answers merely to the momentarily-expressed need. (1995, p. 10/1911, p. 8)

Humans are the unwitting instruments of this blind striving for existence, and both patriotism (which induces them to place the ends of state above their own egoistic goals) and religion (which counsels resignation in the face of the wretchedness of existence) are ruses by which they are induced to serve the ends of Will. This outlook leads Wagner to give the following analysis of religious feeling:

> Its inmost kernel is denial of the world—i.e., recognition of the world *as a fleeting and dreamlike state reposing merely on illusion*—and struggle for Redemption from it, prepared-for by renunciation, attained by Faith. In true Religion a complete reversal thus occurs of all the aspirations to which the State had owed its founding and its organising: what is seen to be unattainable here, the human mind desists from striving-for upon this path, to ensure its reaching by a path entirely opposite. To the religious eye the truth grows plain that there must be another world than this, because the inextinguishable bent-to-happiness cannot be stilled within this world, and hence requires another world for its redemption. What, now, is that other world? So far as the conceptual faculties of human Understanding reach, and in their practical application as intellectual Reason, it is quite impossible to gain a notion that shall not

5 I am thinking in particular of Hans Sachs' famous *Wahn-monologue* at the end of act III, scene 1. The passage Nietzsche cites is from the beginning of act III, scene 2.

6 See Barbera 1994, p. 219 (no. 4). As late as 1873, Nietzsche thought fit to praise this work of Wagner's as 'in the highest sense "edifying"' (Nietzsche to Gersdorff, 2 March 1873, KSB II 3, p. 131).

> clearly show itself as founded on this selfsame world of need and change: wherefore, since this world is the source of our unhappiness, that other world, of redemption from it, must be precisely as different from the mode of cognisance whereby we are to perceive that other world must be different from the mode which shews us nothing but this present world of suffering and illusion. (1995, pp. 23–24/1911, pp. 20–21, emphasis mine)

Religious feeling is awesome in nature—Wagner calls it 'wonder-working' (*wunderwirkend*) and 'sublime' (*erhaben*) (1995, p. 25/1911, p. 21), but is nonetheless illusion for all that. In explicitly associating religious thought with illusion and dream, Wagner goes further than Schopenhauer does in the passages I have quoted, but both the language and the content of this recognizably Schopenhauerian train of thought foreshadow those aspects of *The Birth of Tragedy* 1 that Porter emphasizes.[7]

If Porter's argument that the Dionysian or the metaphysical originates as the compensatory fantasy of needy and suffering human beings is granted, it seems nevertheless that the elaboration of these thoughts in the opening sections of *The Birth of Tragedy* draws considerably on Schopenhauer's treatment of the same theme. It is, moreover, not merely Nietzsche's account of the *origins* of metaphysics that is Schopenhauerian in tenor. His discussion of the *resurgence* of the need for metaphysics in his contemporary era is also redolent of Schopenhauer. According to Nietzsche, this need is provoked anew by the eventual bankruptcy of the optimistic, 'Socratic' belief that science can provide a fully satisfactory explanation of the world (see especially BT 15, 18). The second half of Schopenhauer's chapter 'On Man's Need for Metaphysics' is likewise devoted to an extensive and scathing discussion of the ambitions of science to explain the world:

> Naturalism, or the purely physical way of considering things, will never be sufficient, it is like a sum in arithmetic that never comes out. Beginningless and endless causal series, inscrutable fundamental forces, endless space, beginningless time, infinite divisibility of matter, and all this further conditioned by a knowing brain, in which alone it exists just like a dream and without which it vanishes—all these things constitute the labyrinth in which naturalism leads us incessantly round and round ... In fact, even if a man wandered through all the planets of all the fixed stars, he would still not have made one step in *metaphysics*. On the contrary, the greatest advance in *physics* will only

7 The connection between metaphysical 'knowledge' and dreams is treated at length in Schopenhauer's essay on spirit-seeing (Schopenhauer 1960 [German]/1974 [English]). This discussion inspired Wagner's 1870 centenary essay on Beethoven, which Nietzsche praises in the Preface to BT and in section 16 (KSA 1, pp. 23, 104).

> make the need for a system of *metaphysics* felt more and more, since the corrected, extended, and more thorough knowledge of nature is the very knowledge that always undermines and finally overthrows the metaphysical assumptions that till then have prevailed. (Schopenhauer 1969, vol. 2, pp. 177–178/1949b, pp. 196–197)

Like Nietzsche, Schopenhauer is disdainful of the ambitions of science, and believes that it will eventually refute itself, provoking a return to metaphysical speculation. Not only are there general thematic parallels, but the very terms in which Nietzsche expresses the cultural importance of Socratism echo the cosmic imagery of Schopenhauer's contemptuous dismissal.[8]

An element of continuity with Schopenhauerian ideas is also, I would argue, implied by the imagery of veiling that Nietzsche uses to depict the insight offered by the Dionysian state:

> Now, hearing this gospel of universal harmony, each person feels himself to be not only united, reconciled or merged with his neighbour, but one with him, as if the veil of maya had been torn apart, so that mere shreds of it flutter before the mysterious primordial unity. (BT 1, KSA 1, pp. 29–30; see too BT 15, KSA 1, pp. 98–99; BT 24, KSA 1, p. 150)

Porter points out that the veracity of this vision is far from assured, suggesting that the subjunctive character of the 'as if'-clause and the continued fluttering of the tattered veil imply that the Dionysian vision does not provide immediate insight into the beyond (2000a, pp. 51–52). He concludes that this represents a critique of Schopenhauer; but again, there are Schopenhauerian precedents. We have already seen Schopenhauer speak of 'the veil of the forms of perception' in *The World as Will and Representa-*

8 'For the first time, thanks to this universality, a common network of thought was stretched over the whole globe, with prospects of encompassing even the laws of the entire solar system' (BT 15, KSA 1, p. 100). They also contain echoes of Wagner. Nietzsche characterizes the Socratic instinct for scientific knowledge as a 'sublime metaphysical illusion' (BT 15, KSA 1, p. 99) and comments that without its influence, human energy would have been 'applied instead to the practical, i.e., egotistical goals of individuals and nations'. The 'wars of extinction' that would have ensued would have led to a generalized and suicidal pessimism of the kind which, Nietzsche claims, 'has existed throughout the entire world, wherever art has not appeared in one form or other, especially as religion or science, to heal and to ward off the breath of that pestilence' (BT 15, KSA 1, pp. 100; see also p. 102). Wagner had likewise argued that patriotic or political *Wahn* is still too close to individual egoism to be stable, and will collapse into war unless supplemented by the illusions of faith (1995, pp. 15–19/1911, pp. 12–14). Nietzsche's account of the way science functions as a form of illusion is thereby aligned with Wagner's discussion of religion.

tion II, chapter 17, when wrestling with the thorny issue of human beings' 'inner' experience of the thing-in-itself (1969, vol. 2, pp. 182–183, quoted above). He resorts to this metaphor again in the following chapter, this time to confess the impossibility of an unshrouded view:

> Meanwhile it is to be carefully noted, and I have always kept it in mind, that even the inward observation we have of our own will still does not by any means furnish an exhaustive and adequate knowledge of the thing-in-itself ... in this inner knowledge, the thing-in-itself has indeed to a great extent cast off its veils, but still does not appear quite naked ... Accordingly we have to refer the whole world of phenomena to that one in which the thing-in-itself is manifested under the lightest of all veils, and still remains phenomenon only insofar as my intellect, the only thing capable of knowledge, still always remains distinguished from me as the one who wills, and does not cast off the knowledge-form of *time* even with inner perception. (Schopenhauer 1969, vol. 2, pp. 197, 198/1949b, pp. 220–221)[9]

These passages are taken from the second volume of *The World as Will and Representation,* which was added in the second edition of 1844 and forms a supplement to volume 1. There is no question that such statements are hard to reconcile with the confidence with which the thesis that the world is Will is presented in the first edition of Schopenhauer's work. It is nevertheless evident that the terms of what Porter sees a radical critique of Schopenhauer are available from Schopenhauer himself.

I have, I hope, succeeded in showing that allusions to these particular chapters of *The World as Will and Representation* are prominent at several points in *The Birth of Tragedy*. Nietzsche's selective allusions may fairly be said to emphasize this self-critical moment in Schopenhauer, but do they thereby amount to a wholesale rejection of whatever he may have understood the elder philosopher to stand for? An alternative interpretation is suggested by yet another apologia for the use of metaphysical language—this time from Nietzsche's own notebooks. The passage is from an early draft of Fragment 10[1], which survives labelled by Nietzsche as

9 The imagery of the veil has a long pedigree in German philosophical aesthetics, evoked by Kant Goethe, Schiller, Novalis, Hegel, and others. See Gombrich 1985 for some examples. The implication is always double-edged: a veil conceals as much as it reveals. It is this tradition that Nietzsche taps into with his remarks about the veiling and unveiling in BT 15 and in *The Gay Science* (GS Preface to the second edition 4, KSA 3, pp. 351–352; GS 57, KSA 3, pp. 421–422).

'Fragment of an extended form of "The Birth of Tragedy" written in the first weeks of the year 1871' (KSA 7, pp. 333ff.):

> If I ventured in passing to speak of genius and of appearance as if a knowledge that exceeded every bound stood at my disposal and as if I were able to see out of the pure, great eye of the world, in what follows it will be explained that in using this figurative language [*Bildersprache*] I do not believe that I have stepped beyond anthropomorphic bounds. But who could endure existence without such mystical possibilities? (KSA 14, p. 541)

Porter states that Nietzsche's position in *The Birth of Tragedy* is 'not only that metaphysics is a fictional enterprise worthy of being shattered once and for all *but also that its resurrection is an inescapable and constitutional need deeply implanted in human nature*' (Porter 2000a, p. 9; emphasis mine). Although he recognizes that Nietzsche portrays metaphysical speculation as a matter of human need, his overall discussion of *The Birth of Tragedy* suggests that its argument is weighted heavily towards critique. In the passage above, however, we see Nietzsche underlining in poignant terms a conclusion that we have also seen Schopenhauer and Wagner emphasize: the need for a myth such as the metaphysical provides in order to endure existence. Nietzsche's acknowledgement of this need, together with its Schopenhauerian precedent, raises the possibility that *The Birth of Tragedy* deploys Schopenhauer not in parodic fashion, as a weapon with which to shatter all such illusions, but rather as a means of developing them in a new and superior form.[10]

This interpretation also coheres with Nietzsche's comments about the work of Friedrich August Lange. Porter argues that it was reading Lange that caused Nietzsche to apostatize from Schopenhauer (Porter 2000a, pp. 5, 9–16). Yet, in the same August 1866 letter to Gersdorff in which he praises Lange's *History of Materialism* as 'splendid and highly instructive' (KGB I 2, p. 159), Nietzsche draws a different conclusion: 'You see that even in the face of this most exacting critique our Schopenhauer remains for us, indeed, he almost becomes us even more.' What Lange's arguments show is, according to Nietzsche, that philosophy can only be a form of art, of which none other than Schopenhauer furnishes the highest example:

> If philosophy is art, then even Haym may hide from Schopenhauer; if philosophy should be edifying, then I at least know no philosopher who edifies more than our Schopenhauer. (KGB I 2, p. 160)

My suggestion is therefore that Nietzsche draws upon Schopenhauer in *The Birth of Tragedy* as part of his attempt to foster a new form of Wagnerian

10 Note the echo in Nietzsche's 1873 praise of Wagner's 'On State and Religion' as 'highly "edifying"', quoted in n. 6 above.

Wahn: an acknowledgement and indulgence of the need to find a higher meaning in existence, however illusory that meaning may be. This project may seem opposed to Schopenhauer's goal of presenting a system of metaphysics in the grand style, yet there are sufficient counter-currents in *The World as Will and Representation* to enable Nietzsche to enlist his predecessor in the service of this enterprise. Schopenhauer claims that his philosophy embodies a set of 'truths' (1969, vol. 2, p. 185/1949b, p. 206), yet not in the sense that it presents a system of conclusions derived deductively from true premises, nor because it relies on some form of privileged intuition. Rather, it is true in virtue of providing, in contrast to science, an 'understanding' (*Verständniß*), 'interpretation' (*Auslegung*), or 'deciphering' (*Entzifferung*) of the world of phenomena which is, so he claims, rich, satisfying and complete (1969, vol. 2 pp. 184–186/1949b, pp. 204–205). It is such a humanly satisfying interpretation of existence that, according to the arguments of *The Birth of Tragedy*, only art can provide. Nietzsche picks up on those elements of *The World as Will and Representation* which can be redeployed creatively in order to support this insight. The presentation of Schopenhauer which results from his refashioning is, admittedly, partial and one-sided. It may nevertheless be concluded that *The Birth of Tragedy* extends Schopenhauerian themes and concerns in order to hammer its message home.

The Birth of Tragedy's co-option of Schopenhauer extends further than this, however. Nietzsche does not stop at drawing upon his predecessor's arguments in order to announce the crisis of science; he also dramatizes this crisis and casts Schopenhauer in a leading role. He does so by constructing a narrative which has its beginnings in sixth-century Greece, and which locates Schopenhauer—along with Aeschylus, Sophocles, Euripides, Socrates, Kant, and Wagner—at points along a cultural development that will culminate in a new form of tragic art.[11] This chronology is putatively historical, but insofar as it may be characterized as a form of illusion that aims, via a representation of the past, to generate a constellation of

11 This comment assumes that the 'rebirth' of tragedy Nietzsche envisages in *The Birth of Tragedy* is, indeed, a Wagnerian Renaissance. Although this has sometimes been questioned, it still seems to me the best way to make sense not only of *The Birth of Tragedy* but of the references to Wagner in Nietzsche's notes and letters of the early 1870s. The scope of the rebirth Nietzsche has in mind is, however, far too broad and indeed open-ended to encompass Wagner alone. Although Wagner is identified with the fulfilment of this ideal in *The Birth of Tragedy*, this is compatible with the view that he later retracted this association and, as occurred in *Ecce Homo*, disavowed *The Birth of Tragedy*'s Wagnerianism without disowning the 'hope' that speaks out from the work (EH III BT 1 and 4).

beliefs and attitudes that legitimate a particular form of cultural activity, it might more aptly be termed ideological. Its function is to alert its readers to the climacteric shift taking place in European culture and to raise their hopes for tragedy's rebirth.

Nietzsche's most general verdict on Schopenhauer's significance within this narrative comes in *The Birth of Tragedy* sections 18 and 19, when he is describing the disintegration of the Socratic-optimistic outlook:

> The catastrophe slumbering in the womb of theoretical culture is gradually beginning to frighten modern man ... Meanwhile great natures with a bent for general problems have applied the tools of science itself, with incredible deliberation, to prove that all understanding, by its very nature, is limited and conditional, thereby rejecting decisively the claim of science to universal validity and universal goals. Thanks to this demonstration it has been recognized for the first time that it is a delusion [*Wahnvorstellung*] to believe that we can penetrate to the innermost essence of things by following the chain of causality. The hardest-fought victory was won by the enormous courage and wisdom of Kant and Schopenhauer, a victory over the optimism which lies hidden in the nature of logic and which in turn is the hidden foundation of our culture ... This insight marks the beginning of a culture which I now dare to describe as a tragic culture. Its most important feature lies in putting wisdom in place of science as its highest goal. (BT 18, KSA 1, pp. 117–118)

> Let us recall then, how Kant and Schopenhauer made it possible for the spirit of German philosophy ... to destroy scientific Socratism's contented pleasure in existence by demonstrating its limits, and how this demonstration ushered in an incomparably deeper and more serious consideration of ethical questions and art, one which can be defined as the conceptual formulation of Dionysiac wisdom. In what direction does this mysterious unity of German music and German philosophy point, if not towards a new form of existence, the content of which can only be guessed at from Hellenic analogies? (BT 19, KSA 1, p. 128)

In these remarks, Schopenhauer is lauded (alongside Kant) for having demonstrated the bankruptcy of the Socratic attempt to view the world as amenable to human understanding. The philosophers are not praised for their residual hope for a form of knowledge that transcends the bounds of experience, but rather because of the demonstration their arguments furnish of those very bounds. Although it is Socrates whom Nietzsche dubs 'the vortex and turning point of so-called world history' (BT 15, KSA 1, p. 100), in *The Birth of Tragedy* 18 and 19, Kant and Schopenhauer appear almost as important as actors on the world-historical stage. In finally discrediting Socratism, they clear the way for the replacement of corrosive scientism with a 'new form of existence': a renewed kind of artistic orientation to the world. Schopenhauer is significant in this story not as the last metaphysician, but rather as the philosopher who demonstrates the need for a new myth and who anticipates its form.[12] In doing so, he merits praise as an augur of the rebirth of tragedy. His successor, both in this prophesying and in this anticipating, is Nietzsche himself.

This paper has tried to rehabilitate some claims about *The Birth of Tragedy* which may seem rather traditional: namely, the positive character of its appropriation of Schopenhauer and Wagner, and the importance of the (quasi-)historical structure of its argument. Being traditional does not, of course, amount to being mistaken, and I hope I have shown that such claims can be supported by crediting Nietzsche with a less naive reception of Schopenhauer than has sometimes been suggested. Nietzsche famously warns philosophers to be vigilant about the unnoticed and subtle commitments inherent in the grammar of our language (BGE 2, KSA 5, p. 54; TI '"Reason" in Philosophy' 5, KSA 6, p. 78), but the manner in which the areas and positions of long-running debates come to be defined may occasionally be just as insidious.[13]

References

Barbera, Sandro, 1994, 'Ein Sinn und unzählige Hieroglyphen. Einige Motive von Nietzsches Auseinandersetzung mit Schopenhauer in der Basler Zeit', in: Tilman Borsche / Federico Gerratana / Aldo Venturelli (eds.), Centauren-Geburten. Wis-

12 As Nietzsche emphasizes in BT 16, it is Schopenhauer's analysis of the representational and expressive capacity of music which also provides an intimation of the kind of art by means of which the crisis can be overcome. Schopenhauer's writings suggest that a work of art which combines music with images or action can represent 'the innermost kernel preceding all form, or the heart of things' (Schopenhauer 1949a, p. 311, quoted by Nietzsche, BT 16, KSA 1, p. 106). This is, of course, the kind of representation Nietzsche characterizes as *myth*: 'the symbolic image ... with the highest degree of significance' (BT 16, KSA 1, p. 107). Regrettably, space considerations preclude any further discussion of this aspect of Nietzsche's appropriation of Schopenhauer here.

13 The research for this paper was begun when I was a Junior Postdoctoral Fellow in the Institute of Greece, Rome and the Classical Tradition at the University of Bristol, UK. I am grateful to the Institute Board for funding my research and to the Bristol Classical Seminar for their responses to an early presentation. I also owe thanks to Martin Ruehl, Raymond Geuss, Mike Levene, and Thomas Brobjer for their comments and questions on my initial conference paper, and to Nicholas Jardine and Dawn Phillips for subsequent constructive criticism.

senschaft, Kunst und Philosophie beim jungen Nietzsche, Berlin: Walter de Gruyter, pp. 217–233.
de Man, Paul, 1979, Allegories of Reading: Figural Language in Rousseau, Nietzsche, Rilke, and Proust, New Haven: Yale University Press.
Gombrich, E. H., 1985, 'The Symbol of the Veil: Psychological Reflections on Schiller's Poetry', in: Peregrine Horden (ed.), Freud and the Humanities, London: Duckworth, pp. 75–109.
Lacoue-Labarthe, Philippe, 1971, 'Le detour', in: Poétique, 5, pp. 53–76.
Montinari, Mazzino, 1980, Nietzsche lesen, Berlin: Walter de Gruyter.
Nehamas, Alexander, 1985, Nietzsche: Life as Literature, Cambridge, MA: Harvard University Press.
Oehler, Max, 1942, Nietzsches Bibliothek, Weimar: Gesellschaft der Freunde des Nietzsche-Archivs.
Porter, James I., 2000a, The Invention of Dionysus: An Essay on The Birth of Tragedy, Stanford, CA: Stanford University Press.
Porter, James I., 2000b, Nietzsche and the Philology of the Future, Stanford, CA: Stanford University Press.
Schopenhauer, Arthur, 1864, Aus Arthur Schopenhauer's handschriftlichem Nachlaß. Abhandlungen, Anmerkungen, Aphorismen und Fragmente, ed. Julius Frauenstädt, Leipzig: F. U. Brockhaus.
Schopenhauer, Arthur, 1949a, Die Welt als Wille und Vorstellung, Erster Band, in: Arthur Schopenhauer, Sämtliche Werke. Nach der ersten, von Julius Frauenstädt besorgten Gesamtausgabe neu bearbeitet, ed. Arthur Hübscher, vol. 3, Wiesbaden: Eberhard Brockhaus Verlag.
Schopenhauer, Arthur, 1949b, Die Welt als Wille und Vorstellung, Zweiter Band, in: Arthur Schopenhauer, Sämtliche Werke, ed. Arthur Hübscher, vol. 4, Wiesbaden: Eberhard Brockhaus Verlag.
Schopenhauer, Arthur, 1960, 'Versuch über das Geistersehn und was damit zusammenhängt', in: Arthur Schopenhauer, Parerga und Paralipomena. Kleine philosophische Schriften, Erster Band, in: Arthur Schopenhauer, Sämtliche Werke, ed. Arthur Hübscher, vol. 5, Wiesbaden: Eberhard Brockhaus Verlag, pp. 239–329.
Schopenhauer, Arthur, 1969, The World as Will and Representation, trans. E. F. J. Payne, 2 vols., New York: Dover.
Schopenhauer, Arthur, 1974, 'Essay on Spirit-Seeing and Everything Connected Therewith', in: Arthur Schopenhauer, Parerga and Paralipomena: Short Philosophical Essays, trans. E. F. J. Payne, vol. 1, Oxford: Clarendon Press, pp. 225–309.
Skinner, Quentin, 1969, 'Meaning and Understanding in the History of Ideas', in: History and Theory, 8, pp. 3–53.
Staten, Henry, 1990, Nietzsche's Voice, Ithaca, NY: Cornell University Press.
Wagner, Richard, 1911, 'Über Staat und Religion', in: Richard Wagner, Sämtliche Schriften und Dichtungen. Volksausgabe, Sechste Auflage, vol. 8, Leipzig: Breitkopf und Härtel, pp. 3–29.
Wagner, Richard, 1995, 'On State and Religion', in: Richard Wagner, Art and Politics, trans. William Ashton Ellis, Lincoln: University of Nebraska Press, pp. 5–34.

Translations

The Birth of Tragedy, ed. Raymond Geuss, trans. Ronald Speirs, Cambridge: Cambridge University Press, 1999.

Nietzsche's Musical Conception of Time

Jonathan R. Cohen

1. Time in Music

My title is 'Nietzsche's Musical Conception of Time,' but in order to say something about that, I must first discuss Nietzsche's conception of musical time.

I will approach this topic by way of a passage in which Nietzsche criticizes Wagner. Nietzsche's psychological, political, and cultural criticisms of Wagner are fairly well-known, but his musicological criticisms are not. What I will do is examine closely one passage in which Nietzsche criticizes Wagner for musical reasons, and use that passage (and in particular one sentence in it) as a point of entry for Nietzsche's view of musical time.[1]

The passage appeared originally in 'Assorted Opinions and Maxims' (1879) as section 134. It is reprinted in the anthology *Nietzsche contra Wagner* (1895) as part 1 of the section entitled 'Wagner as a Danger'; however, the later version is slightly shortened, in what I shall argue is a revealing way. Here is the original version, entire, in the Hollingdale translation:

> How modern music is supposed to make the soul move.—The artistic objective pursued by modern music in what is now, in a strong but nonetheless obscure phrase, designated 'endless melody' can be made clear by imagining one is going into the sea, gradually relinquishing a firm tread on the bottom and finally surrendering unconditionally to the watery element: one is supposed to swim. Earlier music constrained one—with a delicate or solemn or fiery movement back and forth, faster and slower—to dance: in pursuit of

1 The musicological discussion in this essay is heavily indebted to Dr Steven Pane, a musicologist at my university with whom I've been studying nineteenth-century music the past few years in order to better understand Nietzsche's comments about music. I have repeatedly offered Dr Pane co-authorship of the articles which have resulted but he has so far always refused, saying that the only keyboard he wants anything to do with is the one with 88 keys.

> which the needful preservation of orderly measure compelled the soul of the listener to a continual self-possession: it was upon the reflection of the cooler air produced by this self-possession and the warm breath of musical enthusiasm that the charm of this music rested. —Richard Wagner desired a different kind of movement of the soul: one related, as aforesaid, to swimming and floating. Perhaps this is the most essential of his innovations. The celebrated means he employs, appropriate to this desire and sprung from it—'endless melody'—endeavours to break up all mathematical symmetry of tempo and force and sometimes even to mock it; and he is abundantly inventive in the production of effects which to the ear of earlier times sound like rhythmic paradoxes and blasphemies. What he fears is petrifaction, crystallization, the transition of music into the architectonic—and thus with a two-four rhythm he will juxtapose a three-four rhythm, often introduce bars in five-four and seven-four rhythm, immediately repeat a phrase but expanded to two or three times its original length. A complacent imitation of such an art as this can be a great danger to music: close beside such an over-ripeness of the feeling for rhythm there has always lain in wait the brutalization and decay of rhythm itself. This danger is especially great when such music leans more and more on a wholly naturalistic art of acting and language of gesture uninfluenced and uncontrolled by any higher plastic art: for such an art and language possesses in itself no limit or proportion, and is thus unable to communicate limit and proportion to that element that adheres to it, the all too feminine nature of music. (AOM 134)

'Endless melody' (sometimes 'infinite melody'; the German is *unendliche Melodie*) is defined by contemporary musicologists as melody which 'avoids, or bridges, caesuras and cadences' (Sadie 1980, p.121). Caesuras are the rests that come at the end of completed musical phrases, and cadences are the harmonic resolutions at the ends of phrases by which the music returns to the tonic, or home key. Endless melody, then, is music which just keeps going, without resolving in the way in which we are accustomed. The result is (i) harmonically, a loss of a sense of home key and harmonic resolution to it, (ii) rhythmically, a loss of a sense of regular rhythm and the sense of resolution created when a phrase fills out its allotted measures, and (iii) structurally, a loss of distinction between aria and recitative: unlike traditional 'number' opera, in which choral parts, solos, and narrative sections are distinct, the music in classical Wagnerian operas flows along endlessly.

Thus, for example, *Tristan und Isolde*—considered the *locus classicus* for endless melody—has rests, but not caesuras; that is, the rests don't represent resolutions (see, e.g., the Prelude, Fig. 1). For comparison, Bizet's *Carmen*—an appropriate foil given Nietzsche's deployment of it against Wagnerian opera in *The Case of Wagner* (1888)—features classic cadences, making it always easy to tell when the phrase is done (see, e.g., the Overture, Fig. 2).

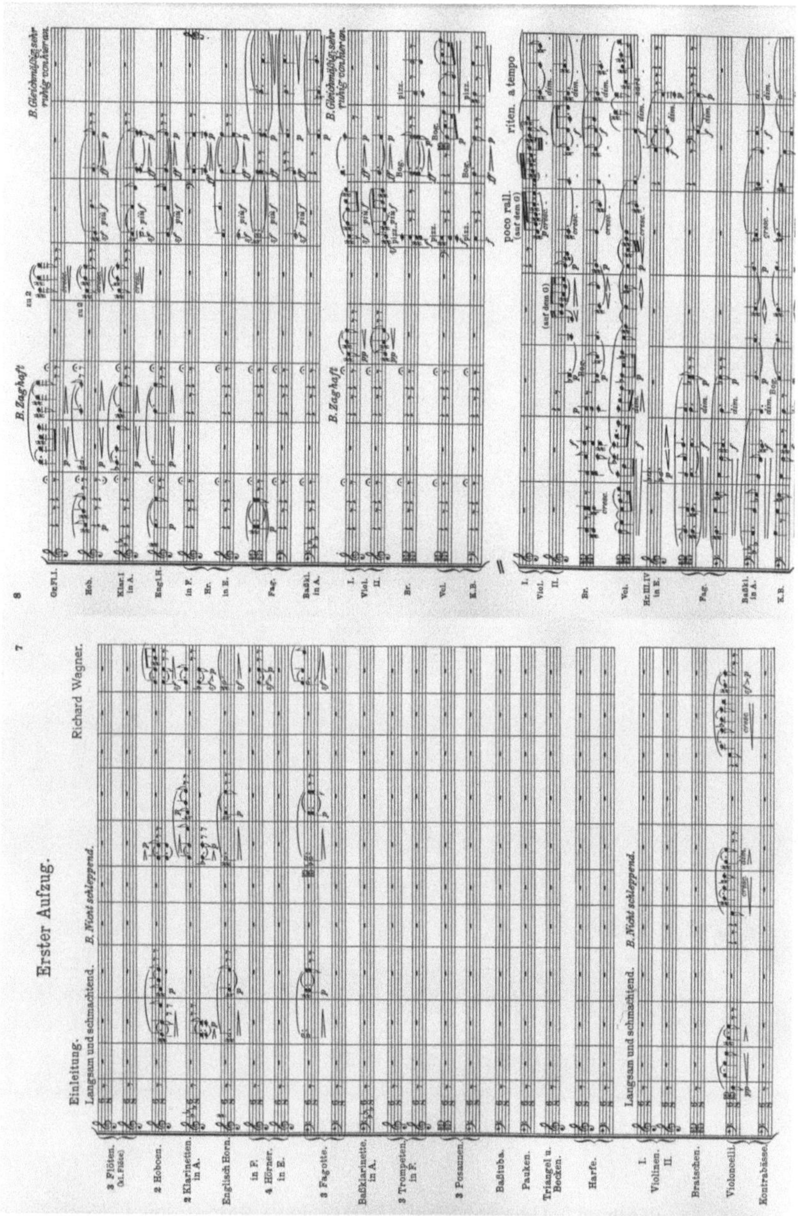

Figure 1. Richard Wagner, *Tristan und Isolde*, Prelude, measures 1–19 (New York: Dover Publications, 1973)

Figure 2. Georges, Bizet, *Carmen*, Prelude, measures 1–16
(New York: Dover Publications, 1989)

The contrast is clear: whereas traditional music features a finite rhythmic structure which repeats, a strong sense of home key, unmistakable cadence, etc., endless melody has none of these.[2] Clear too, I think, is Nietzsche's point that endless melody invokes in us the sensation of floating, while more traditional cadences are appropriate for dancing, even if they don't always make us want to get up and dance right then and there. Nietzsche's criticism of Wagner on this head is that Wagner's music leads us to an abandonment of ourselves. We give ourselves over to pure feeling, losing our sense of structure. While Nietzsche seems often to be promoting such a loss—'I am no man; I am dynamite' (EH 'Why I Am a Destiny' 1)—at least as often he in fact praises structure, and even argues that structure is necessary for a flourishing and creative life:

> Every morality is, as opposed to laisser aller, a bit of tyranny against 'nature'; also against 'reason'; but this in itself is no objection, as long as we do not have some other morality which permits us to decree that every kind of tyranny and unreason is impermissible. What is essential and inestimable in every morality is that it constitutes a long compulsion ... one should recall the compulsion under which every language so far has achieved strength and freedom—the metrical compulsion of rhyme and rhythm ... What is essential 'in heaven and on earth' seems to be, to say it once more, that there should be obedience over a long period of time and in a single direction: given that, something always develops, and has developed, for whose sake it is worth while to live on earth; for example, virtue, art, music, dance, reason, spirituality—something transfiguring, subtle, mad, and divine. (BGE 188, trans. Kaufmann)

Nietzsche's praise of structure takes different forms at different times in his career; while this passage from *Beyond Good and Evil* (1885) evinces a certain monomania, *The Gay Science* (1882) displays an appreciation of multifarious structures. There Nietzsche praises what he calls 'brief habits', a habit that 'nourishes' one for a time but then is discarded and replaced with the next one. 'Enduring habits I hate,' he says, but 'Most intolerable, to be sure, and the terrible par excellence would be for me a life entirely devoid of habits, a life that would demand perpetual improvisation' (GS 295). This last phrase describes endless melody almost exactly.

---

2 The reader may want to pause at this point long enough to listen to recordings of the relevant pieces, in order to have the music in his/her ears. It only takes a minute for the point to become obvious. I have chosen to focus on the orchestral beginnings of both operas, since in both cases the beginning sets the tone for the rest. Since opera is ultimately vocal music, however, the reader may want to hear the contrast also in vocal passages from the two works; if so, I recommend the Transfiguration from *Tristan* (and see Fig. 3) and the act II 'Chanson' from *Carmen*.

At any rate, our issue here is not whether Nietzsche is right about the necessity of structure for life to flourish, nor about the justice of his criticism of Wagner. (For example, we might defend Wagner, at least within the context of *Tristan*, by pointing out that music which never resolves is perfectly appropriate for a story about unfulfilled love.) Rather, we will return to the issue of Nietzsche's conception of musical time by analysing the aspect of endless melody which Nietzsche criticizes in 'Assorted Opinions and Maxims' 134, namely, rhythm.

This is already an idiosyncratic way for Nietzsche to approach the question, for it is not at all obvious that rhythm *is* the defining characteristic of endless melody. As just noted, current musicology defines endless melody in terms of lack of caesura and cadence, and consequently lack of resolution. However, the musicological analysis is ambivalent, since resolution has both a harmonic and a rhythmic component. And thus Bryan Magee, for example, can ignore rhythm entirely and account for the effect of *Tristan* on the listener solely in terms of harmonics:

> The first chord of *Tristan* ... contains within itself not one but two dissonances, thus creating within the listener a double desire, agonizing in its intensity, for resolution. The chord to which it then moves resolves one of these dissonances but not the other, thus providing resolution-yet-not-resolution ... And this carries on throughout a whole evening. (Magee 2001, pp. 208–209)

For that matter, Wagner himself introduced the term 'endless melody' (in his essay 'Zukunftsmusik', written in 1860, at about the same time he was composing *Tristan*) in neither rhythmic nor harmonic terms. For him, the term 'melody' connotes music which is expressive and significant; the rest of what is included in a piece of music—harmonies, connecting passages, etc.—is formulaic and says nothing. So for Wagner, the point about endless melody is that it describes music which is always saying something and has no gratuitous padding. Thus he avoids cadences primarily because they are formulaic (Sadie 1980, p. 121).

For Nietzsche, however, musical formulae, if successful, are to be cherished, representing as they do the fruit of many years of work by many hands on problems of musical composition. As noted above, Nietzsche finds traditions acceptable if they allow one to flourish, and some sort of structure is necessary if one is to flourish. So Nietzsche looks at the music itself and asks about its *effect*—does it allow one to flourish? In the case of endless melody, Nietzsche does not explore it harmonically, as critics such as Magee do, but rather turns the conversation to rhythm. In other words, even if the musicologists and Wagner himself disagree, Nietzsche *makes* endless melody be about rhythm, and thus by the same token about *time*.

The crux, then, of the criticism of endless melody in 'Assorted Opinions and Maxims' 134 is this sentence in the middle of the passage:

> What [Wagner] fears is petrifaction, crystallization, the transition of music into the architectonic—and thus with a two-four rhythm he will juxtapose a three-four rhythm, often introduce bars in five-four and seven-four rhythm, immediately repeat a phrase but expanded to two or three times its original length.

Rhythms are described here in terms of time signatures. A time signature consists of two numbers, the top one indicating the number of beats per measure, and the bottom one indicating the denominator of the fraction defining the musical note which counts as a single beat. Thus 2/4 time means two beats per measure, with a quarter note counting for a single beat; 6/8 mean six beats per measure, with an eighth note counting for a single beat; and so on. The image one gets from the sentence just quoted, then, is that endless melody consists of a jumble of incongruent time signatures which produces rhythmic chaos.

However, when we actually look at Wagner's scores we find nothing of the kind. The *Tristan* Prelude, for example, is in 6/8, while the Transfiguration is in 4/4—both perfectly traditional time signatures which hold sway in the score for a perfectly traditional length of time (Figs. 1 and 3). The image of a jumble of time signatures in fact describes some later composers such as Stravinsky. In *Les Noces,* an irregular jumble of 3/8 and 2/8 time signatures does indeed produce a sense of floating, which surely could not be danced to (see Fig. 4).[3] However, Wagner does *not* use a jumble of time signatures. And Nietzsche must have known this, since he was familiar with (at least) the piano score of *Tristan.*[4] So what is Nietzsche talking about in this sentence?

3 Again, the reader is encouraged to find and listen to the first minute of *Les Noces* (and again, that's all it will take to understand the point) before reading on. It should be noted that Stravinsky does sometimes write pieces featuring a jumble of time signatures which are nonetheless (at least titled as) dances, such as the 'Russian Dance' from *Petruschka.* As I will argue below, Nietzsche's point is not really about time signatures but rather about the rhythm of the piece as the listener experiences it, and the 'Russian Dance', which I think is indeed danceable in a way, proves his point quite nicely.

4 See EH 'Why I Am so Clever' 6, and Storr 1994, p. 215.

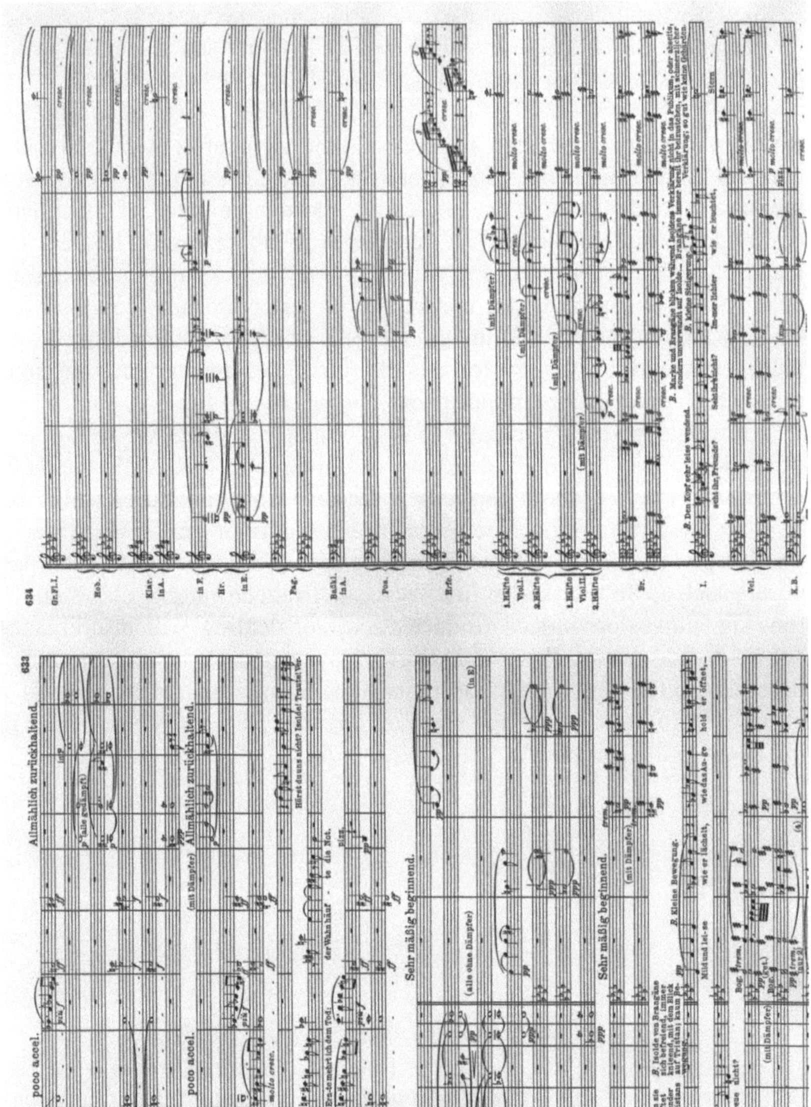

Figure 3. Richard Wagner, *Tristan und Isolde*, Transfiguration, measures 1–8 (New York: Dover Publications, 1973)

Figure 4. Igor Stravisky, *Les Noces*, Part I, scene I, measures 1–18
(London: J. & W. Chester Music Ltd., 1922)

Part of the difficulty is that Hollingdale's translation is wrong.[5] The original German for the sentence under examination is as follows:

> [Wagner] fürchtet die Versteinerung, die Krystallisation, den Übergang der Musik in das Architektonische,—und so stellt er dem zweitactigen Rhythmus einen dreitactigen entgegen, führt nicht selten den Fünf- und Siebentact ein, wiederholt die selbe Phrase sofort, aber mit einer Dehnung, dass sie doppelte und dreifache Zeitdauer bekommt.

The words *zweitactigen, dreitactigen, Fünf- und Siebentact* mean literally just 'two-beating, three-beating, five- and seven-beat'. So in fact all Nietzsche is doing is simply giving the numbers of beats in a measure. If these were the top numbers of time signatures, bottom numbers (indicating the length of note taking one beat) would be necessary, and since the most common bottom number is four, Hollingdale supplies 'four' each time. But Hollingdale has assumed Nietzsche means the two, three, five, and seven to indicate time signatures, whereas this is not necessarily so. In Hollingdale's version of the passage, the criticism of endless melody is that it consists of a jumble of time signatures, and, in the case of 5/4 and 7/4, unusual ones at that. But in fact Nietzsche is complaining about a jumble of clusters of beats: now two, now three, now five, now seven.

So the anachronistic confusion of Wagner with the later Stravinsky turns out to be a translator's error. However, correcting the error does not make the passage all that much clearer. Even retranslated, the problematic sentence still cries out for explication: where and how *does* Wagner combine two beats with three, and five with seven, in a piece in which the time signature remains constant?

At one time I had hoped to find a particular passage in *Tristan* which would manifest some explicit combination of two, three, five, and/or seven beats, but so far I have not been able to find one that does. Apparently, then, the reason Nietzsche lists two, three, five, and seven as the number of beats in Wagnerian endless melody is to expressly leave out four, six, and eight, the most common numbers used in time signatures, and the actual numbers in the time signatures for the Prelude and Transfiguration in *Tristan* (Figs. 1 and 3). But what is Nietzsche getting at?

---

5 Criticizing the Hollingdale translation, as I am about to do, seems impertinent, if not sacrilegious, in the context of this conference of which Reg Hollingdale was a founder and guiding light. The last time I saw him was the last time this conference was held in Cambridge, four years ago, and he died less than a month afterwards. So I feel bound to express here my respect for his work—I *lived with* his translation of *Human, All Too Human* while writing my dissertation—and to temper my criticism with gratitude.

What he seems to be pointing to is the way Wagner's melodies don't reside comfortably within their measures—they 'overflow' their measures, as it were. Traditional composers situate their melodies nicely within musical measures, acknowledging the downbeat and accepting the natural caesura at the end of a phrase. Thus the Overture to *Carmen* features classic 'four-by-four' structure consisting of four measures of four beats each, and because the musical phrases sit comfortably within their measures, they also *sound like* four measures of four beats each (Fig. 2). Wagner, on the other hand, denies the listener the satisfaction given by the traditional structure. The opening phrase of the *Tristan* Prelude, though set in 6/8 time, in fact starts with five beats of rest leading up to one beat of music at the end of the first measure, then two full measures of six beats each, then four beats of music in a fourth measure before two beats of rest (see Fig. 1).[6] This is what Nietzsche seems to be referring to when he says that Wagner puts two beats in a space where the measure makes one expect three or vice versa, or introduces a phrase of five or seven beats even though measures never (ordinarily) accommodate those numbers of beats. His point is that there is no match between the number of beats in the musical phrase and the number of beats in the underlying measures.

In addition (so the rest of the crucial sentence quoted above continues), when a phrase seems to repeat, Wagner deliberately extends it so that the second hearing's length does not match the first. Thus while the part played by the woodwinds in the Prelude's opening theme maintains its length (in measures 3–4, 7–8, and 11–12), the cello's part does not (compare measures 1–3 and 5–7 with 9–11—the theme starts earlier in measure 9 than it does in measures 1 and 5). In other words, despite the putative regularity of the 6/8 time signature *in the score*, in fact 'endless melody' *as played and heard* is entirely irregular. The internal rhythm of endless melody does not match the structure set up by the musical measures and thus overrides the listener's own internal sense of structure.

Thus when Nietzsche says Wagner combines differing numbers of beats and repeats the same phrase at different lengths, he is referring not to the time signature (which remains constant) but to the successive length of musical phrases which one would expect to be matched both to each other and to the downbeat. Wagner, Nietzsche asserts, deliberately mismatches

6 It's true that this musical phrase totals seventeen beats (1 + 6 + 6 + 4), and that seventeen equals two plus three plus five plus seven, but I think that's only coincidence: (1) The seventeen aren't broken up into sets of two, three, five, and seven beats, (2) later phrases in *Tristan* contain more than seventeen beats (or fewer), and (3) Nietzsche does not list the four numbers together as a sum but rather in two pairs.

so as to detach the listeners from their secure anchors and set them adrift on a sea of endless melody. This Nietzsche objects to as making impossible the sort of keeping-one's-feet-on-the-ground which he regards as necessary for the dance of life.[7]

What does all this tell us about Nietzsche's view of musical time? One ordinarily thinks of musical time in terms of the time signature or of the tempo meted out by the conductor or the metronome. But this is to take the view of the musicians, whereas Nietzsche in this passage takes the perspective of *the listener*. In this way, Hollingdale's mistranslation is actually instructive. It is indeed natural to assume that when Nietzsche begins throwing around numbers, he must be talking about time signatures—those are the numbers of which music seems to be made. As we have just seen, however, the only way to make sense of 'Assorted Opinions and Maxims' 134 is to understand it in terms of the music's effect on the listener. Whatever it looks like in the score, endless melody is played and heard in such a way as to provoke chaos in the listener.[8]

The only feature of musical time which matters, then, in Nietzsche's view, is the *perceived* rhythm of the musical phrases. The score and its time signature represent time 'in itself', as it were; to the listener, however, there is only time *as perceived*—i.e., the number and frequency of beats in the musical phrase itself as played and heard. In Wagner's music, not only are there irregular beats, by this standard—the two, three, five, and seven Nietzsche lists—but even then, these combinations recur in Wagner's music irregularly, so that the listener has no purchase, no structure. The melodic rhythm differs from the harmonic rhythm, and both differ from the underlying rhythm the musicians are counting out. The musicians are (presumably) counting out a stable six beats per measure, but to Nietzsche the beats heard by the listener are the only things that matter. They have an effect on the listener, not only during the time the music is actually being played but, Nietzsche clearly worries, afterwards as well. The jumble of rhythms cultivate a jumble in the soul. Thus musical time is a matter of the perception of, and the effect on, the listener. Its own intrinsic features—what's written in the score and what's counting in the musicians' heads—might as well not exist.

7 The metaphor of dance as being necessary for life can be found in the first volume of HA as section 278. And criticism of endless melody as incompatible with dance can be seen again—without the musicological details—in book II of GS (see especially sections 80, 84, and 86).

8 This marks Nietzsche's break from Wagner in yet another way—in his early period, as Wagner's ally, he took the point of view of the composer; now, as Wagner's enemy, he speaks solely as a listener.

2. Time as Music

When we turn to Nietzsche's view of time itself, it seems to me we can do no better than extrapolate from his view of musical time.

In the first volume of *Human, All Too Human* (1878), Nietzsche admits that there could be things-in-themselves, but denies that they could matter to us (HA I 9). And once one has shown that they do not matter to us, one has in effect 'refuted' them (HA I 21). This should apply, then, to time: the possible existence of time-in-itself, though undeniable, is irrelevant to us. But this is as much as to say that there *is* no time-in-itself for us—the only time that matters is time as we experience it. What this means is captured nicely by the musical criticism of endless melody in 'Assorted Opinions and Maxims' 134: the time signature in the score ('time-in-itself') is irrelevant to us; all that matters is the music's rhythm as we experience it ('time for us'). The musical analysis of time serves both to concretize the abstract metaphysics and to provide one of its most telling illustrations.

Music continues to be a useful avenue for discussing time, both as metaphor and as foremost example, in the later works. In his later period, Nietzsche denies categorically the existence of things-in-themselves, and so time-in-itself is no longer acceptable even as a mere supposition.[9] Musically, the result is an entirely intuitive analysis of music's effect. *The Case of Wagner* (1888) is the newly written late text employing this method, but Nietzsche also repackaged several earlier passages about Wagner in *Nietzsche contra Wagner.* And when it came time to revise 'Assorted Opinions and Maxims' 134 for this purpose, the sentence we have focused on in this essay was excised. The reason, I suggest, is that by then Nietzsche has moved beyond the half-hearted positivism of his middle period in which 'Assorted Opinions and Maxims' is situated to a full rejection of the thing-in-itself. The crucial sentence, however, envisages a contrast between an underlying 'real' musical time—the time recorded in the musical score—and musical time as perceived by the listener. The later Nietzsche rejects this distinction outright, and thus when 'Assorted Opinions and Maxims'

9 Although most of the passages in which Nietzsche explicitly denies the existence of things-in-themselves are in the notebooks (many of them included by Nietzsche's sister in *The Will to Power*), the view can be seen clearly at such published *loci* as GM I 13, TI 'How the "Real World" Finally Became a Fable', and TI 'The Four Great Errors' 3.

134 is re-used in *Nietzsche contra Wagner* he wants to draw no attention to the existence of the musical score and so leaves out this sentence.[10]

In both versions, Nietzsche's criticism of endless melody employs a view of musical time which provides insight as to the nature of our experience of time as a whole. Time, that is, is far more important than just another instance of a supposed thing-in-itself. The implication of this passage, that certain experiences of time are deleterious, suggests that, for Nietzsche, we each have our own internal rate of living, our own tempo, derived, presumably, from our internal physiological rhythms—breathing rate, heart rate, metabolism, etc.[11] A structured tempo connotes a body functioning well; presumably this is the condition we must be in in order to flourish creatively. Our internal tempo can change, as the tempo of a piece of music changes from section to section, and then time itself is indeed different for us than it was. However, internal chaos, lack of consistent tempo, means nothing gets done.

Our experience of music, then, can help or harm us. It can help structure our internal rate of time—either directly or by providing a contrasting rhythm to serve as a beneficial tonic—or it can harm it. Nietzsche himself finds Wagner always harmful:

> My objections to the music of Wagner are physiological objections ... My 'fact' is that I no longer breathe easily once this music begins to affect me; that my foot soon resents it and rebels; my foot feels the need for rhythm, dance, march; it demands of music first of all those delights which are found in good walking, striding, leaping, and dancing. But does not my stomach protest, too? my heart? my circulation? my intestines? Do I not become hoarse as I listen? (GS 368 = NCW 'Where I Offer Objections')

Nietzsche believes this music to be harmful because there is in it a deliberate undermining of temporal structure. And with no time-in-itself to fall back on, such undermining can be utterly destructive. It requires great strength to resist it and maintain one's own tempo. And thus *The Gay Science* 368 concludes with a Wagnerian responding to Nietzsche's criticism, 'Then you really are merely not healthy enough for our music?' In Nietzsche's shocked silence we hear the unspoken retort that it is the Wagneri-

10 It is also possible, of course, that he simply wanted to avoid requiring the reader of NCW to engage in the sort of involved interpretation to which that sentence has driven us in this essay. (Interestingly, the canard about the 'all too feminine' in music which closes the AOM version is left behind as well in NCW—one can only speculate why.)

11 For a similar view of the relation between musical and physiological rhythms, see Langer 1953, pp. 126–129, 328–330.

ans who are unhealthy, and that resisting this music is precisely a proof of health in Nietzsche's view.[12]

3. Time as the Music of our Lives

Nietzsche is often categorized as an existentialist, or a proto-post-modernist, or something of the sort. But he seems to me best categorized (if categorization be necessary) as a post-Kantian. Most of his views can be explained as 'like Kant—but with a twist'. So it is, in my view, with time. I will close by briefly characterizing Kant's understanding of time and the twist Nietzsche applies to Kant's conception.

For Kant, time is a form of sensibility. That is, rather than being a feature of the external world, time is a feature of our minds. Our minds are constructed—the contemporary metaphor of hardwiring is convenient—so as to arrange sense experience in a sequential order. We experience events in the world as happening before, simultaneously, or after each other. We cannot say how they 'really' happen, since what they are in themselves is not accessible to us. All we can say is that we experience them as happening in a regular, sequential order, and that is reality for us.

Though this seems to be a radically subjectivist position, Kant insists that it does not deny the objectivity of time. Since the hardwiring of our minds is not subject to our wills, time is still out of our control, and thus confronts us as a brute fact. To be sure, since Kant says time is a feature of *our* minds, which might well be different from those of other rational creatures, he should probably have described time as intersubjective rather than objective. Still, time is the same for all of us, and functions equally for all of us as a brute fact we must adapt ourselves to, *just as if* it were a feature of the external world.

Kant's conception of time is parallel to his conception of space—space is a form of sensibility by which our minds organize sense experience into a three-dimensional world. In the case of space, it is somewhat easier to imagine how other beings might perceive the world differently. For example, if one covers an eye one loses the parallax effect of having two eyes, and one now sees the world two-dimensionally. One can thus get a sense of what the world might seem like to a creature that has no visual parallax

12 This line too is missing from the NCW version—why? I suspect that by then Nietzsche had become nervous about another way to interpret his silence and lack of retort—perhaps his recurrent, debilitating illnesses actually gave the Wagnerian's gibe at him some credence.

(frogs, say), and also get some very vague sense of how there could come to be creatures that perceived in four dimensions. However, there is no similar way to conceive of how time might appear to creatures whose brains were hardwired differently from ours. In a way, this only proves Kant's point that for us the world simply *is* this way, i.e., that three-dimensional space and uni-directional time are indeed features of reality *for us*. But it does make it hard to understand the other side of Kant's position, that time is ideal.

Nietzsche, I think, can help here. On the one hand, by denying the existence of things-in-themselves, Nietzsche blocks the contrast between how time might be in itself and how we perceive time. That is, Kant must maintain that time is a hard and fast feature of reality, yet also say that there might well be other ways to perceive it. But what is the 'it' that other creatures are perceiving differently? While we might have an inkling of what that might mean in the case of space, in the case of time it's quite mysterious. For Nietzsche, however, time *is* our perception, and there's no time-in-itself that other creatures might have a different perception of, so the difficulty disappears.

But there is more: while Nietzsche follows Kant in asserting that it is our minds that structure reality, rather than reality impressing itself directly on our blank mental wax tablets, he makes one crucial adjustment. Kant assumes that our minds all function the same way, that we are hardwired in the way he describes. Nietzsche, however, asserts that our minds are all different. For Nietzsche, it is the individuality of perception that is crucial, not its intersubjectivity. While he does not deny that our perceptions can and do overlap—allowing us to live in some sort of concert with each other—he emphasizes our perspectival differences. He agrees that it is our minds which structure our reality, but sees the differences between our minds as sufficient to make our realities perspectival rather than intersubjective (and thus make it necessary to use 'realit*ies*' in the plural).

The result of Nietzsche's line of thought is that the best way to describe our perceptions of time is to resort to the realm and language of music.[13] Each musical piece sets its own tempo—that is, *it determines its own temporal reality*. There is no time-in-itself against which to compare these various tempos—they *establish* temporal reality for the world of that piece of music. We too live, think, and function at our own tempo—we

13 It is surely instructive that, whereas Nietzsche was obsessed with music and wrote about it extensively throughout his career, Kant did not appreciate music very much, and his otherwise magisterial aesthetic theory does not work very well in the case of music (see Higgins 1991, pp. 55–67).

establish the reality of time *for us*. There is no absolute time to measure ourselves against, or by which to criticize our individuality. At the same time, we can compare our own inner tempo, and challenge it, with that of others. Thus the experience of music can be a tonic for us, giving our souls rest, or perhaps a new rhythm to live by. It is either beneficial or harmful depending on its complementarity or conflict with the music of our lives. Wagner's use of endless melody to destroy the listener's sense of time, then, constitutes the most pernicious form of his nihilism.[14]

References

Higgins, Kathleen, 1991, The Music of our Lives, Philadelphia: Temple University Press.

Langer, Susanne, 1953, Feeling and Form, New York: Scribner's.

Magee, Bryan, 2001, The Tristan Chord: Wagner and Philosophy, New York: Metropolitan.

Sadie, Stanley (ed.), 1980, The New Grove Dictionary of Music and Musicians, vol. 20, London: Macmillan.

Storr, Anthony, 1994, 'Nietzsche and Music', in: Royal Institute of Philosophy Supplement, 37, pp. 213–227.

Translations

Beyond Good and Evil, trans. Walter Kaufmann, New York: Vintage Books, 1966.

The Gay Science, trans. Walter Kaufman, New York: Vintage Books, 1974.

Human, All Too Human, trans. R. J. Hollingdale, Cambridge: Cambridge University Press, 1986.

Nietzsche contra Wagner, in: *The Portable Nietzsche*, ed. and trans. Walter Kaufmann, New York: Viking Press, 1954.

14 In this context, an old joke passed on to me by Dr Pane—'A Wagnerian opera starts at 8:00, three hours pass, you look at your watch, and it's 8:15'—takes on an uncanny double meaning.

Index rerum et nominum

H

I

Bei Fragen zur Produktsicherheit wenden Sie sich bitte an:
If you have any questions regarding product safety,
please contact:

Walter de Gruyter GmbH
Genthiner Straße 13
10785 Berlin
productsafety@degruyterbrill.com